# THE VERY LARGE TELESCOPE INTERFEROMETER CHALLENGES FOR THE FUTURE

A JENAM 2002 Workshop
Porto, Portugal
3–5 September 2002

*Edited by*:
PAULO J.V. GARCIA
*Centro de Astrofísica da Universidade do Porto,*
*Porto, Portugal*

ANDREAS GLINDEMANN
*European Southern Observatory*
*Garching, Germany*

THOMAS HENNING
*Max-Planck-Institut für Astronomie*
*Heidelberg, Germany*

FABIEN MALBET
*Laboratoire d'Astrophysique de Grenoble*
*Grenoble, France*

Reprinted from *Astrophysics and Space Science*
Volume 286, Nos. 1–2, 2003

KLUWER ACADEMIC PUBLISHERS
DORDRECHT / BOSTON / LONDON

Library of Congress Cataloging-in-Publication Data

The Very Large Telescope Interferometer-challenges for the future / edited by Paulo J.V. Garcia ... [et al.].
p. cm.
Proceedings of a workshop held as part of JENAM 2002 organized by Sociedade Portuguesa de Astronomia and held on Sept. 5–7, 2002 at the Departamentos de Matemática Aplicada e Matemática Pura da Universidade do Porto.
ISBN 1-4020-1518-6 (alk. Paper)
1. Very Large Telescope Interferometer – Congresses I. Garcia, Paulo J.V.

2003055495

---

Published by Kluwer Academic Publishers,
P.O. Box 17, 3300 AA Dordrecht, The Netherlands.

Sold and distributed in the U.S.A. and Canada
by Kluwer Academic Publishers,
101 Philip Drive, Norwell, MA 02061, U.S.A.

In all other countries, sold and distributed
by Kluwer Academic Publishers,
P.O. Box 322, 3300 AH Dordrecht, The Netherlands.

*Printed on acid-free paper*



Printed in the Netherlands

TABLE OF CONTENTS

## INTERFEROMETRIC SCIENCE



## THE VLTI FUTURE

*Astrophysics and Space Science* publishes original contributions, invited reviews and conference proceedings over the entire range of Astronomy and Astrophysics. It publishes observational and theoretical papers as well as those concerned with the techniques of instrumentation. Observational papers can include data from ground-based, space, and atmospheric facilities. *Astrophysics and Space Science* has published, and continues to publish, landmark papers in its field and is an indispensable source of information for professional astronomers, astrophysicists and space scientists.



---

*Astrophysics and Space Science* is published bimonthly.
Subscription prices (2003, 6 volumes of 4 issues): EUR 2849.00/USD 2853.00 (print *or* electronic access); for print *and* electronic access, please add 20%.
Periodicals postage paid at Rahway, N.J. USPS No. 035–140.
U.S. Mailing Agent: Mercury Airfreight International Ltd., 365 Blair Road, Avenel, NJ 07001.

Published by Kluwer Academic Publishers, Van Godewijckstraat 30, P.O. Box 17, 3300 AA Dordrecht, The Netherlands, and 101 Philip Drive, Norwell, MA 02061, U.S.A.

*Postmaster*: Please send all address corrections to *Astrophysics and Space Science*, c/o Mercury Airfreight International Ltd., 365 Blair Road, Avenel, NJ 07001, U.S.A.

*Printed on acid-free paper*

# FOREWORD

The workshop on *The Very Large Telescope Interferometer: Challenges for the Future*, which was part of JENAM 2002 organised by Sociedade Portuguesa de Astronomia, took place at the Departamentos de Matemática Aplicada e Matemática Pura da Universidade do Porto (FCUP) from the 5th to the 7th of September 2002.

The Very Large Telescope Interferometer (VLTI) is the major interferometric facility in the world, operated by the European Southern Observatory. It is a unique facility as it is available to the community and has a sensitivity that will bring into reach vast classes of objects in the fields of planet searches, star formation and evolution and extragalatic astrophysics. The VLTI was open to the community on a shared risk basis in March 2002. This workshop was therefore an ideal place to obtain a broad view of optical interferometry and its scientific prospects.

The workshop started with a presentation of the basics of interferometry theory/practice, then of the VLT, Keck and LBT interferometers. The VLTI instruments MIDI and AMBER were introduced to its users as well as specific calibration and modelling tools. The bulk of the workshop addressed astrophysics being currently done with interferometers with very high quality reviews in the fields star formation, imaging, pulsating stars, Mira stars, active stellar shells and extragalactic astronomy. The workshop featured some talks presenting fresh VLTI data using the VINCI commissioning instrument. Finally the future of the VLTI was discussed in the context of long range science perspectives and second generation instrumentation.

We would like to thank all referees for their excellent work towards improving the quality of the workshop papers.

Financial support for the workshop participants and proceedings came from research grants from the Centro de Astrofísica da Universidade do Porto, Fundação para a Ciência e Tecnologia, Programa Operacional de Ciência, Tecnologia e Inovação do Quadro Comunitário de Apoio III and from the European Southern Observatory. The European Space Agency contributed towards the JENAM 2002 Abstract Book.

We are grateful to the CAUP system managers (Manuel Monteiro and Júlio Carreira), to the popularisation unit (Filipe Pires, Nelminha Silva, Ricardo Reis, Elsinha Silva and Jorge Martins) and to Darren O'Brien who provided crucial support on many occasions.

*Astrophysics and Space Science* **286:** vii–xvi, 2003.

We acknowledge the efficient support from Kluwer during the editing of the proceedings. Cover images credits are from the European Southern Observatory (ESO). The book cover was designed by Richardo Reis.

Last, but not least, a special word of thanks must go to Mário Monteiro and the rest of the JENAM LOC (Vitor Costa, João Fernandes, Jorge Filipe Gameiro, João Lima, Catarina Lobo, Carlos Martins and Margarida Serote Roos) for the impressive amount of work done in maintaining all the JENAM 2002 backbone.

*Paulo J.V. Garcia*
*Andreas Glindemann*
*Thomas Henning*
*Fabien Malbet*

# COMMITTEES

## Workshop Scientific Organising Committee:

| | |
|---|---|
| Malcolm Fridlund | *ESA, The Netherlands* |
| Paulo J.V. Garcia | *Porto, Portugal* |
| Andreas Glindeman | *ESO, Germany* |
| Chris Haniff | *Cambridge, United Kingdom* |
| Thomas Henning | *Heidelberg, Germany* |
| Fabien Malbet | *Grenoble, France* |
| Didier Queloz | *Genève, Switzerland* |
| Andreas Quirrenbach | *Leiden, The Netherlands* |
| Leonardo Testi | *Arcetri, Italy* |

Vitor Costa (CAUP & ISEP)
João Fernandes (OAUC)
Jorge Filipe Gameiro (CAUP & DMA-FCUP)
João Lima (CAUP & DMA-FCUP)
Catarina Lobo (CAUP & DMA-FCUP)
Carlos Martins (CAUP & Cambridge)
Mário Monteiro (CAUP & DMA-FCUP, Chair)
Margarida Serote Roos (CAAUL)

# LIST OF PARTICIPANTS

Acke, Bram (Bram.Acke@ster.kuleuven.ac.be)
Instituut voor Sterrenkunde, Celestijnenlaan 200B, 3001 Leuven, Belgium

Bacciotti, Francesca (fran@arcetri.astro.it)
Osservatorio Astrofisico di Arcetri, L.go E. Fermi 5, 50125 Firenze, Italy

Balazs, Lajos (balazs@konkoly.hu)
Konkoly Observatory, Budapest, Box 67, H-1525, Hungary

Bhattacharya, Nandini (bhattach@tnw.tudelft.nl)
Dept. of Applied Sciences, Delft University of Technology, PO BOX5046, 2600 GA, Delft, The Netherlands

Boden, Andy (bode@ipac.caltech.edu)
Infrared Processing and Analysis Center, California Institute of Technology, 1201 East California Boulevard, Pasadena, CA 91125, USA

Boone, Frederic (frederic.boone@obspm.fr)
LERMA, Observatoire de Paris, 61 Av. de l'Observatoire, 75014 Paris, France

Butcher, Harvey (butcher@astron.nl)
ASRON, PO BOX 2, 7990 AA, Dwingeloo, Netherlands

Chelli, Alain (alain.chelli@obs.ujf-grenoble.fr)
Laboratoire d'Astrophysique de l'Observatoire de Grenoble, Universite Joseph Fourier, 38041, Grenoble Cedex 9, France

Collin-Zahn, Suzy (suzy.collin@obspm.fr)
Observatoire de Paris, 5 place Jansen, 92195 Meudon, France

Correia, Serge (scorreia@eso.org)
European Southern Observatory, Karl-Schwarzschild-Str. 2, 85748 Garching, Germany

Costa, Vitor (vcosta@astro.up.pt)
Centro de Astrofísica da Universidade do Porto, Rua das Estrelas, 4150-762 Porto, Portugal

Cunha, Margarida (mcunha@astro.up.pt)
Centro de Astrofísica da Universidade do Porto, Rua das Estrelas, 4150-762 Porto, Portugal

Daigne, Gerard (daigne@observ.u-bordeaux.fr)
Observatoire Aquitain des Sciences de l'Univers (OASU), 2, rue del'Observatoire BP 89, 33270, Floirac, France

Delplancke, Francoise (fdelplan@eso.org)
European Southern Observatory, Karl-Schwarzschild-Str. 2, 85748 Garching, Germany

Derie, Frederic (fderie@eso.org)
European Southern Observatory, Karl-Schwarzschild-Str. 2, 85748 Garching, Germany

Diamond, Philip J. (pdiamond@jb.man.ac.uk)
Jodrell Bank Observatory, University of Manchester, Macclesfield, Cheshire SK11 9DL, UK

Dias, João (jdias@oal.ul.pt)
CAAUL, Observatório Astronómico de Lisboa, Tapada da Ajuda, 1349-018 Lisboa, Portugal

Dougados, Catherine (catherine.dougados@obs.ujf-grenoble.fr)
Laboratoire d'Astrophysique de l'Observatoire de Grenoble, Universite Joseph Fourier, 38041, Grenoble Cedex 9, France

Eckart, Andreas (eckart@ph1.uni-koeln.de)
I. Physikalisches Institut, University of Cologne, 77, Zulpicher Str., 50937 Cologne, Germany

Folha, Daniel (dfmf@astro.up.pt)
Centro de Astrofísica da Universidade do Porto, Rua das Estrelas, 4150-762 Porto, Portugal

Fridlund, Malcolm (malcolm.fridlund@esa.int)
Astrophysics Division – Space Science Department of ESA, ESTEC, Postbus 299, 2200 AG Noordwijk, The Netherlands

Gameiro, Filipe (jgameiro@astro.up.pt)
Centro de Astrofísica da Universidade do Porto, Rua das Estrelas, 4150-762 Porto, Portugal

Garcia, Paulo J. V. (pgarcia@astro.up.pt)
Departamento de Física da Faculade de Engenharia e Centro de Astrofísica da Universidade do Porto, Portugal

Gil, Carla (Carla.Gil@obs.ujf-grenoble.fr)
Laboratoire d'Astrophysique de l'Observatoire de Grenoble, Universite Joseph Fourier, 38041, Grenoble Cedex 9, France

Glindeman, Andreas (aglindem@eso.org)
European Southern Observatory, Karl-Schwarzschild-Str. 2, 85748 Garching, Germany

Gusev, Alexander (alexander.gusev@ksu.ru)
Dept. of Astronomy and Geophysics, Kazan University, Kremlevskaya Str., 18, 420008 Kazan, Russia

Hanniff, Chris (cah@mrao.cam.ac.uk)
Astrophysics Group, Cavendish Laboratory, Madingley Road, Cambridge, CB3 0HE, UK

Henning, Thomas (henning@mpia-hd.mpg.de)
Max-Planck-Institut für Astronomie, Königstuhl 17, 69117 Heidelberg, Germany

Herbst, Tom (herbst@mpia-hd.mpg.de)
Max-Planck-Institut für Astronomie, Königstuhl 17, 69117 Heidelberg, Germany

Hron, Josef (hron@astro.univie.ac.at)
Institute for Astronomy, University of Vienna, Türkenschanzstr. 17, A-1180 Vienna, Austria

Kitiashvili, Irina (irina.kitiashvili@ksu.ru)
Dept. of Astronomy and Geophysics, Kazan University, Kremlevskaya Str., 18, 420008 Kazan, Russia

Klockner, Hans-Rainer (hrklock@astro.rug.nl)
Keypten Intitute, Landleven 12, 9747 AD Groningen, The Netherlands

Koechlin, Laurent (koechlin@ast.obs-mip.fr)
UMR 5572 d'Astrophysique, Observatoire Midi-Pyrénées – CNRS, 14 avenue Édouard Belin, 31400 Toulouse, France

Krautter, Joachim (j.krautter@lsw.uni-heidelberg.de)
Landessternwarte Königstuhl, 69117 Heidelberg, Germany

Lampens, Patricia (patricia.lampens@oma.be)
Koninklijke Sterrenwacht van België, Ringlaan 3, 1180 Brussel, Belgium

Lane, Ben (ben@gps.caltech.edu)
Department of Geological and Planetary Sciences, MS 150-21, California Institute of Technology, Pasadena, CA 91125, USA

Malbet, Fabien (fabien.malbet@obs.ujf-grenoble.fr)
Laboratoire d'Astrophysique de l'Observatoire de Grenoble, Universite Joseph Fourier, 38041, Grenoble Cedex 9, France

Marengo, Massimo (mmarengo@cfa.harvard.edu)
Harvard-Smithsonian Center for Astrophysics, 60 Garden Street, Cambridge, MA 02138, USA

Meinser, Jeffrey (meisner@strw.leidenuniv.nl)
NEVEC, Leiden Observatory, University of Leiden, PO Box 9513, 2300 RA Leiden, The Netherlands

Mouillet, David (mouillet@bagn.obs-mip.fr)
Laboratoire Astrophysique Observatoire Midi-Pyrenees, 57, Avenue d'Azereix, BP 826, 65008, Tarbes cedex, France

Mourard, Denis (denis.mourard@obs-azur.fr)
Observatoire de la Cote d'Azur, Dept. Fresnel UMR6528, 2130 Route del'Observatoire, 06460, Caussols, France

Niedzielski, Andrezej (aniedzi@astri.uni.torun.pl)
Torun Center for Astronomy (TCfA), N. Copernicus University, ul. Gagarina 11, 87-100 Torun, Poland

Novais, Liliana (00a017@astro.ma.fc.up.pt)
Rua da Igreja n° 163, 4480-754 Vila do Conde, Portugal

Nurnberger, Dieter (nurnberg@iram.fr)
European Southern Observatory, Alonso de Cordova 3107, casilla 19001, Santiago 19, Chile

Paresce, Francesco (fparesce@eso.org)
European Southern Observatory, Karl-Schwarzschild-Str. 2, 85748 Garching, Germany

Pascucci, Ilaria (pascucci@mpia-hd.mpg.de)
Max-Planck-Institut für Astronomie, Königstuhl 17, 69117 Heidelberg, Germany

Percheron, Isabelle (percheri@strw.leidenuniv.nl)
NEVEC, Leiden Observatory, University of Leiden, PO Box 9513, 2300 RA Leiden, The Netherlands

Perez-Torres, Miguel (torres@ira.cnr.it)
Istituto di Radioastronomia/CNR, via P. Gobetti 101, 40129 Bologna, Italy

Perrin, Guy (guy.perrin@obspm.fr)
LESIA, Observatoire de Paris, 5 place Jules Janssen, 92190, Meudon, France

Petrov, Romain (petrov@unice.fr)
UMR 6525, Parc Valrose, 06108 Nice Cedex 2, Nice, France

Pinheiro, Fernando (fjgp@astro.up.pt)
Centro de Astrofísica da Universidade do Porto, Rua das Estrelas, 4150-762 Porto, Portugal

Preibisch, Thomas (preib@mpifr-bonn.mpg.de)
Max-Planck-Institut für Radioastronomie, Auf dem Hügel 69, 53121 Bonn, Germany

Przygodda, Frank (przygodd@mpia-hd.mpg.de)
Max-Planck-Institut für Astronomie, Königstuhl 17, 69117 Heidelberg, Germany

Quirrenbach, Andreas (aquirrenbach@ucsd.edu)
University of California, San Diego, Mail Code 0424, La Jolla, CA 92093, USA

Richichi, Andrea (arichich@eso.org)
European Southern Observatory, Karl-Schwarzschild-Str. 2, 85748 Garching, Germany

Santos, Nuno (nuno.santos@obs.unige.ch)
Observatoire de Genève, 51 ch. des Maillettes, 1290 Sauverny, Switzerland

Schilizzi, Richard (schilizzi@jive.nl)
Joint Institute for VLBI in Europe, PO BOX 2, 7990 AA, Dwingeloo, The Netherlands

Schoeller, Markus (mschoell@eso.org)
European Southern Observatory, Alonso de Cordova 3107, casilla 19001, Santiago 19, Chile

Smith, Kester (kester@mpifr-bonn.mpg.de)
Max-Planck-Institut für Radioastronomie, Auf dem Hügel 69, 53121 Bonn, Germany

Stee, Phillipe (philippe.stee@obs-azur.fr)
Observatoire de la Cote d'Azur – Site de Grasse-Roquevignon, Avenue Copernic, 06130, Grasse, France

Swings, Jean-Pierre (swings@astro.ulg.ac.be)
Institut d'Astrophysique, Universite de Liege, Avenue de Cointe 5, B-4000 Liege, Belgium

Takami, Michihiro (takami@herts.ac.uk)
Department of Physical Sciences, University of Hertfordshire, College Lane, Hatfield Herts AL10 9AB, UK

Thiébaut, Eric (thiebaut@obs.univ-lyon1.fr)
Observatoire de Lyon/CRAL, 9 avenue Charles André 69561 Saint Genis Laval cedex, France

Verhoelst, Tijl (Tijl.verhoelst@ster.kuleuven.ac.be)
Institute of Astronomy, K.U. Leuven, Celestijnenlaan 200 B, 3001 Leuven, Belgium

Wilkin, Francis (f.wilkin@astrosmo.unam.mx)
Instituto de Astronomia, UNAM, Apto. POSTAL 3-72 (Xangari), 58089 Morelia, Michoacan, Mexico

Woltjer, Lodewijk (ulla.dw@bluewin.ch)
Observatoire de Haute Provence, F04870 St. Michel l'Observatoire, France

# INTERFEROMETRY PRIMERS

# INTERFEROMETRY THEORY PRIMER

DAVID MOUILLET
*Observatoire Midi-Pyrénées. 57, av Azereix. BP826. 65008 Tarbes Cédex, France*

**Abstract.** As an introduction to the following talks, this presentation reviews the theoretical background of interferometric measurements. Starting from a general formalism, I then emphasize the relationship between those measurements and the observed object properties.

**Keywords:** interferometry, theory, tutorial

## 1. Interferometric information

### 1.1. WAVEFRONT COHERENCE IN PUPIL PLANE

The distribution of the light intensity in the sky is commonly obtained in the optical and the near IR range, by collecting this light in a so-called *pupil plane*, focussing it and detecting the energy in the focal plane or *image plane*. Each position in this image plane is illuminated by the sources from a given direction in the sky, the scale being defined by the optic's focal length. The electric field propagation from the source to the optics and from the optics to their focal plane are respectively described by two successive Fourier transforms under the usual assumptions of Fraunhoffer diffraction theory:

$$E(\theta) \longrightarrow \hat{E}\left(\tfrac{\rho}{\lambda}\right) . Pupil\left(\tfrac{\rho}{\lambda}\right) . Atm\left(\tfrac{\rho}{\lambda}\right) \longrightarrow E_{det}(\theta)$$

where

- the conjugated variables are respectively i/ the direction in the sky $\theta$ and ii/ the vector distance in pupil plane, in wavelength units, given by $\rho/\lambda$;
- the optics are described by the $Pupil$ function: its modulus indicates the optical transmission and its phase indicates some delays of the light propagation, non homogeneous over the pupil (due to unperfectly flat mirrors for instance). Similarly, the effect of atmosphere (transmission but also and mainly non constant optical path length) can be described by a complex $Atm$ function. This function is also highly time dependent.

Now, let us consider how is obtained the light intensity in the detector plane ($I_{det}(\theta)$), that is practically measured in the optical and IR range with quadratic detectors. In particular, how does it depend on the intensity in the source plane ($Obj(\theta)$), and how do the pupil and atmosphere modify it? This intensity $I_{det}(\theta)$ (resp. $Obj(\theta)$) is the average square modulus of the electric field $E_{det}(\theta)$ (resp.

*Astrophysics and Space Science* **286:** 3–15, 2003.

$E(\theta)$). Under assumptions that are reasonable in astronomy and that includes the previous assumptions for the Fraunhoffer diffraction (source of limited size, at large distance, and spatially incoherent), the Van Cittert-Zernike theorem ensures that the Fourier transform of the square modulus operation in one space translates into the autocorrelation of the conjugated fields. We derive the following relationships between the source and detector plane intensities on the one hand, and the autocorrelation of the pupil plane quantities on the other:

$$Obj(\theta) \longrightarrow \widehat{Obj}\left(\tfrac{\rho}{\lambda}\right) \, . \, MTF\left(\tfrac{\rho}{\lambda}\right) \longrightarrow I_{det}(\theta) = Obj(\theta) \star PSF(\theta),$$

where,

- $\widehat{Obj}\left(\frac{\rho}{\lambda}\right)$, the Fourier transform of the source plane intensity is not an intensity but an autocorrelation:

$$\widehat{Obj}\left(\frac{\rho}{\lambda}\right) = \hat{E}\left(\frac{\rho}{\lambda}\right)^{\star} * \hat{E}\left(\frac{\rho}{\lambda}\right) = \iint_{pupil} \hat{E}\left(\frac{r}{\lambda}\right)^{\star} \hat{E}\left(\frac{r-\rho}{\lambda}\right) dr$$

- the effect of the pupil and atmosphere is described by the *modulation transfer function*

$$MTF\left(\frac{\rho}{\lambda}\right) = \left\langle \text{Autocorrelation}\left\{Pupil\left(\frac{\rho}{\lambda}\right) \, . \, Atm\left(\frac{\rho}{\lambda}\right)\right\} \right\rangle_{time}$$

- This filtering appears in the detector plane as a convolution of the intrinsic source image by the instrumental point spread function $PSF(\theta) = \widehat{MTF}\left(\frac{\rho}{\lambda}\right)$

In other words, the information defining the source image is described by this autocorrelation function of the electric field in the pupil plane. According to the previous formula, it is given by the relative values of the electric field on two points of the pupil separated by a given distance, averaged over the time and over the pupil (for all pairs of points with such a separation). This quantity is also called the *wavefront coherence* in the pupil plane. It is a function of the considered spatial frequency $\rho/\lambda$ and may also depend on wavelength. Only part of the initial intrinsic information is selected and transmitted by the instrument and atmosphere, as described by the modulation transfer function.

This formalism is general and appropriate for both classical imaging or interferometry. In interferometry, the pupil shape may be of any type and the way of measuring the signal may differ from the direct detection of an image in the focal plane. In this context, the coherence properties of the electric field in pupil plane is widely used; the simplicity of the impact of instrumental effects as a multiplication of the object intrinsic signal depending on its spatial frequency, rather than an image convolution, makes also the discussion of the signal in the pupil plane very convenient.

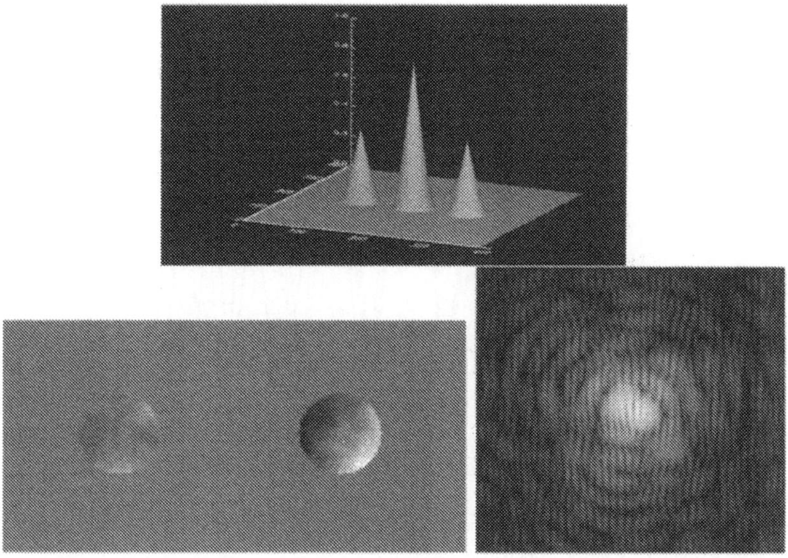

*Figure 1.* Top: example of transfer function corresponding to a pupil constituted of two circular apertures. On top of the low frequency peak transmitted by each individual aperture with a cut-off frequency defined by the individual aperture diameter $D/\lambda$, the system transmits the information at higher frequency (secondary peaks) centered around the frequency defined by the apertures separation or baseline $B/\lambda$. Bottom: image in log scale of a point-like source through this system (right), affected by random turbulent phase delays over the pupil, represented on the left.

## 1.2. Measuring the Visibilities

### 1.2.1. *Coding the interferometric information*

The problem is to measure the correlation of the electric field taken in 2 positions of the pupil plane: $< E(P_1, t) \, . \, E^\star(P_2, t) >$ (normalized to the target flux). In the optical range, the use of heterodyne techniques (that make it possible to numerically compute such correlations) dramatically limits the spectral bandpass, and consequently the sensitivity, of the system (the frequency bandpass is directly defined by the electronics sampling frequency, which remains very small compared to the electric field frequency, typically $10^{14-15}$Hz).

Alternatively, this quantity can be derived from the more easily obtained intensity after the beam combination,

$$I_{12} = < \left\| E(P_1, t) + E^\star(P_2, t + \tau) \right\|^2 >$$

where the delay $\tau$ should be controlled and modulated by the instrument set-up. This modulation can be i/ temporal (a rapidly moving element in the path modulates the optical length of one arm, such as a Michelson set-up) or ii/ spatial (as in the Young experiment: depending on the pixel position in the detector plane, the optical length from collected points $P_1$ and $P_2$ is not identical; this optical path difference, OPD, can be translated into a delay: $\tau = OPD/c$).

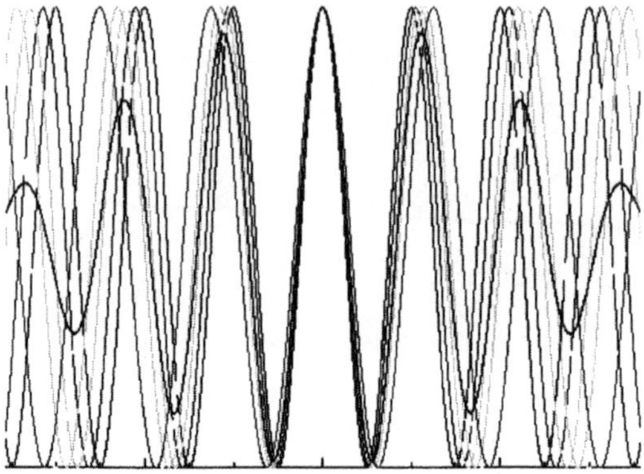


*Figure 2.* Illustration of the fringe pattern formed in the Young experiment. The fringe separation is defined by the wavelength and the separation of the 'Young holes' (or output pupil) that form the fringe pattern (independently from the entrance pupil or baseline for which it measures the visibility). Each individual thin curve corresponds to a given wavelength. Their average (in thick line) shows the expected fringe pattern for a broad band observation: the fringe contrast decreases on a typical scale OPD $\simeq R.\lambda$, where $R$ is the spectral resolution.

If (at least partially) coherent, the electric fields $E(P_1)$ and $E(P_2)$ will interfere constructively or destructively, depending on $\tau$: $I_{12}(\tau)$ shows then a fringe pattern.

$$I_{12} = I_1 + I_2 + 2\mathcal{R}e\left\{< \ E\left(P_1, t\right) \ . \ E^{\star}\left(P_2, t+\tau\right) \ >\right\},$$

where $I_1$ and $I_2$ refer to the intensity of the electric field propagated from $P_1$ and $P_2$ respectively down to the detector (see Figure 2).

Around the zero OPD position (or white fringe), the fringe contrast measures the visibility modulus and its position the visibility phase (an offset of one fringe corresponds to a source offset of $\lambda/B$ and a phase shift of $2\pi$). Due to the chromaticity of the fringe separation, a broad band fringe pattern gets blurred far from the white fringe. In other words, we compare electric fields which get temporally de-correlated: $|< \ E\left(P_1, t\right) \ . \ E^{\star}\left(P_2, t+\tau\right) \ >|$ gets significantly smaller than $|< \ E\left(P_1, t\right) \ . \ E^{\star}\left(P_2, t\right) \ >|$. The number of fringes of the pattern is limited to spectral resolution R=$\lambda/\Delta\lambda$ ( the coherence length of the electric field is $R.\lambda$). Finally, this number of fringes may also be limited (for a given source) to the windowed field ($B/D$ fringes for a monomode instrument).

The principle mentioned above makes it possible to derive the mutual coherence of two points in the entrance pupil from the contrast of the fringe pattern obtained after beam combination. This principle can be extended to more than 2 apertures, with dedicated pixels for each pair of apertures. When using a spatial coding of the fringe patterns, the patterns related to different pairs of apertures may be detected on a single line of pixels as long as the corresponding fringe separations are non redundant in order to be able to associate the measurement of each fringe pattern with the relevant pair of apertures. This condition is ensured

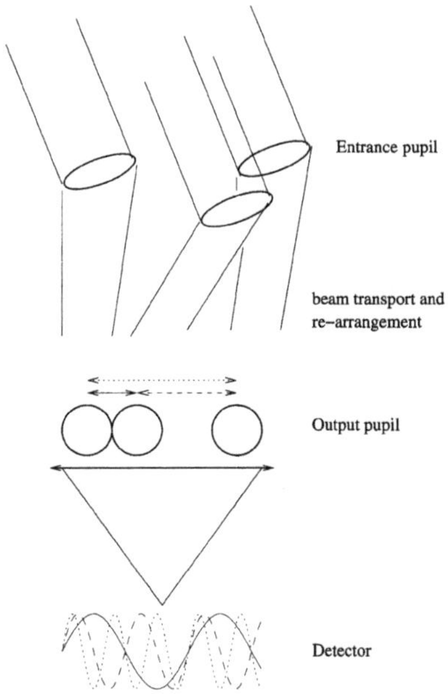


*Figure 3.* Illustration of the spatial coding of 3 fringe patterns on a single line of pixels with non-redundant fringe separation. The fringe separation of each fringe pattern is defined by the separation each corresponding pair of apertures in the output pupil. The corresponding measured visibility is related to the source image spatial frequency defined by the separation of telescopes (in the entrance pupil).

when the configuration of the pupil used to recombine the various apertures (or output pupil) is non-redundant.

Finally, in the particular case where the output pupil, that determines the configuration of the fringe pattern on the detector, is homothetic to the entrance pupil (telescope array configuration), the obtained fringe pattern is a direct image of the observed source, in the sense that it is the convolution of the source by the instrument PSF.

In principle, a number of methods for the interferometric signal coding are ideally equivalent. In practice, the choice is not trivial and should be optimized according to the array configuration and to the main astrophysical goals. They indeed do not imply the same number of physical pixels, or pixel read-outs (and corresponding noise). Also, a temporal modulation of the fringes assumes that the MTF remains stable during the fringe scan: the atmosphere consequently impose very short exposure times. Alternativelly, when using a spatial modulation of the fringe (the complete fringe pattern being simultaneously detected on different pixels), the various beams are not transmitted through same optics: the system transmission for each beam down to the detector has to be well calibrated and stable for an accurate visibility measurement.

1.2.2. *Effect of atmospheric turbulence*
Instantaneously, for 2 pupil points, the propagation through the atmosphere induces a random delay for one point compared to the other (differential piston). As a consequence, the instantaneous fringe pattern is translated: the corresponding visibility phase is shifted by a random value. If averaged over time (with an integration time longer than the atmospheric coherent time, which is typically a few ms in visible and scales as $\lambda^{6/5}$), the random phase delay varies significantly: the fringe pattern is blurred; in the complex plane, the visibility vector rotates and is attenuated after the average. Not only the visibility phase is modified but also the modulus is attenuated. In other words again, the transfer function is attenuated.

In case of extended sub-apertures (larger than the turbulent coherent length, that is typically a few 10 cm in visible), the visibility measurement for a given spatial frequency results (even instantaneously) from the spatial average over each pair of pupil points with the appropriate separation. This spatial average induces a similar degradation on the transfer function*. In the general case, both spatial and temporal averages degrade the signal: it reduces the transfer function and makes it difficult to calibrate accurately.

Dealing with the atmospheric perturbation is very difficult and remains often the main limitation in various observing modes. This limitation motivates the use of various specific devices or observing procedures:

- *short exposure times* within an atmosphere coherent time to freeze the turbulence: this is critical for accurate visibility measurements when scanning a fringe pattern (temporal coding); this is also important when using a spatial coding of the fringe in order to preserve a high fringe contrast, but this reduces the limiting magnitude
- *real-time correction* of the atmosphere perturbation: fringe tracker and/or adaptive optics. For extended apertures, the adaptive optics acts with a fast deformable mirror and aims at correcting for the wavefront perturbation over each aperture. The fringe tracker, for each baseline (or pair of apertures), adjusts the OPD and aims at stabilizing the corresponding fringe pattern. Those corrections in fast servo-loop also require bright stars (typically H or $K \simeq 11$–$13$) to acquire high signal-to-noise measurements but they make possible much longer interferometric integrations, which may be very valuable in particular to get higher spectral resolution.
- *spatial filtering* through monomode fibers may reject the residuals from adaptive optics: only the coherent part of the incoming flux on each aperture is transmitted. The atmospheric variations of the incoming wavefronts are then translated into calibrable photometric variations. This method which makes possible more accurate visibility measurements limits the field of view to the resolution of a single aperture.

* This degradation due to the spatial average has motivated pupil masking techniques in order to create a non redundant pupil.

– *differential measurements*: the atmospheric OPD is (in first approximation) achromatic. The relative visibility (modulus and phase) obtained simultaneously at 2 distinct wavelengths can be measured much more accurately than each individual visibility. This mode requires an appropriate instrument for such simultaneous measurements; it is well suited for spectrally dependent target structures more than for continuum measurements.

## 2. Generic Interferometer Purposes

Generally speaking, the interferometer will aim at collecting the light over a pupil of any shape (this pupil plane is also called the frequency plane or u-v plane as common notation for the frequency coordinates) and at estimating the wavefront coherence over this pupil for each possible available spatial frequency (i.e. for each possible pair of points separation available within this pupil). Not going into the discussion of practical problems with this instrumentation (see M. Schoeller, this volume), I draw hereafter in general terms a non-exhaustive list of limitations or challenges for interferometers that conditions their design or discussions:

- *u-v coverage or selection of transmitted spatial frequencies of the target:* Unlike connected pupils (in classical imaging) that provide the complete image information up to the cut-off frequency (or resolution) defined by the aperture diameter, a diluted pupil may provide information for much higher frequencies (defined by the largest baseline), but this information is generally incomplete. The usual way to represent the frequency plane or u-v coverage, is to plot the pupil autocorrelation (optical transfer function). It is possible to improve a sparse u-v coverage in different ways.
  i/ a large number of apertures provides simultaneously a large number of baselines ( $n(n-1)/2$ ), each of them independent if the array geometry is non redundant.
  ii/ alternatively (or additionally), it may be possible to move such apertures so as to modify the array and measure sequentially more and more frequencies
  iii/ finally, for a given array geometry, the earth rotation itself modifies the projected baselines (or aperture separations as seen from the source).
  In each case, the light will have to be propagated to the combination laboratory, and the optical path length to be accurately controlled and equalized: this implies possibly long delay-lines (up to a significant fraction of the baselines).
- *transfer function:* on the available frequency window defined by the array geometry, any instrumental defects (such as non constant phase delays in time or spatially over the pupil, non homogeneous polarization effects, vibrations...) induces a coherence loss, i.e. an attenuation of the transfer function. The ideal instrument will have
  – a high transfer function (measured as the fringe contrast in case of an unresolved source for which the wavefront is intrinsically perfectly coherent)

not to loose the interesting signal down to the noise level; this requirement sets very severe specifications on the quality of the whole interferometric opto-mechanics,
   - and known to a very accurate level, that will drive the final accuracy of the measured target visibility. This implies accurate calibration procedures of this transfer function (which is a complete issue by itself, see Percheron, this volume) and a high instrumental stability between these calibrations.

- *sensitivity and speed:* other sources of noise (detector and photon noise) definitely also limit the signal quality. This sensitivity issue is often critical due to the number of relay optics involved in the complete interferometer. Also, the small structures of the targets (close binaries, stellar surfaces, circumstellar ejecta, ...) may often involve a small fraction only of the total flux and be variable on short timescales. The sensitivity on moderate exposure times for a single array configuration is in many cases a critical issue. This need drives the motivation for the interferometry of large coherent apertures, i.e large telescopes equipped with adaptive optics
- *spectral coverage and resolution:* as for any other instrument, the target interferometric information may be interesting and complementary in various spectral ranges and up to high spectral resolution (see Petrov and Przygodda, this volume). Additionally, one will note that the observing wavelength directly impacts the measured spatial frequency $B/\lambda$ for a given geometric baseline. For instance, for an achromatic source, the simultaneous observations at various wavelengths may replace advantageously some time consuming array re-configurations in order to complete the u-v coverage. On contrary, in the cases of object geometries that strongly depend on wavelengths (for instance a point-like object in continuum that is extended in an excited emission line), simultaneous observations may provide very accurate self calibration.
- *polarization:* here again, the interest for measurements in various polarization states of the target light may be interesting in interferometry also. And here also, if theoretically possible, the number of optics usually involved here make this measurements very delicate and involve specific calibration procedures

This list is very general: a single instrument may probably not be optimized for all of these items. After a generation of pioneering instruments that have explored a number of instrumental capabilities and that have provided interesting astrophysical results, the up-coming instruments dedicated to a wide community will significantly improve some of the mentioned goals. The case of VLTI is a good example. The common infrastructure itself is very powerful and designed for *a wide range of use* (see Glindemann, this volume): from the visible to mid-IR, combining either medium-sized or large telescopes, with an increasing number of delay-lines. The instrument plan aims at covering various astronomical purposes with complementary instruments. Each focal instrument is focussed on a specific observational niche.

In order to maximize the return from such instruments, to validate some program feasibility and finally to optimize the use of the interferometer, the observer himself definitely needs to identify *which part* of the general interferometric signal is critical to constrain his target, what specific accuracy and calibration are required. The generic approach is to model the target properties, and simulate the expected interferometric signal (at various frequencies). Varying the model parameters makes possible to discuss the optimal frequency range where to constrain them, the amplitude of the effect and required SNR. These requirements will be compared to the available instrument capabilities and configurations. The following sections illustrate in some examples such an approach, as a basic reference frame, discussing in particular the relation between the object information and the observed visibility modulus and phase.

## 3. Which Visibilities Tell Me What Object Properties?

### 3.1. VISIBILITY MODULUS

The visibility modulus at a given spatial frequency indicates the fraction of the object image involved in flux variations on the typical $\lambda/B$ angular scales. This baseline, that does not replace more complete numerical simulations, explains easily as a first approach what are the relevant baselines in order to characterize the object properties (see examples in Fig. 4). From such measurements, it will in particular be possible to derive the information relative to the *object structure*: whether most of the object intensity is extended or not, on what scale, with what orientation and aspect ratio (if it is more extended in one direction than in another), whether sharp edges are present (implying non-zero visibilities up to high frequencies, much larger than the global object typical size)...

I will not illustrate here, in a general paper, many cases of such information content, but I will only point out a few remarks:

- *number of measurements:* very often, the interesting astrophysical information is not only a 'typical size' of the source in a given direction. This means that a single measurement with one baseline $B$ will very often not be enough and will certainly not provide a complete image at the resolution $\lambda/B$ ! This statement is already clear to most of the readers but has to be emphasized: even in the case of relatively simple models, the structure will require probably various appropriate configurations to be unambiguously identified, and will imply a priori information (with at least as many measurements as parameters to be constrained)
- *accuracy of the measurements:* here again, obviously, the ability to distinguish between similar models, or in other words the propagated uncertainties onto the determination of the model parameters, is determined directly from the visibility measurements accuracy. This accuracy comes from the source

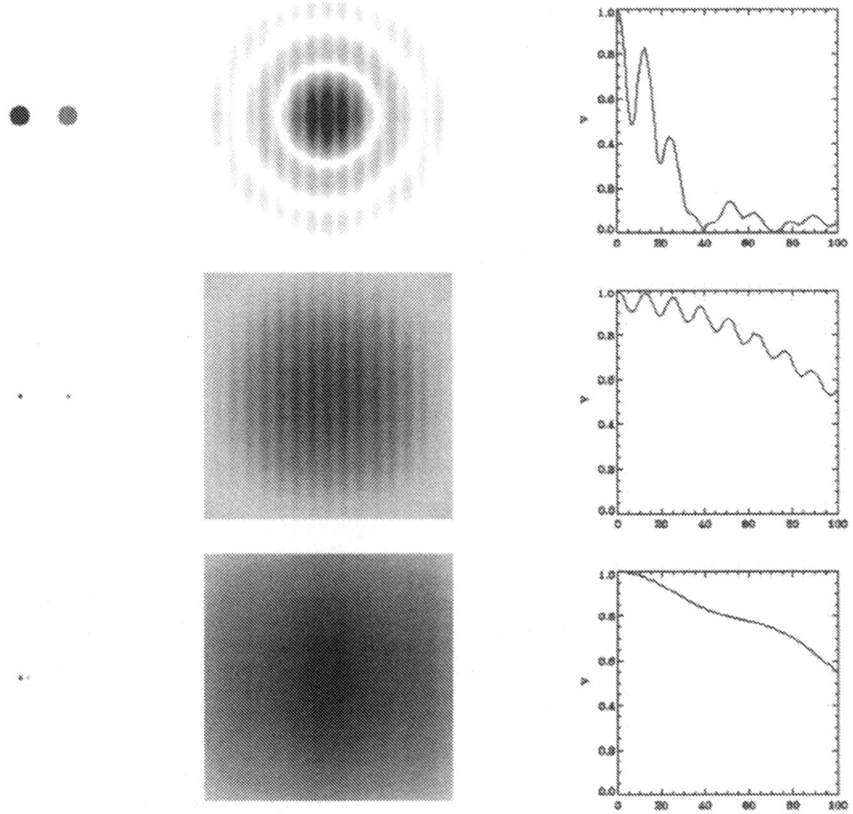

*Figure 4.* Examples of visibility modulus corresponding to the image of binaries, with arbitrary spatial scale units. Left: assumed object image (with various stellar radius, binary separation and contrast); middle: the Fourier transform of the image provide the expected visibility modulus, on each point for the corresponding baseline; right: extracted visibility curve for baselines aligned with the binary separation. The 3 examples from top to bottom illustrate the impact of the each star diameter (general decrease of visibilities), the binary separation (the closer the separation – bottom –, the larger the fringe separation), and binary flux ratio (from 3 for the top example to 20 for the two others) as the amplitude of the visibility oscillation.

brightness, the collecting surface, but also very importantly from the instrument transfer function stability and calibration accuracy. The requirement for a very high accuracy is in particular obvious in the case of high dynamic range detections (for instance a binary with a high brightness ratio, but also small spots on a stellar surface, or gap/inhomogeneities over a disk...): as a rule of thumb, the amplitude of the visibility modulus signature of a structure is roughly (at the appropriate frequency) the fraction of the flux involved in this structure.

- *measurement of low visibilities:* the performance of an interferometer (sensitivity, accuracy) are usually given for unresolved sources (high visibility). Observations of resolved sources with low visibilities are more difficult to i/ detect and track the fringes, and ii/ estimate the visibility (SNR $\propto V^2$). Such observations may however be interesting in some cases (to maximize the variation of visibility according to the model parameters, as in the case of constraining a stellar diameter) or even unavoidable in other cases (to detect and characterize small structures of a larger source). When observing with more than two apertures and when an interesting but very low visibility is expected with the largest baseline, 'bootstraping' techniques may be very useful. It consists of using an array configuration where the apertures creating the largest baseline are also involved in shorter baselines. On the shorter baselines, the fringes are more easily detected, the optical path can be efficiently monitored so as to make possible longer and coherent integrations for the low visibility measurement. A similar technique can also be applied with a single baseline but simultaneous observations at different wavelengths.
- complementary observations: finally, as for any instruments, the measurements of the visibility modulus may be astrophysically interesting for various wavelengths, polarizations and at different time.

### 3.2. VISIBILITY PHASE

Information on the visibility modulus only (even for a large number of baselines and high SNR) is not enough to reconstruct the target image. The information on the visibility phase is critical for such reconstruction; such a reconstruction is possible when this information is obtained for many spatial frequencies (to be compared to the target complexity), as demonstrated with radio observations.

For a given spatial frequency (or a given baseline), the visibility phase is relative to the position of this frequency component of the target image: a $2\pi$ shift corresponds to a source offset of $\lambda/B$ compared to the instrument axis. This offset induces the incoming wavefront to be slightly tilted so that the optical path length is slightly longer (by a wavelength) on one arm compared to the other. Note that any other source of OPD will produce the exactly similar measurement as an instrumental or atmospheric visibility phase, in addition to the intrinsic astrophysical signature. It is very difficult, in practice, to control accurately the complete optical path in the interferometer and to manage the random delays due to the turbulent atmosphere so that the critical issue is to find a relevant phase reference (or in other words to control/calibrate the phase of the turbulent and instrumental transfer function).

Various options can be explored (and are actually considered in the VLTI: see Delplancke, and Petrov, this volume) in order to retrieve the astrophysical visibility phase information:

- *external star as a phase reference:* by observing simultaneously and through the same optical path (as far as possible) an external astrophysical source, it may be possible to detect and measure any phase shifts suffered by the target of interest. In other words, the external source is used as a reference. The method requires the presence of such a bright phase reference source in the vicinity of the target (within the piston isoplanetic patch, which is typically in units of 10 arcsec, depending on the wavelength); the exact simultaneity is strictly necessary for the calibration of turbulent effects; additionally, the required field of view implies dual-beam optics (and differential delay lines so as to monitor the optical path lengths for both targets) but the instrumental phase should be very similar and calibrated for both beams. In order to measure absolutely the separation of both sources, an internal metrology of the 2 paths is necessary.
- *differential measurements at various wavelengths:* in case that only one target is observed, the visibility in one wavelength, where the object structure is known a priori, can be used as a reference for other wavelengths. Here again, this method requires the simultaneity; it relaxes interestingly the problem of phase delays to 'only' their chromatic part (which still are to be calibrated accurately). It applies efficiently to objects with a variable position or structure with wavelength (variation of photo-center with wavelength, structure in an emission line...)
- *closure phase:* in the case of simultaneous observations with 3 (or more) apertures, the addition of each visibility phase corresponding to the 3 baselines forming a closed loop of telescopes cancels out the optical path lengths variations affecting any individual aperture (such a variation affects 2 baselines, with opposite signs). The resulting value, the closure phase, is a quantity that is intrinsic to the astrophysical source, and is in particular non zero for non centro-symmetrical sources.

This information on the phase is the basis of *astrometry* at the resolution of the interferometer. In such context, the visibility modulus may or may not be simultaneously used. In a pure astrometric instrument, the specification on the modulus of the transfer function (or instrument contrast) is very much relaxed: it may even vary in some range as long it remains high enough to make possible the fringe detection with a good SNR.

As mentionned in the introduction of this section, the visibility phases, when obtained for a large number of baselines and combined with visibility modulus measurements, make possible to reconstruct a high angular resolution *image* of the target. As for radio interferometry, the reconstruction is directly conditioned by the obtained u-v coverage (see sect 2, first item), and by the measurements SNR; it also takes benefit from any a priori information on the target to regularize the inverse problem (see for instance the presentations of C. Haniff and E. Thiébault, this volume).

## 3.3. The case of nulling interferometry

Nulling interferometry is an observing mode with specific objectives and instrumental requirements. Its interests and instrumental requirements correspond closely to those of coronography and the frontier between them is not well defined. The purpose is not to measure a visibility phase or modulus but to produce an interferometric pattern that is destructive for an on-axis source and constructive for a much fainter interesting target. The discrimination between the extinguished and transmitted source is obtained on the scale of the interferometer resolution $\lambda/B$. This capability is primarily motivated by the search for extrasolar planets but also concerns a number of other high dynamic range observations.

We already mentioned that measuring the signature associated to a high dynamic source requires very accurate visibility measurements (with an accuracy better than the brightness ratio between the faint structure of interest and the total flux). A specific requirement for this mode is to obtain a transfer function which is not only very stable and/or calibrated to that accuracy but also that this function is very high (close to unity). In other words, the coherent part only of the incoming on-axis light will be rejected: a rejection rate of a factor 20 would for instance require a instrumental visibility of $> 0.95$. This requirement is very severe.

The specific interest for this mode is to strongly reduce the photon noise associated with the bright on-axis source, which is a fundamental limit otherwise. Eventually, if other sources of noise are well controlled and reduced (in particular in space without the problems related to atmosphere), the photon noise will be the dominant limitation and such a mode will improve the high dynamic detection capabilities (see Fridlund, this volume).

## References

Delplancke, F.: this volume, 'PRIMA science'.
Fridlund, M.: this volume, 'GENIE – the Darwin demonstrator'.
Glindemann, A.: this volume, 'The VLTI interferometer – a status report'.
Haniff, C.A.: this volume, 'Imaging stars and their environments'.
Percheron, I.: this volume, 'The VLTI calibrators program'.
Pétrov, R.: this volume, 'The NIR VLTI instrument AMBER for users'.
Przygodda, F.: this volume, 'MIDI – interferometry in the 10 micron band'.
Schoeller, M.: this volume, 'Interferometry practice primer'.
Thiébault, E.: this volume, 'Imaging with AMBER-VLTI: the case of microjets'.

# OPTICAL INTERFEROMETRY – A PRACTICE PRIMER

MARKUS SCHÖLLER
*European Southern Observatory, Casilla 19001, Santiago 19, Chile*

**Abstract.** The existing and planned interferometers differ largely in the size of telescope area, baseline length and wavelength at which they operate. In this article we focus on the common principle between all of them and discuss the three main ingredients of an interferometer: the light collectors, the delay lines and the beam combiner. We take a look into the different approaches taken, e.g. beam travel in vacuum or air, the use of spatial filters, interferometer layouts etc. Furthermore, we describe the subsystems which are helping in the formation of fringes, namely adaptive optics, fringe trackers and phase reference systems.

**Keywords:** optical interferometry

## 1. Introduction

There are ten optical interferometers in operations today: CHARA (ten Brummelaar, 2002), COAST (Haniff et al., 2002), GI2T (Mourard et al., 2002), IOTA (Traub, 2002), Keck (Boden, 2002), MIRA (Nishikawa, 2002), NPOI (Mozurkewich), PTI (Lane, Colavita and Boden, 2002), SUSI (Tango, 2002) and VLTI (Glindemann et al., 2002). Several more went out of service. Although they are very different in their appearance (e.g. they have telescope sizes between a few cm and more than 10 m), the wavelength regime they operate in (somewhere between 400 nm and 13 $\mu$m), the modes of operation (pure visibility measurements, closure phases, phase referencing, wide and narrow angle astrometry, nulling), they all utilize the same underlying physical principles and use more or less the same kind of hardware. In this article we want to describe the basic technology that is needed to make an optical interferometer work.

When we talk about optical interferometers in this paper, we think of interferometers which work at visible, near- or mid-infrared wavelengths and make a direct detection of fringes. We would like to separate these types of interferometers from e.g. heterodyne interferometers like the ISI (Danchi et al., 2002), which mix the signal of each telescope with a local oscillator and form the fringes electronically, or the intensity interferometer (Hanbury Brown, Davis and Allen, 1967), which was determining correlations of photon arrival statistics instead of making fringes.

For a good overview about the status of optical interferometry in general, the reader is referred to (Quirrenbach, 2001) and for the historic context to (Shao and Colavita, 1992).

*Astrophysics and Space Science* **286:** 17–31, 2003.

In this article we will discuss first the key ingredients that are needed to make an interferometer work. This is followed in section 3 by the devices which make the life of an interferometrist easier and the limiting magnitude of the interferometer fainter. In the next section we describe the insight that has to be put into the effort of designing the interferometer, before we discuss in section 5 briefly dual feed systems and nulling. We conclude in section 6 with what challenges optical interferometry will face in the future.

## 2. Key Ingredients

An interferometer can essentially consist out of the following components: two telescopes and a beam combiner. Furthermore, optical trains are needed to tie the components together, i.e. transport the light from the two telescopes towards the beam combiner. If the telescopes are on the same mount – as for the Large Binocular Telescope (Herbst, 2002) – or the telescopes move to keep the optical path difference to 0 – like the intensity interferometer did (Hanbury Brown, Davis and Allen, 1967) or as proposed for the 20/20 (Angel, 2002) – then this is all it needs. Most interferometers have telescopes which are fixed to the ground, and thus need delay lines to compensate for the optical path difference (OPD) between the arms of the interferometer, while the source moves across the sky. Instead of using a delay line, in the I2T (Labeyrie, 1975) and the first version of the GI2T the beam combination table was moved.

### 2.1. Telescopes

The main task of the interferometric telescopes as in any single dish operation is to collect photons. A rigid design is needed to avoid any introduction of optical path difference due to vibrations, as for all components of the interferometer. As a golden rule, all telescopes of an interferometer should be of the same type, with the same orientation of mirrors and the same coatings on all conjugating optical surfaces to control differential polarization.

Smaller interferometers employ usually siderostats, since they are cost effective and very sturdy. Their drawbacks of complicated field rotation and changing elliptic aperture shapes are of minor concerns for interferometers. Interferometers with larger apertures usually use alt-az mounted regular telescopes equipped with a Coudé train to inject the light beams into the interferometer. Examples of interferometric telescopes can be found in Figure 1.

### 2.2. Delay lines

The delay lines have to compensate the optical path difference (OPD) for all baselines in an interferometer. To do this, they have to cancel the OPD at any given time, i.e. follow the sidereal delay while the object moves across the sky.

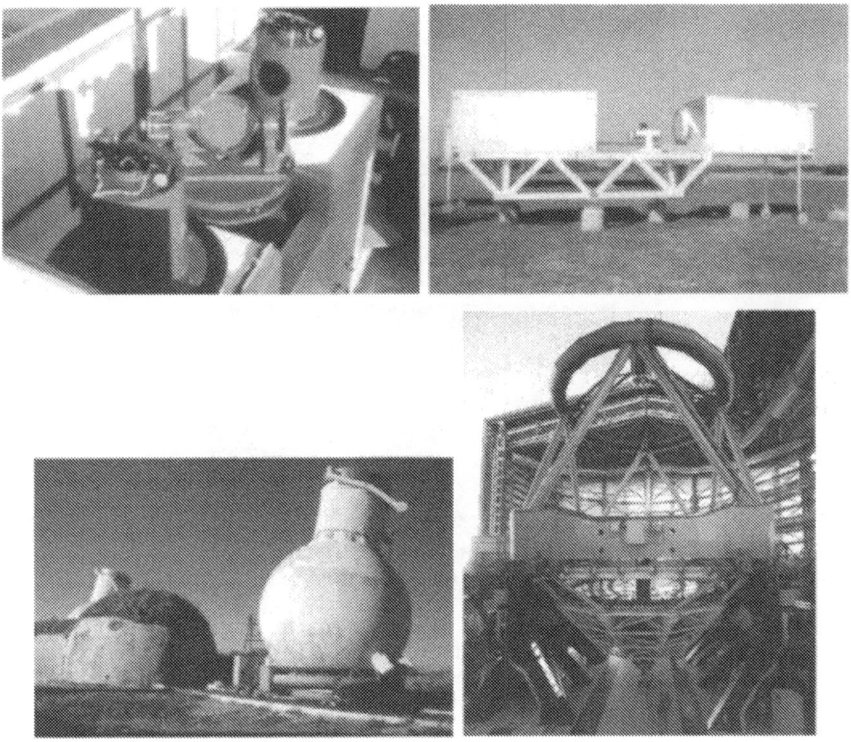

*Figure 1.* Interferometric telescopes: on the top one can see the siderostats used by SUSI (left) and COAST (right), on the bottom left the special interferometric design of the GI2T telescopes and on the bottom right one of the VLTI's 8 m Unit Telescopes.

An interferometer with N telescopes or better with N arms needs delay lines for the first N-1 arms. The last arm needs an optical device which has the same mirrors under the same angles as the delay lines. Many interferometers put another delay line onto this arm for convenience and more flexibility.

Apart from the actual number of delay lines used, there are also different ways in which the total OPD can be canceled. The VLTI implements only one moving delay line per arm with a sufficiently long compensation range. The Keck uses so-called fast delay lines which compensate the sidereal motion, while at the same time compensating most of the OPD with long delay lines in which mirrors can be moved and clamped to give various compensation ranges. The GI2T has one delay line with a short range of compensation length and moves its telescopes into a position so that this range is enough to compensate the overall OPD.

The optics of delay lines are typically either a roof mirror or a cat's eye. The cat's eye has the advantage that light which enters the delay line under an angle $\alpha$ leaves it under the same angle, while it leaves a roof mirror under the angle $-\alpha$.

*Figure 2.* The delay lines of three different interferometers: COAST (top left), PTI (top right) and VLTI (bottom).

Thus, if a cat's eye is moving slightly misaligned, the beam is still reflected back into the direction where it came from.

Delay lines can be moved in various ways, with linear motors, voice coils or piezos. Typically they consist of a composition of these components, e.g. a linear motor for the coarse and a piezo for the fine positioning. This mix of devices makes it possible to achieve the positioning requirements of the delay lines with an accuracy of tenths of nanometers over a stroke length of tenths of meters.

Figure 2 shows various implementations of delay lines.

### 2.3. Beam combination

A beam combiner brings the beams from the telescopes close enough together so they can interfere. If the beam combination scheme requires the coding of the fringes in time, a device to modulate the OPD between the beams is required. Further, some type of detector to record data is needed.

There are two different ways to combine beams in a beam combiner: in a multiaxial beam combiner the beams are placed adjacent to each other and form a fringe pattern in space. In a coaxial beam combiner, the beams are added on each other, for example with a beam splitter. The fringes are produced in time by modulating the OPD between the two beams. Figure 3 illustrates the two combin-

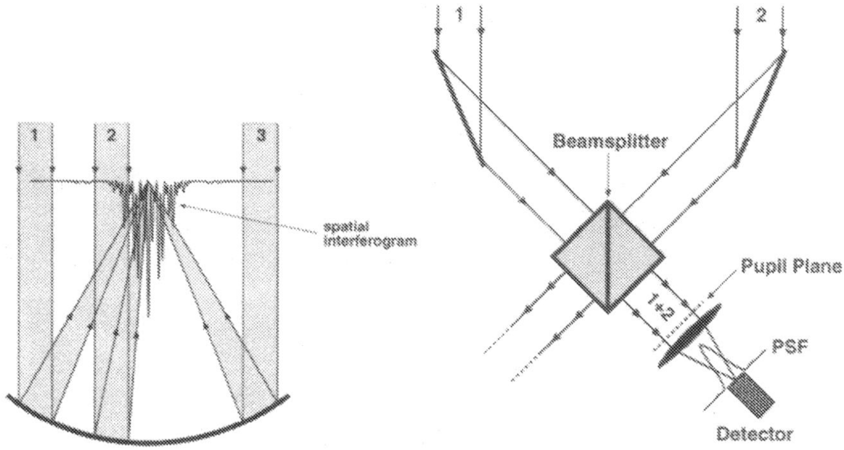


*Figure 3.* Fringe combination schemes: multiaxial (left) and coaxial (right). An OPD modulating device has to be used in one of the arms of the coaxial beam combiner to form fringes.

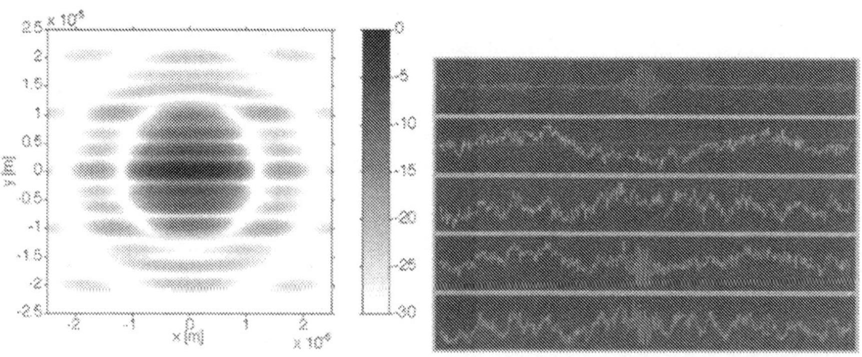

*Figure 4.* The fringes resulting from multiaxial and coaxial beam combination: on the left one can see a simulation of fringes similar to the ones which will be produced by the AMBER instrument on VLTI. The right panel shows fringes as observed by the VINCI instrument on VLTI. The lower two scans show raw fringes in the two interferometric channels, the two scans above give the photometric signals, and the upper row shows the compensated and filtered fringe.

ation schemes and in Figure 4 one can see how the data resulting from the beam combination looks like.

The actual implementation of a beam combiner is done usually in bulk optics, which take space and are susceptible to misalignments. Another type of beam combiners is made out of single mode fibers, which are not only combining the light within the fibers, but also serve as spatial filters (Coude du Foresto, Maze and Ridgway, 1993). Lately, progress in the production of integrated optics has allowed

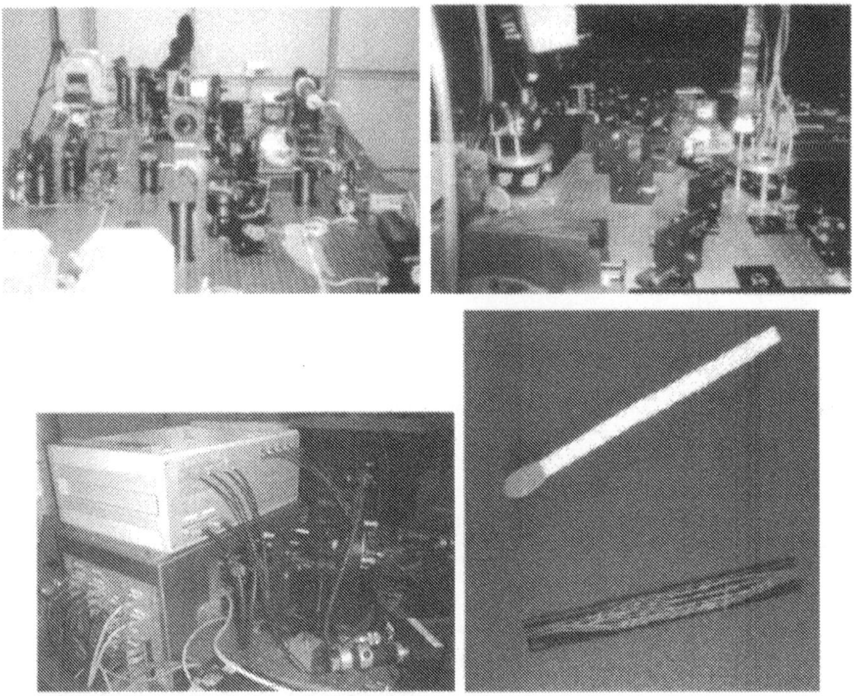

*Figure 5.* Different types of beam combiners: in the upper row one can see the bulk optics assemblies of SUSI (left) and COAST (right). In the lower row the fiber optics beam combiner box of VINCI on VLTI (named MONA, left) and the integrated optics beam combiner IONIC (right) are shown.

to build beam combiners of the size of a coin (Malbet et al., 1999), (Berger et al., 2001). Different types of beam combiners can be found in Figure 5.

When using more than two telescopes, one has to make a choice how to combine the signals on the different baselines. The easiest way is to split the light of the N beams into N-1 parts and interfere all beams pairwise. Another approach is to put all beams onto one detector and encode the different baselines on different spatial or temporal frequencies. By doing this one has to take special care to avoid cross talk between the different baselines, which could bias the signals significantly. A mix of the two schemes is used at the NPOI, where six telescopes are combined in three beam combiners, which each combine four telescopes. The complexity rises for the pairwise scheme with the number of baselines, for the all-on-one scheme only with the number of telescopes. The pairwise combination also has drawbacks when it comes to closure phases, since these are affected by drifts between the beam combiners.

The signal-to-noise ratio (SNR) between pairwise and all-on-one combination is similar, since in one case the light has to be distributed between the beam combiners, in the other case it has to be split between the different encoder frequencies.

It is not clear if it is better to combine e.g. eight telescopes all at once at a lower SNR, or to combine four telescopes twice and getting all baselines and closure phases sequentially. Beside the SNR one also has to take calibration issues into account. The debate about how to combine many telescopes is far from over.

## 3. Adaptive Optics, Spatial Filtering, Fringe Tracker, and Variable Curvature Mirror

Since the light of two (or more) telescopes has to be combined, it is necessary to bring the light of them onto the same spot in an image plane. Tip-tilt sensors for each telescope are employed to ensure the image stability needed. If the telescopes are larger than the diameter of the atmospheric turbulence cells (the Fried Parameter $r_0$), then the telescopes have to be equipped with adaptive optics to ensure the maximum flux being available for the beam combination. Finally, a fringe tracker stabilizes the fringes, which are otherwise moved backwards and forwards in OPD by the atmospheric piston.

### 3.1. TIP-TILT AND ADAPTIVE OPTICS

While in imaging systems tip-tilt and adaptive optics (AO) are used to make sharper images, in optical interferometry they are used similar as in spectrographs. Instead of putting as much light as possible onto a slit, here the goal is to put as much light as possible onto a spatial filter.

The main tradeoff for these systems is between costs (we are talking about as many AO systems as telescopes) and performance.

One also has to consider the optimum position for the wavefront sensor. While a sensor at the telescope is not able to correct errors which result from the optical path between the telescopes and the beam combiner, a sensor in the laboratory leads to sensitivity loss and very likely to alignment problems. A possible solution is a higher order AO system at the telescope and a slower tip-tilt corrector in the laboratory.

### 3.2. SPATIAL FILTERING

Wavefront errors lead to a degradation of the measured visibility. The interfering wavefronts have to have exactly the same shape to give the maximum degree of coherence. A spatial filter ideally removes inhomogeneities of the individual wavefronts (and thus makes them similar), at the cost of reducing the amount of light which is available for beam combination. One can also say that a spatial filter acts as a low pass filter for the wavefront.

There are two different implementations of a spatial filter: single mode fibers and pinholes. While single mode fibers are removing the wavefront errors nearly perfectly, pinholes can be effective in the regime of a small ratio of telescope diameter over $r_0$, and are usually used at wavelengths where no single mode fibers exist.

### 3.3. Fringe tracking

A fringe tracker stabilizes the fringes within a fraction of a wavelength so that the scientific instruments can integrate much longer than the coherence time of the atmosphere would allow.

Although this looks very good at a first glimpse, it is not so obvious that it actually helps a lot. If the fringe tracker is able to track the fringes on the source, this means that there is enough signal in a coherence time interval to measure the phase of the fringes with high accuracy. The science instrument should then also be able to detect and measure the fringe without any problems. But there are possible scenarios where a fringe tracker helps a lot:

- Fringe tracking is done in a wide band while the science instrument uses high spectral resolution.
- The source can be tracked at the fringe tracker wavelength while not at the scientifically interesting wavelength (wavelength bootstrapping).
- Fringes are tracked on two shorter baselines while the science instrument is integrating on a long baseline (baseline bootstrapping).
- Fringes are tracked on an off-axis source in dual feed mode.

### 3.4. Variable curvature mirror (VCM)

The distance from the telescope entrance pupil to the beam combination instrument is variable in an interferometer due to the moving delay line. Without an active element in the optical train no pupil imaging is possible. Pupil imaging is necessary if one wants to put the telescope pupil on a cold stop, e.g. when observing in the thermal infrared, or in the case of homothetic mapping where the entrance pupil of the interferometer has to be mapped onto the exit pupil of the interferometer in all three coordinates to achieve a larger field of view. The variable curvature mirror sits e.g. in an image plane within the delay line and changes its radius of curvature so the entrance pupil is relayed to a given position (Ferrari et al., 2002).

## 4. Optical Path, Array Layout, Alignment and Interferometer Control

### 4.1. Beam transport

An issue which is always taking much attention is how the beams are actually transported from the telescopes to the beam combiner. The light can be transported in air, in vacuum or in fibers.

A travel in air introduces additional wavefront errors due to internal turbulence. To keep these errors to a minimum, a stable environment is required. The two main concepts are the wine cellar approach, where the environment is built underground and left undisturbed, and the building-in-building, where the inner building is also kept undisturbed, while the outer building is kept under thermal control. Another problem which is apparent when transporting the beams through air is the differential refractive index for different wavelengths. Figure 6 shows air exposed tunnels and vacuum pipes.

The use of vacuum solves the problem of internal wavefront errors and is mandatory if the beams are traveling over ground in pipes which are heated during day time, resulting in huge turbulence. They can also reduce somewhat the problem of differential refraction. The disadvantage of a vacuum system is, besides the cost of the investment itself, that entrance windows have to be used and the optics are not easily accessible anymore.

Another way to transport the beams, which was proposed by (Froehly, 1982) is to inject the light at the telescope into a fiber and then transport it towards the beam combination laboratory. Although issues like light injection and dispersion within the fiber have to be taken into account when implementing a fiber transport, other infrastructure related problems like light ducts, underground tunnels, or a large number of mirrors, which have to be set under the exact same angles, disappear. A beam transport within fibers is especially attractive for the implementation of an interferometer on an existing telescope array. The OHANA array (Perrin et al., 2002) will transport beams within fibers.

### 4.2. Array Layout

The layout of an interferometric array has to be optimized to give as good a *uv*-coverage as possible. The shape of the landscape at the observatory has also to be taken into account. Figure 7 shows some examples of possible interferometer layouts.

### 4.3. Aligning

Since an optical interferometer has a huge number of optical surfaces, which might be coupled together in various ways, some thoughts have to be put into the alignment strategy. There is a tendency to have a system with light sources and degrees of freedom everywhere to be able to align the optical system. The alternative is an undisturbed and well engineered system which does not require any frequent alignment. A well designed interferometer has an image alignment and a pupil alignment device and everything else is static or adaptive.

### 4.4. Controlling an Interferometer

Since interferometers can be highly complex systems, one has to invest some effort into the software and hardware which tie the pieces together.

*Figure 6.* Beam transport in interferometers can run through open air or within evacuated pipes. The left column shows the air tunnels of PTI, SUSI and VLTI (the first two building-in-building, the latter a wine cellar). The right column shows the pipes of PTI and the pipes of CHARA outside and inside the interferometric building (both from top to bottom).

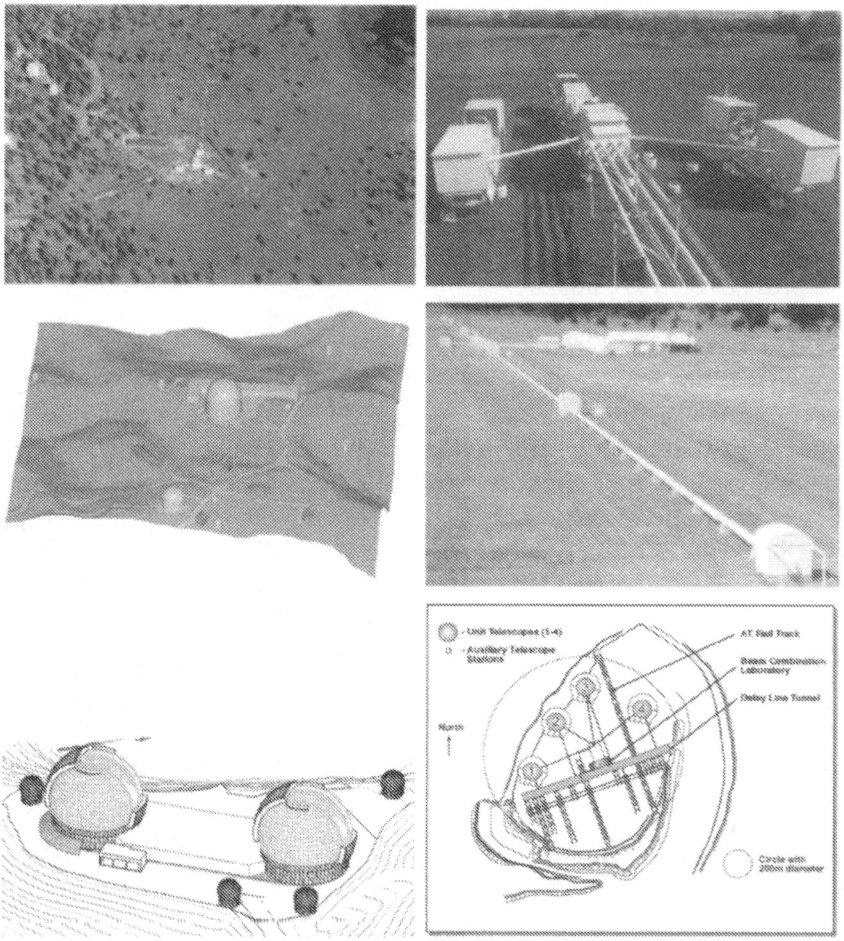

*Figure 7.* Different strategies for layouts of optical interferometers: the arrays of NPOI, COAST and CHARA (top left, top right and middle left) resemble the shapes of a Y, while SUSI (middle right) is linear, KECK (bottom left) has two perpendicular baselines with their outriggers and the VLTI (bottom right) a field of stations which can populate various shapes.

As an example, we present here the sequence which is foreseen for the 10 $\mu m$ instrument MIDI on the VLTI 8 m Unit Telescopes, chopping while a fringe tracker is in use. The following cycle has to be repeated with about 2 Hz:

At the start, MIDI and the fringe tracker are on source and MIDI is integrating while the fringe tracker is tracking. Next, both the fringe tracker and the adaptive optics system open their respective loops to not get confused when they do not get any flux from the observed object. The telescope now chops off source and MIDI integrates a sky fringe to determine the rapidly changing sky background. Once this is done, the telescope chops back on source. The adaptive optics can now again close the loop and once this is achieved, the fringe tracker can follow. After all this is accomplished, MIDI will again integrate on source.

ESO's VLTI will be the first general user interferometer. It will be operated in the framework of the VLT data flow system (Ballester et al., 2001).

## 5. Dual Feed and Nulling

### 5.1. Dual Feed Systems

The field of view of an interferometer is usually very small, much smaller than the field of view of the individual telescopes. A dual feed system allows two objects within the telescope field of view to be picked up and being sent independently through the interferometer where they are both combined individually.

The dual feed system makes three kinds of observations possible: observations of faint objects while fringe tracking on a source within the isoplanatic patch, imaging due to direct access to object phase information via phase referencing and narrow angle astrometry.

A dual feed system requires the following components:

**The star separator** allows the separation of two objects within the field of view at the telescope focus and sends two beams into the interferometer.

**A differential delay line** compensates the OPD between the two objects.

**The metrology** measures the internal optical path of the interferometer.

**A fringe tracker** stabilizes the fringes on one object.

Note that the optics of the interferometer have to be designed in such a way that they can cope with two beams per telescope.

### 5.2. Nulling

Nulling is not an interferometric technique but an imaging technique which can also be used on interferometers. Nulling introduces a wide band achromatic phase shift of $\pi$ in one arm which results in destructive interference. In an interferometer the result is e.g. a wide fringe maximum of a planet in the sharp minimum of its parent star instead of a sharp planet maximum in a shallow star minimum.

## 6. Future Challenges

Optical long baseline interferometry has made huge progress in the last thirty years, delivering science on several facilities, with increasing aperture diameters and baseline lengths. Yet, there are several technologies which still have to be mastered.

Beam transport in fibers has not been successfully implemented in an optical interferometer up to now. If these fibers could coherently amplify the number of photons, subsequent losses could be overcome.

Water vapor very likely has the largest influence on accurate measurements at 10 $\mu$m (Meisner and le Poole, 2002). The studies on humidity dependence of the refractive index of air and the consequences on the longitudinal dispersion are just starting.

Alternative concepts for delay lines have to be developed when larger OPDs have to be compensated on baselines much longer than the ones used today.

With a larger number of telescopes and instruments on the same interferometer, intelligent ways have to be found to switch beams. Active integrated optics may be of help here.

The NPOI uses the most complex beam combiner to combine six telescopes. The combination of a larger number of telescopes in bulk optics will require a lot of space. Integrated optics will have to deliver solutions which use less space and allow simple layouts.

As with other fields in astronomy, interferometry will also gain in new detector technology. For most applications with no inherent field of view, only small area detectors are needed. Detectors with inherent spectral resolution as in superconducting tunnel junctions would also be of help for the photon starved interferometers.

The current implementation of interferometers leads to a limited field of view, which has the size of an Airy disk as defined by the largest telescope used in the array. Accordingly, the field of view gets smaller on larger telescopes. There are three possible solutions for this problem. Optical interferometers have to exploit the mosaicing technique as used in today's radio interferometry. Alternatively one can implement homothetic mapping (Beckers, 1990) or a hypertelescope with a densified pupil (Pedretti et al., 2000). The first concept requires very complex optomechanics. The latter method results in images that have different angular scales superimposed.

The nulling technique has to be developed beyond the laboratory stage, where the deepest nulls have been achieved with lasers and at optical wavelengths.

Phase referencing techniques require highly sophisticated technology. It has to be shown that all the subcomponents can be controlled to routinely deliver science.

Although fringe sensing and adaptive optics have a similar underlying principle, there is no equivalent to an artificial guide star. If someone could produce a method which overcomes this obstacle, the door to faint science with interferometers would be wide open.

Finally, the new generation of interferometers has to deliver the science that has been advertised over the last decade.

## Acknowledgements

I would like to thank the whole interferometric community for fruitful discussions over the years, especially the people in the VLTI group at ESO. I would also like to thank all optical interferometers for making the images used in this paper public. The anonymous referee gave several valuable comments which helped to improve this paper.

## References

Angel, J.R.P.: 2002, Why optical interferometers should have moving elements, ideally two big ones. *Proc. of the SPIE* **4838**(#17), 2002.

Ballester, P., A.M. Chavan, B. Cotton et al.: 2001, Data flow system for the Very Large Telescope Interferometer. *Proc. of the SPIE* **4477**, 225.

Beckers, J.M.: 1990, The VLT interferometer. III – Factors affecting wide field-of-view operation. *Proc. of the SPIE* **1236**, 379.

Berger, J.P., P. Haguenauer, P. Kern et al.: 2001, Integrated optics for astronomical interferometry. IV. First measurements of stars. *A&A* **376**, L31.

Boden, A.F.: 2002, The Keck Interferometer. *these proceedings*.

Coude du Foresto, V., G. Maze and S. Ridgway: 1993, Stellar interferometry with infrared single-mode fibers. *ASP Conf. Ser.* **37**, 285.

Danchi, W.C., C.H. Townes, W. Fitelson et al.: 2002, Status and progress on the upgraded infrared spatial interferometer. *Proc. of the SPIE* **4838**(#05).

Ferrari, M., G.R. Lemaitre, S.P. Mazzanti et al.: 2002, Variable curvature mirrors: implementation in the VLTI delay lines. *Proc. of the SPIE* **4838**(#39).

Froehly, C.: 1982, Coherence and interferometry through optical fibers. *Proc. ESO Conf., Science Importance of High Angular Resolution at Infrared and Optical Wavelengths* **285**.

Glindemann, A., J. Argomedo, R. Amestica et al.: 2002, The VLTI Interferometer – a status report. *these proceedings*.

Hanbury Brown, R., J. Davis and L.R. Allen: 1967, The stellar interferometer at Narrabri Observatory. I. A description of the instrument and the observational procedure. *Monthly Notices of the Royal Astronomical Society* **137**, 375.

Haniff, C.A., J.E. Baldwin, A.G. Basden et al.: 2002, Progress at COAST 2000-2002. *Proc. of the SPIE* **4838**(#03).

Herbst, T.: 2002, Interferometry with the Large Binocular Telescope. *these proceedings*.

Labeyrie, A.: 1975 Interference fringes obtained on VEGA with two optical telescopes. *ApJ* **196**, L71.

Lane, B.F., M. Colavita and A.F. Boden: 2002, PTI status report. *Proc. of the SPIE* **4838**(#08).

Malbet, F., P. Kern, I. Schanen-Duport et al.: 1999, Integrated optics for astronomical interferometry. I. Concept and astronomical applications. *A&AS* **138**, 135.

Meisner, J.A. and R.S. le Poole: 2002, Dispersion affecting the VLTI and 10 micron interferometry using MIDI. *Proc. of the SPIE* **4838**(#126).

Mourard, D., D. Bonneau, P. Stee et al.: 2002, Status report on the GI2T interferometer. *Proc. of the SPIE* **4838**(#02).

Mozurkewich, D., D.J. Hutter and J.A. Benson: 2002, NPOI progress report. *Proc. of the SPIE* **4838**(#07).

Nishikawa, J.: 2002, Status report on Mitaka optical and infrared array (MIRA-1.2). *Proc. of the SPIE* **4838**(#12).

Pedretti, E., A. Labeyrie, L. Arnold et al.: 2000, First images on the sky from a hyper telescope. *A&AS* **147**, 285.
Perrin, G., O. Lai, J. Woillez et al.: 2002, OHANA: the Optical Hawaiian Array for nanoradian astronomy on top of Mauna Kea. *Proc. of the SPIE* **4838**(#15).
Quirrenbach, A.: 2001, Optical Interferometry. *Annual Review of Astronomy and Astrophysics* **39**, 353–401.
Shao, M. and M.M. Colavita: 1992, Long-baseline optical and infrared stellar interferometry. *Annual Review of Astronomy and Astrophysics* **30**, 457–498.
Tango, W.J.: 2002, Status of the SUSI program. *Proc. of the SPIE* **4838**(#04).
ten Brummelaar, T.A.: 2002, Update on the CHARA array. *Proc. of the SPIE* **4838**(#09).
Traub, W.A.: 2002, Recent progress at IOTA with 3 telescopes. *Proc. of the SPIE* **4838**(#06).

# THE VLTI AND OTHER INTERFEROMETERS

# THE VLTI – A STATUS REPORT

A. GLINDEMANN, J. ALGOMEDO, R. AMESTICA, P. BALLESTER, B. BAUVIR, E. BUGUEÑO, S. CORREIA, F. DELGADO, F. DELPLANCKE, F. DERIE, PH. DUHOUX, E. DI FOLCO, A. GENNAI, B. GILLI, P. GIORDANO, PH. GITTON, S. GUISARD, N. HOUSEN, A. HUXLEY, P. KERVELLA, M. KIEKEBUSCH, B. KOEHLER, S. LÉVÊQUE, A. LONGINOTTI, S. MÉNARDI, S. MOREL, F. PARESCE, T. PHAN DUC, A. RICHICHI, M. SCHÖLLER, M. TARENGHI, A. WALLANDER, M. WITTKOWSKI and R. WILHELM
*European Southern Observatory, 85748 Garching, Germany*

**Abstract.** The Very Large Telescope (VLT) Observatory on Cerro Paranal (2635 m) in Northern Chile is approaching completion. After the four 8-m Unit Telescopes (UT) individually saw first light in the last years, two of them were combined for the first time on October 30, 2001 to form a stellar interferometer, the VLT Interferometer. The combination in pairs of all four UTs was completed in September 2002. In this article, we will describe the subsystems of the VLTI and the planning for the following years.

**Keywords:** interferometry, very large telescopes, interferometric instrumentation, very high angular resolution, adaptive otics, astrometry, phase referenced imaging, VLTI

## 1. Introduction

On October 30, 2001 at about 1am, the two 8-m Unit Telescopes Antu and Melipal of the Paranal Observatory with a baseline of 102m were combined for the first time as a stellar interferometer observing fringes on the star Achernar (see Figure 1), only six months and twelve days after the VLTI produced the first fringes with two siderostats. In the first year of commissioning, more than one hundred different objects were observed to verify the performance and the scientific potential of the VLTI. We found that all specifications were met or exceeded (Schöller et al., 2002 and Koehler et al., 2002).

Recently, on September 15/16 and 16/17, 2002, the combination in pairs of all four Unit Telescopes was completed using a total of five different baselines (see Figure 2). Only the combination MELIPAL – YEPUN could not be provided due to the current configuration of delay lines in the interferometric tunnel.

The first period of a total of 150 hours shared risk science operations with the test instrument VINCI and the siderostats started in October 2002. 40 proposals from the community were received, representing about 10% of all proposals submitted to ESO for the VLT observatory. During 2002, all science data between First Fringes in March 2001 and September 2002 have been released through the archive

*Astrophysics and Space Science* **286:** 35–44, 2003.

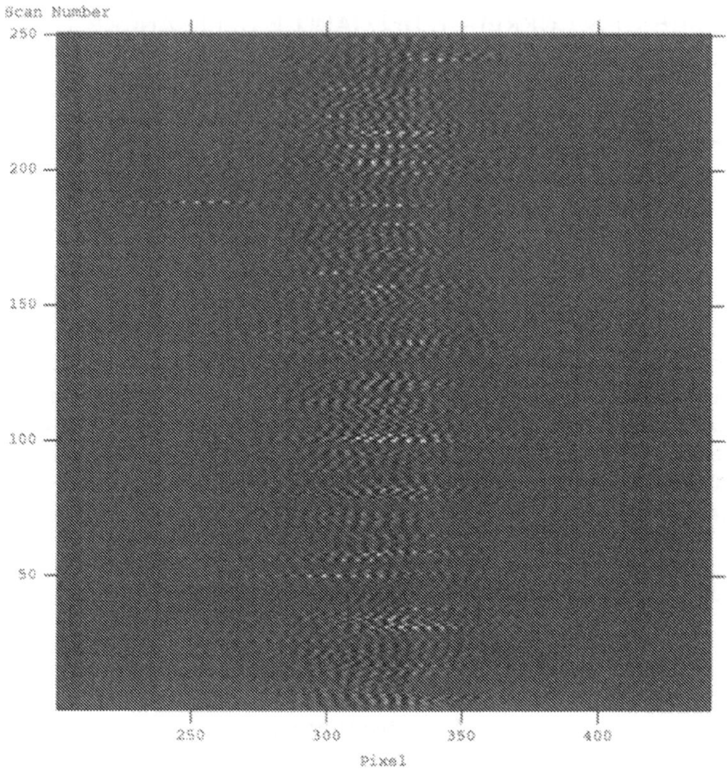


*Figure 1.* First fringes of Achernar with VINCI and the two 8-m Unit Telescopes ANTU and MELIPAL on October 30, 2001. Each horizontal line represents the interferometric fringes in the K-band registered during a single scan with scan numbers increasing from bottom to top. Due to atmospheric turbulence, the fringes are slightly shifted sideways between individual scans. The shift is typically less than one fringe (2.2 $\mu$m), illustrating the stability of the VLTI.

resulting in first scientific results which are described in Kervella et al., 2002; Wittkowski and Hummel, 2002; Richichi and Wittkowski, 2002, and Segransan et al., 2003. A summary of the scientific results of the first year can be found in Paresce et al. (**?**).

The mid infrared instrument MIDI was delivered to Paranal in October 2002 and it is awaiting first fringes every minute. In the course of 2003, the fringe sensor unit FINITO and the near infrared instrument AMBER will arrive, the UTs will be equipped with the adaptive optics system MACAO, and four 1.8-m Auxiliary Telescopes (AT) will be integrated. Three more Delay Lines for a total of six will complete the first phase of the VLTI.

In 2005, the dual feed facility PRIMA will extend the capabilities of the VLTI to faint objects (K = 16–19) and it will allow for high precision astrometry. Last but

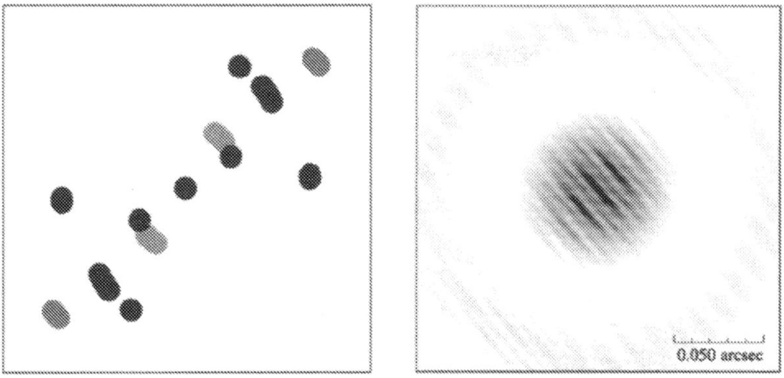


*Figure 2.* The result of observing Achernar in the K-band with all four Unit Telescopes on September 15/16 and 16/17, 2002. The uv coverage is shown on the left, and the point spread function (PSF) on the right. The width of the central fringe, i.e. the angular resolution limit, is about 3 by 15 milliarcsec due to the distribution of baselines used during these observations. On the largest scale, the image is enveloped by the Airy disk of a single 8-m Unit Telescope. Its first minimum at 57 milli arcsec off the center can be clearly seen.

not least, a joint ESO/ESA project for GENIE, the DARWIN ground demonstrator planned for 2006, will open the door for planet hunts with the VLTI.

The interferometric array of the VLT observatory is displayed in Figure 3. It is unique in offering the possibility to combine four 8-m UTs with a maximum baseline of 130m, and to combine a maximum of eight 1.8-m ATs if the Delay Line tunnel is equipped with eight Delay Lines. The ATs can be moved to 30 different stations with a maximum baseline of 200m providing an excellent uv-coverage.

## 2. VLTI Instrumentation

MIDI was designed and built by a European consortium led by the Max-Planck-Institute for Astronomy in Heidelberg. It will operate in the N-band (8–12 $\mu$m). The details of the instrument are described in Leinert et al. (2002). MIDI was delivered to Paranal in October 2002; first light with the Unit Telescopes is scheduled for December 2002. Regular science operations are planned to start in October 2003.

The near-infrared science instrument, AMBER, will operate between 1 and 2.5 $\mu$m, at first with two telescopes, with a spectral resolution up to 10000. The European consortium in charge of designing and manufacturing this instrument is led by the Universities of Nice and Grenoble (Petrov et al., 2002 and Malbet et al., 2002). AMBER was designed for three beams to enable imaging through phase closure techniques. It is planned to start commissioning AMBER with the Siderostats in the third quarter of 2003.

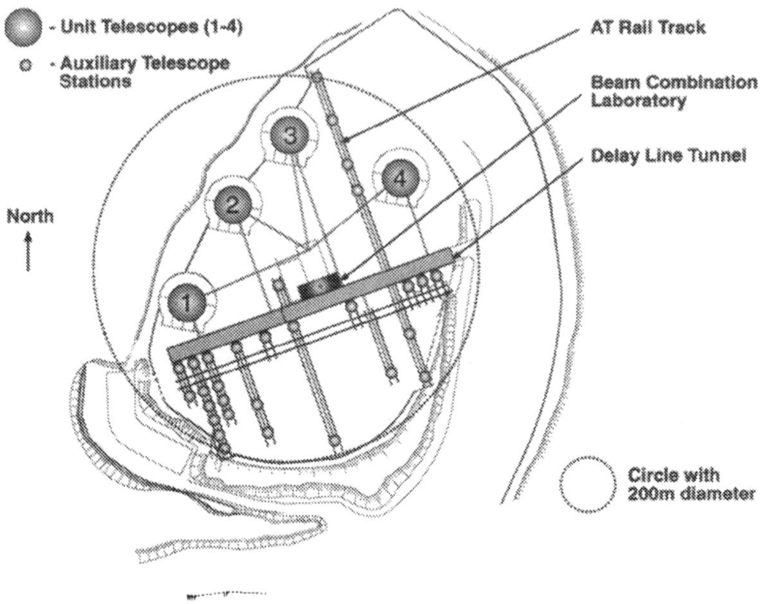


*Figure 3.* The layout of the VLTI. The four 8-m Unit Telescopes (UT) and the 30 stations for the 1.8-m Auxiliary Telescopes (AT) are displayed. The AT stations are connected by rail tracks on which the ATs can be relocated. Also shown are the Delay Line tunnel and the beam combination laboratory. The Delay Line tunnel has room for eight Delay Lines allowing the operation of eight ATs and a total of 28 baselines for each exposure. The longest baseline with two ATs is 200m (indicated by the circle with 200 m diameter). The longest baseline with two UTs is 130 m.

The VLTI fringe sensor unit FINITO is manufactured at the Observatory of Torino (OATo) (Gai et al., 2001). FINITO operates in the H-band using fibers as spatial filters. The closed loop system of the fringe tracker consists of FINITO as fringe sensor unit, and of a piezoelectric element in the Cat's-eye of the Delay Line as actuator. FINITO can manage up to three beams, thus providing fringe tracking for AMBER in closure phase mode. The delivery to Paranal is planned for the first quarter of 2003.

The adaptive optics system MACAO will have a 60-actuator bimorph mirror and a curvature wavefront sensor in the visible. MACAO is an in-house development (Arsenault et al., 2002). MACAO is essential for all near-infrared instrumentation including FINITO when observing with the Unit Telescopes. This means that also a mid-infrared instrument like MIDI needs adaptive optics in order to improve the limiting magnitude by using FINITO.

The first two 1.8-m Auxiliary Telescopes built by AMOS in Liège will be ready for the VLTI at the end of 2003, the third and fourth in 2004 (Flebus et al., 2002). The telescopes are relocatable on 30 stations providing baselines between 8 and 200m. Using three telescopes with AMBER and, thus, three baselines at the same time will allow the application of closure phase techniques eliminating the influence of atmospheric turbulence on fringe position. Each AT will be equipped with a tip-tilt system correcting for the fast image motion induced by atmospheric turbulence. One should note that the ATs are available exclusively for the VLTI, forming an observatory that is operated independent of the UTs.

## 3. PRIMA and GENIE – The Next Phase of the VLTI

PRIMA and GENIE will complete the suite of first generation instruments around 2005. GENIE is a joint ESO/ESA project providing the ground demonstrator for DARWIN as a science instrument for the VLTI. The concept of GENIE is currently under discussion (Gondoin et al., 2002). The goal is to use GENIE for planet detection with the VLTI.

The phase referenced imaging and micro-arcsec astrometry (PRIMA) facility is the third VLTI instrument. It is a dual feed system adding a faint object imaging and an astrometry mode to the VLTI (Quirrenbach et al., 1998 and Glindemann et al., 1999). PRIMA is the key to access: 1) higher sensitivity, with a limiting magnitude of K = 19, 2) imaging of faint objects with high angular resolution (<10 milli arcsec), and 3) high precision astrometry (<10 $\mu$arcsec over a 10 arcsec field).

As a detector for PRIMA, either one of the two scientific instruments MIDI or AMBER can take advantage of the fringe stabilisation provided by PRIMA, or a dedicated PRIMA detector is used for high precision astrometry (Paresce et al., 2002). PRIMA enables simultaneous interferometric observations of two objects – each with a maximum size of 2 arcsec – that are separated by up to 1 arcmin, without requiring a large continuous field of view. Then, the sensitivity of the VLTI is improved by using a bright guide star for fringe tracking similar to the guide star in adaptive optics for wavefront sensing – in one of the two ÔfeedsÕ, allowing to increase the exposure time on the science object in the other feed up to 10-30 minutes depending on the position in the sky.

PRIMA can be subdivided in four sub-systems: 1) star separator, an opto-mechanical system in the Coude focus of UTs and ATs to pick two objects within the 2 arcmin field of view and send the light to the Delay Line tunnel, 2) fringe sensor unit to provide the signal for the fringe tracker, 3) laser metrology system to measure $OPD_{int}$, and 4) differential Delay Lines to correct the differential OPD for two objects that are separated up to 2 arcmin with a baseline up to 200 m – then, the maximum differential OPD is 130 mm.

In order to optimise the scientific output, a phased implementation plan was drafted. In the first phase, contracts for manufacturing the star separator systems for the ATs, and for two fringe sensor units will be placed. The second fringe sensor unit provides the high precision astrometry detector. Two main Delay Lines instead of differential Delay Lines will be used. The laser metrology system will be an in-house development with support from the Institute of Microtechnology in Neuchâtel, Switzerland (Lévêque et al., 2002). This phase is dedicated to astrometry with the ATs.

Call for Tenders were issued for the star separators and the fringe sensor units, and the first kick-off meeting for the manufacturing of the fringe sensor unit took place in July 2002. Without a laser metrology system, the fringe pattern can be stabilised over 10–100 sec, and the expected limiting magnitudes on the ATs are about K=13 and N=5.Phase information that is required for imaging and for astrometry becomes available as soon as the laser metrology system is installed. The limiting magnitudes with the ATs are $K \approx 16$ and $N \approx 8$. The Strehl ratio in the reconstructed image can be as good as 30% in the K-band and 80% in the N-band depending on the uv coverage. PRIMA shall be operational in 2005.

In the second phase, upgrades of the laser metrology allow for reaching the final goal of 5nm rms over 30min and, thus, 10$\mu$arcsec astrometric accuracy. Equipping also the UTs with star separators, increases the limiting magnitudes by 3 reaching $K \approx 19$ and $N \approx 11$. The second phase of PRIMA shall be operational by 2007.

## 4. VLTI Operations and Fringe Tracking

The final goal of the VLTI is to produce images with a few milli arcseconds resolution. Figure 2 shows the point spread function when observing Achernar with all four UTs in September 2002. The 'image' of Achernar is very similar to the theoretical point spread function since its diameter of 2 milli arcsec is smaller than the VLTI resolution limit fo 3 milliarcsec.As a first step towards this goal, FINITO will provide fringe tracking in the H-band. Without a dual feed facility the reference star has to be found within the 2 arcsec field of view of the VLTI. Most likely, the science object itself has to be used as a reference star. This gives only a slight advantage for the accuracy of the visibility measurement, but it gives a considerable advantage for AMBER in spectroscopic mode if the light of e.g. the K-band is dispersed over many pixels.

The performance of FINITO is displayed in Figure 4. The limiting magnitude* depends – like for adaptive optics systems – on the required performance. For fringe tracking with the VLTI, the specifications for the residual OPD is 70nm in order to loose less than 2% of contrast in the K-band. Then, the limiting correlated magnitude is H = 11. If the acceptable residual OPD is 150nm (corresponding to a contrast loss of 8%) the limiting magnitude is H = 14. Even with a reference star of H = 16 the residual OPD of 250 nm still reduces the contrast by less than 25%. In the N-band at a wavelength of 10$\mu$m a residual OPD of 250nm means only a 2% loss in contrast. However, if the science object is very red the H-magnitude might become fainter than 16, although the N-magnitude is still manageable by MIDI.

The calculated performance of FINITO relies on MACAO, the adaptive optics system. MACAO is specified to deliver a Strehl ratio of 50% in the K-band for a

* All limiting magnitudes are given for the UTs.

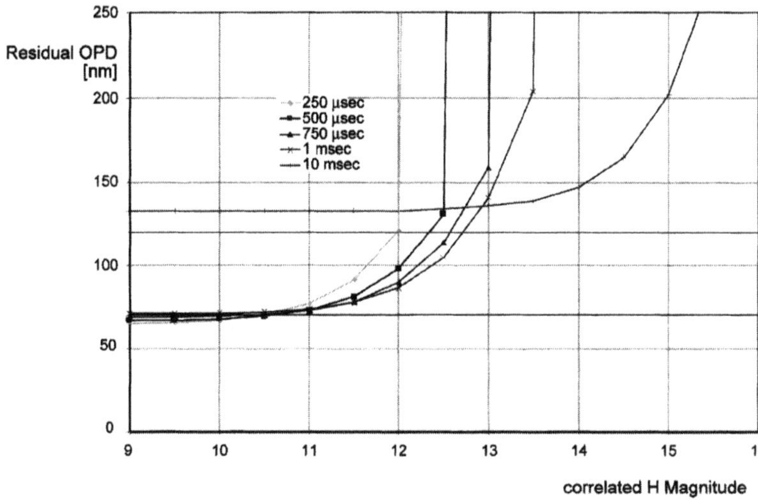


*Figure 4.* Residual OPD for fringe tracking with FINITO as function of correlated H-magnitude with UTs. The curves represent the closed loop performance for different exposure times on the FINITO detector, taking into account the specified performance of the fringe sensor unit FINITO and of the piezoelectric actuator in the Delay Lines, and the delays in signal transfer between these two elements. The performance of the adaptive optics system MACAO is also taken into account. The assumption for the atmospheric coherence time was 50msec in the K-band which is twice as long as the median coherence time at Paranal.

guide star brighter than V = 13. The Strehl ratio is reduced to 25% with a V = 16 guide star. This on-axis performance has to be corrected for anisoplanacy: the K-band Strehl ratio is approximately reduced to 50% of its on-axis value if the guide star is 30 arcsec off-axis. Since the wavefront sensor is installed in the Coudé focus of the UTs, the guide star can be picked in a field of 2 arcmin. All these numbers were calculated for typical Paranal seeing conditions.

The arrival of PRIMA in 2005 will add one more parameter to the performance of the VLTI: the guide star for the fringe tracking can also be picked in the 2arcmin field of view of the Coudé focus. In addition, the PRIMA fringe sensor unit is specified to have a magnitude limit that is at least one magnitude fainter than FINITO.

The main difference compared to the former scenario is the consideration of the isoplanatic angle for fringe tracking introducing a random OPD of typically 400nm if the guide star is 25arcsec off axis. This means a loss in contrast of 50% in the K-band, and, thus, a loss of sensitivity since the correlated magnitude is reduced. The situation in the N-band is more relaxed; a residual OPD of 400nm reduces the contrast by only a few %. Now, one has to evaluate whether a fainter star closer to the science object introduces less residual OPD than a brighter star further away. The combination of a loss in Strehl due to a MACAO guide star far off-axis with a large residual OPD due to a fringe tracking guide star far off-axis can substantially reduce the performance.

## 5. Second Generation Instruments

The main limitations of the first generation instrumentation are the small field of view of one Airy disk (250 resp. 57milli arcsec in the K band for ATs resp. UTs), and the restriction to two (MIDI) resp. three (AMBER) beam combination. While phase referenced imaging – delivering contrast and phase for every baseline individually – as in PRIMA can cope with only two beams, the closure phase technique requires more than three beams to reconstruct unambigously contrast and phase of individual baselines. However, both techniques benefit from an instrument combining more beams (6–8) allowing for more efficient observing and producing instantly an excellent image quality (Glindemann et al., 2001). The VLTI infrastructure can comfort instruments combining up to 8 beams. Thus, there is a need for a second generation instrument with a multi way beam combiner. The question of how to combine the beams – with integrated optics or with bulk optics – is intimately related to the second important topic which is an enlarged field of view.

There are two different schemes to increase the field of view: mosaicing and homothetic mapping. Mosaicing an image means to scan the object in steps of one Airy disk and to put the individual images together to form the 'large' image, like in radio interferometry. Homothetic mapping relies on reimaging the interferometric array into the entrance pupil of the instrument (Beckers, 1990), thus forming on the detector a regular image of the object displaying a superposed fringe pattern with a fringe spacing as small as 2 milliarcsec for a baseline of 200m at 2 $\mu$m. Taking images for many different array configurations one can then superpose the Fourier transforms of these images and reconstruct the complete image with a resolution down to 2 milliarcsec. One should note that the detector pixels should not be larger than 0.5 milliarcsec in order to scan the fringe at four points over one period. The required detector size is then 2000$\times$2000 pixels for a 1 arcsec field of view. Using mosaicing for enlarging the image makes fibers and integrated optics ideally suited for guiding and combining the beams. Optical fibers have proven their usefulness (Coudé du Foresto et al., 1998) and, recently, integrated optics showed some very promising results (Kern et al., 2002). These techniques would help to reduce the size of interferometric instruments. The conclusion is that the most important feature of the second generation instrumentation is its ability to combine many beams, improving the image quality and the observing efficiency. Both closure phase and dual feed imaging would profit from using many beams at the same time. Large fields of view are very interesting but should be implemented through mosaicing rather than homothetic mapping.

## 6. The Overwhelmingly Large Array – La OLA

With extremely large telescopes like OWL lurking above the horizon interferometry only makes sense if it delivers an angular resolution that is at least a factor of 10 higher. This means baselines of a few kilometers. Then, the optical delays that have to be compensated are of the order of kilometers, too. However, rather than building Delay Line tunnels that are kilometers long one should combine moving cat's eyes like in the VLTI with static Delay Lines. Again, the technical progress in integrated optics would be extremely helpful when delivering fast optical switches. One could then continuously observe fringes while the static Delay Lines are being switched on with optical switches. The static Delay Lines and the beam transport could be built with bulk optics or with fibers. First experiments with a fiber interferometer with 500m long fibers were successful (Delage et al., 2001). In order to avoid any intensity loss at all one could use fibers with phase preserving amplification of light as in fiber lasers. If the amplification could be triggered with only a few photons there would be virtually no limit for the length of the fibers.The details of such an overwhelmingly large array such as number and size of the telescopes have to be discussed in more detail than can be done here. The possibilities range from a large number of 4m telescopes to a modest number of 8m telescopes. The boldest and most ambitious approach, however, is clearly to copy the VLTI concept by combining several OWL telescopes surrounded by an array of movable auxiliary telescopes with a diameter of e.g. 8 m.

## References

Arsenault, R. et al.: 2002, *Proc. SPIE* **4839**: *Adaptive Optical System Technologies II*, in press.
Beckers, J.M.: 1990, *Proc. SPIE* **1236**: *Advanced Technology Optical Telescopes IV*, 379–389.
Coudé du Foresto, V., Perrin, G., Ruilier, C., Mennesson, B., Traub, W. and Lacasse, M.: 1998, *Proc. SPIE* **3350**, 856.
Delage, L. and Reynaud, F.: 2001, *OPTICS EXPRESS* **9**, 267–271.
Flebus, C. et al.: 2002, *Proc. SPIE* **4838**: *Interferometry for Optical Astronomy II*, in press.
Gai, M. et al.: 2001, *Scientific Drivers for ESO Future VLT/VLTI Instrumentation*, 328–330.
Glindemann, A. and Lévêque, S.: 1999, *VLT Opening Symposium*, 468–473.
Glindemann, A. et al.: 2001, *Scientific Drivers for ESO Future VLT/VLTI Instrumentation*, 279–288.
Gondoin, Ph. et al.: 2002, *Proc. SPIE* **4838**: (see above), in press.
Kern, P. et al.: 2002, *Proc. SPIE* **4838**: (see above), in press.
Kervella, P., Schöller, M., van Boekel, R., di Folco, E., Correia, S., Glindemann, A., Paresce, F., Richichi, A., Tarenghi, M. and Wittkowski, M.: 2002, *Conf. Proc., Exotic Stars as Challenges to Evolution* (IAU 187).
Koehler, B. et al.: 2002, *Proc. SPIE* **4838**: (see above), in press.
Leinert, Ch. et al.: 2002, *Proc. SPIE* **4838**: (see above), in press.
Lévêque, S. et al.: 2002, *Proc. SPIE* **4838**: (see above), in press.
Malbet, F. et al.: 2002, *Proc. SPIE* **4838**: (see above), in press.
Paresce, F. et al.: 2002, *Proc. SPIE* **4838**: (see above), in press.
Petrov, R. et al.: 2002, *Proc. SPIE* **4838**: (see above), in press.

Quirrenbach A., Coudé du Foresto V., Daigne G., Hofmann K.-H., Hofmann R., Lattanzi M., Osterbart R., Le Poole R., Queloz D. and Vakili F.: 1998, *Proc. SPIE* **3350**, 807–817.
Richichi, A. and Wittkowski, M.: 2003, these proceedings.
Schöller, M. et al.: 2002, *Proc. SPIE* **4838**: (see above), in press.
Segransan, D. et al.: 2003, *A&A Lett.*, in press.
Wittkowski, M. and Hummel, Ch.: 2002, *Proc. SPIE* **4838**: (see above), in press.

# INTERFEROMETRY WITH THE LARGE BINOCULAR TELESCOPE

TOM HERBST
*Max-Planck-Institut für Astronomie, Heidelberg Germany*

**Abstract.** The Large Binocular Telescope (LBT) will be the largest single telescope in the world when it is completed in 2005. The unique structure of the telescope incorporates two, 8.4 meter diameter primary mirrors on a 14.4 meter center-to-center mounting. This configuration provides the equivalent collecting area of a 12 meter telescope, and when combined coherently, the two optical paths offer very interesting possibilities for interferometry. Two initial interferometric instruments are planned for the LBT. A group based at the University of Arizona is constructing LBTI, a pupil-plane, nulling beam combiner operating in the thermal infrared N band. This instrument will search for and measure zodiacal light in candidate stellar systems for the Terrestrial Planet Finder (TPF) and Darwin missions. Expansion ports can accomodate additional instruments. A second group, based in Heidelberg, Arcetri, and Köln, is building LINC-NIRVANA, a near-infrared Fizeau-mode beam combiner. This type of observation preserves phase information and allows true imagery over a wide field of view. Using state-of-the-art detector arrays, coupled with advanced adaptive optics, LINC-NIRVANA will deliver the sensitivity of a 12 m telescope and the spatial resolution of a 23 m telescope, over a field of view up to 2 arcminutes square.

**Keywords:** interferometry, nulling, fizeau, infrared

## 1. The Large Binocular Telescope

The *Large Binocular Telescope* (LBT, Hill and Salinari, 2002), currently nearing completion on Mount Graham in Arizona, is an ambitious and innovative undertaking to construct the world's largest single telescope. The LBT promises observational breakthroughs in our understanding of the origin and evolution of our universe and its contents. It also offers a unique opportunity to search for planets outside the Solar System, and will be an exciting tool to study how stars and planets form. The LBT project is a joint undertaking by the University of Arizona, the German LBTB consortium, the Italian astronomical community (represented by the Arcetri Astrophysical Observatory), the Ohio State University, and the Research Corporation in Tucson.

Figure 1 shows the configuration of the Large Binocular Telescope. The two, 8.4 meter diameter mirrors are joined on a single, common telescope mount. The secondary mirrors are fully adaptive, with 672 actuators each. In order to facilitate diagnostics and efficient operation, the secondaries are concave. This necessitates a Gregorian configuration, with a real, prime focus approximately a meter below the secondary mirrors. The fast, f/1.14 primaries keep the overall telescope length small, and hence reduce the size and cost of the enclosure.

*Astrophysics and Space Science* **286:** 45–53, 2003.

*Figure 1.* Configuration of the Large Binocular Telescope. The human figure near the right-hand primary mirror gives a sense of scale. Image courtesy of European Industrial Engineering, S.r.l.

Having both mirrors on a single mount lets the LBT operate in three modes:

1. As two 8.4 meter telescopes with separate focal planes and instruments. The mount can point the telescopes to fields 1-2 arcminutes apart, and multi-object instruments can either double the effective integration time or double the number of objects observed.
2. As a 12 meter telescope, with light from both mirrors combined incoherently in a single focal plane. In practice, this mode is not very useful, since in the background limit, identical instruments mounted on each individual telescope can achieve the same result with a simpler, more robust design.
3. As an interferometer with light from the two mirrors combined coherently.

LBT supports a total of ten focal stations, including two prime, two 'direct' Gregorian (below the primary mirrors) and a total of six 'bent' Gregorian foci (on the central instrument platform). The two interferometric instruments, LBTI (see section 2) and LINC-NIRVANA (see section 3), will occupy two pairs of these six, central, focal stations. Reconfigurable swing-arms (see Figure 1) allow rapid switching of instruments to respond to different science needs and changing observing conditions.

### 1.1. LBT Status

Figure 2 shows the status of the LBT at the end of 2002. The actual telescope structure was produced by Ansaldo (now Ansaldo Camozzi) in Milano, Italy. Pre-erection and tests took place in the factory in late 2001. Disassembly and shipping to Arizona occupied much of the first half of 2002, and now all parts are either in the summit enclosure or the base camp. The first primary mirror blank was cast at the University of Arizona Mirror Lab in January, 1997, with the second one following in May 2000. The Mirror Lab declared grinding, polishing, and figuring of the first mirror complete in early December 2002. Shaping of the second primary mirror will begin after installation of a new polishing machine in early 2003. Construction of the enclosure on Mt. Graham began in earnest in mid-July 1996. By the end of the construction season in late 1997, most of the important concrete foundation and fixed steelwork was in place. At this writing (December 2002), the enclosure is essentially complete, and re-assembly of the telescope is underway. Adaptive optics are at the core of LBT operation, particularly for interferometry. A prototype of the adaptive secondaries, with 332 actuators, is currently being commissioned on the MMT.

First light for the LBT, defined as operation with a single mirror and one facility instrument, is scheduled for June, 2004. Second light, with both mirrors operating as a single telescope, will take place approximately one year later. Delivery of the two interferometric instruments is scheduled for late 2005.

## 2. LBTI

Within the context of the NASA Origins program, the University of Arizona is developing a beam combiner, which will sit at the central, combined focus of the LBT. This instrument, called LBTI (Large Binocular Telescope Interferometer, Hinz, 2002), is a fully cryogenic instrument devoted to near and mid- infrared observations (2–20 $\mu$m). In particular, LBTI will support nulling interferometry in a program to search for extrasolar planets and zodiacal light.

### 2.1. Nulling interferometry with LBTI

The relatively short-baseline pupil configuration and the simple, symmetric optical path of the LBT enable efficient *nulling* interferometry. In this mode, the beam combiner optics overlay the two telescope pupils with a $\pi$ phase shift. This produces a null in sensitivity along the pointing direction. The glare from a bright star can thus be suppressed, and if the neighbouring maxima in transmission are properly aligned, the fainter glow of zodiacal light or planets can come through. The signal indicating such extended emission or companions would be a modulation in the unsuppressed radiation as the projected baseline changes due to the

*Figure 2.* LBT progress at the end of 2002. (Clockwise from upper left) Assembly of the azimuth platform within the observing chamber. Completed enclosure on Mt. Graham. Aspherization of a Zerodur blank for the 672 actuator adaptive secondaries. Both LBT mirrors in the casting lab – figuring of the first primary (top) ended in late 2002.

earth's rotation. Because of improved contrast, most nulling experiments operate at mid-infrared wavelengths around 10 $\mu$m.

Figure 3 shows the LBTI instrument mounted at the central, bent focus of the Large Binocular Telescope. The instrument consists of three main subsystems. The Universal Beam Combiner (UBC) brings the radiation from the two optical trains to a common axis on the midline of the telescope. There, the Nulling Interferometer for the LBT (NIL) creates and overlays the two pupil images with the necessary $\pi$ phase shift. Finally, the Nulling-Optimized Mid-IR Camera (NOMIC) serves as the detector system. Potential future instruments can be placed immediately above NIL and selected with a rotatable mirror.

*Figure 3.* (top) LBTI mounted at the central, shared focus. (bottom) Detail of the instrument, showing the beam combiner (UBC), nulling interferometer (NIL), and the mid-IR camera (NOMIC)

## 2.2. LBTI GOALS

The Large Binocular Telescope Interferometer team has established a series of well-focused goals for the instrument. They are:

1. To carry out a sensitive nulling survey of nearby stars (NIREST - see below) to search for zodiacal light.
2. To develop nulling interferometry techniques, including active phase control and observing strategies.
3. To provide a testbed for multi-pair nulling interferometry techniques, to help verify optical systems for the TPF/Darwin space interferometry missions.
4. To provide a combined-beam focal station suitable for future expansion, including pupil plane and Fizeau interferometers.

### 2.3. The NIREST survey

The Arizona team has proposed a major science program for LBTI. NIREST is the Nulling InfraRed survey of Extra-solar Systems for TPF. This survey of approximately 80 nearby stars is essential as a ground-based 'precursor' of the TPF/Darwin missions. It will attempt to characterize and understand a sample of nearby sun-like stars to be examined by the space interferometer. A survey of approximately 80 stars will identify zodiacal disks, which signal the existence of a planetary system for followup study with TPF/Darwin. On the other hand, NIREST will also weed out those systems with very intense zodiacal emission, which can overwhelm the signal of terrestrial planets. Given a significant sample, NIREST can also address the question of zodiacal emission strength as a function of spectral type and age of the central star.

LBTI will also be sensitive enough to detect giant planets with masses in the range 5-10 $M_{jupiter}$. In addition to being interesting themselves, such planets would dominate the dynamics of their host systems, potentially protecting earth-like planets from bombardment and thereby increasing their habitability.

### 2.4. LBTI status

The implementation plan for LBTI is strongly coupled to that of the telescope – it makes little sense to bring an interferometer to Mt. Graham before the second mirror. The Arizona team successfully passed an initial NASA review in November 2001 and is now well into the design phase. Current planning calls for completion of the instrument and lab verification in late 2005, with integration, testing, and commissioning on the mountain accomplished by autumn, 2006. The instrument should be available to the LBT and NASA partner communities soon thereafter.

## 3. LINC-NIRVANA

LINC-NIRVANA (Herbst et al., 2002; Ragazzoni et al., 2002) is a near-infrared image-plane beam combiner with Multi-Conjugate Adaptive Optics (MCAO). It will be built by a consortium of three Institutes: the Max-Planck-Institut für Astronomie (MPIA) in Heidelberg, the Osservatorio Astrofisico di Arcetri in Firenze, and the University of Köln.

### 3.1. Fizeau interferometry with LINC-NIRVANA

LINC-NIRVANA combines the radiation from the two 8.4 m primary mirrors of the Large Binocular Telescope in 'Fizeau' mode. In a Fizeau interferometer, the wavefronts interfere in the focal plane, not in the pupil plane, as with essentially all current interferometric instruments (including the LBTI nuller). Unlike their pupil-plane cousins, Fizeau interferometers are true imaging devices. Perhaps the simplest way to think of a Fizeau interferometer is as a very large telescope with diameter equal to the baseline, but with a mask corresponding to the configuration of component telescopes placed in the entrance pupil. The fourier transform of this pupil layout is the point spread function (PSF), and all objects within the field display this PSF.

Figure 4 shows the design concept. The fundamental features of LINC-NIRVANA are:

- Ambient temperature fore-optics
- Cold cryostat with detector systems for NIR
- Mult-Conjugate Adaptive Optics for atmospheric correction
- Optical bench for rigidity, easy access, etc.
- Stiff, lightweight superstructure
- Light-tight enclosure

The telescope focal planes lie just inside the instrument enclosure. At this location, the Ground-Layer Wavefront Sensors (GLWS) measure turbulence above the telescope using natural guide stars in an annular region outside the science field. The GLWS send their correction signals to the adaptive secondaries. The central two arcminutes of the field not sampled by the GLWS continues further into the instrument and is collimated by a five-element lens. The optical path is folded twice before forming a scaled down image of the telescope entrance pupil. For MCAO, these folding optics will be 349-actuator deformable mirrors.

A dichroic beam splitter above the cryostat separates the visible wavelength wavefront sensing channel from the near-infrared science channel. The NIR pupil lies just inside the window of the cryostat, where a cold mask suppresses excess thermal background radiation. A reflective camera inside the dewar focuses and combines the radiation on the near-infrared science and fringe-tracking detectors. The visible light reflected from the dichroic is folded to the side toward a three element camera lens, which forms an f/20 focus. A 45° folding mirror directs the light upward into the Mid-High Layer Wavefront Sensors (MHWS). These devices measure the turbulence at mid and high altitudes, and send correction signals to the deformable mirrors located in the collimator lens area.

### 3.2. LINC-NIRVANA SCIENCE PERFORMANCE AND GOALS

Table I lists the anticipated spatial resolution, field of view, and point source sensitivity for LINC-NIRVANA. The sensitivity calculation conservatively assumes that the light from both telescopes is distributed within an 8.4 m Airy disk (i.e.

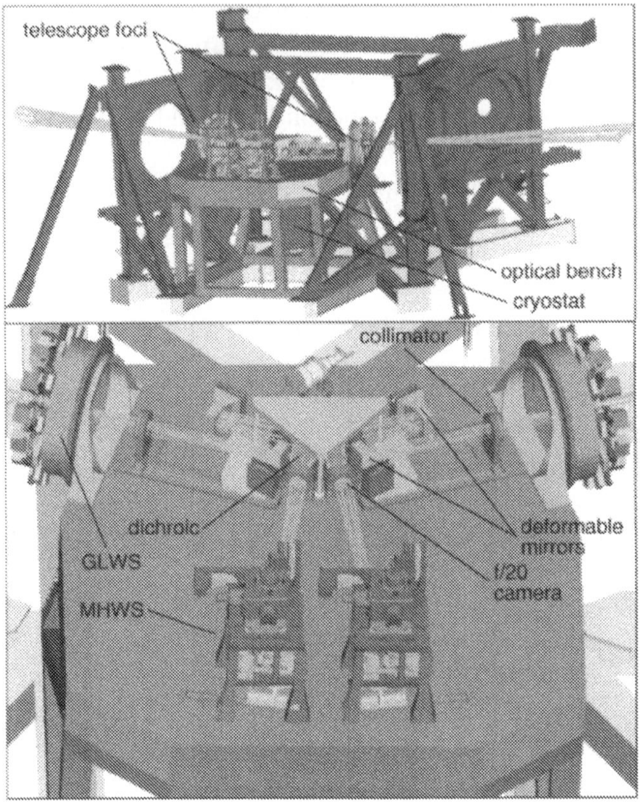


*Figure 4.* (top) LINC-NIRVANA at one of the bent shared foci. (bottom) Closeup of the optical path, including the ambient temperature collimator lenses, the cold pupil stop, and the cryogenic camera. Note the additional folds in the optical path to accomodate the deformable mirrors. The Mid-High Layer Wavefront sensor uses visible light reflected off the dichroic.

no fringes). Optimal extraction techniques geared to the interferometric PSF may improve the point-source sensitivity.

Because it is a true imager, LINC-NIRVANA enables a broad variety of 'conventional' science programs brought to the next level by its enhanced spatial resolution and sensitivity. Key programs being pursued by the LINC-NIRVANA team include pushing supernova cosmology studies to beyond redshift 3, following the evolution of complex structures in star-forming regions, and detecting Jupiter-mass planets around stars within 100 pc due to their reflex, astrometric wobble.

### 3.3. LINC-NIRVANA STATUS

As with LBTI, the LINC-NIRVANA implementation schedule is closely tied to that of the telescope and other instruments. As of December 2002, the optical design was complete, and work on the mechanics and software is well under-

TABLE I

Spatial resolution, field of view, and sensitivity of LINC-NIRVANA

| Band | Wavelength ($\mu$m) | Angular Resolution ($\lambda/B$, mas) | Field of View (arcsec)[a] | Sensitivity[b] |
|---|---|---|---|---|
| J | 1.25 | 11 | 10 | 28.0 |
| H | 1.65 | 15 | 15 | 26.6 |
| K′ | 2.12 | 20 | 20 | 26.0 |

[a] Limited by detector size – future expansion possible.
[b] Point source detection, $5\sigma$ in one hour, assuming sky backgrounds of J=15.2, H=13.6, and K′=13.3.

way. A concept review took place in November 2001, and a formal Preliminary Design Review (PDR) is scheduled for March 2003. Given the combination of a complex MCAO system and interferometry, the LINC-NIRVANA team plans to implement the instrument in stages, beginning with single-star AO interferometry, then bringing the various subsystems (GLWS, MHWS) online one by one.

## 4. Conclusions

The unique optical configuration of the Large Binocular Telescope lends itself readily to interferometric use. Two innovative instruments, LBTI and LINC-NIRVANA, will exploit this opportunity and provide the LBT community with unparalleled nulling and imaging interferometry capability.

## Acknowledgements

The author thanks Phil Hinz for updated information regarding the LBTI project at the University of Arizona.

## References

Angel, J.R.P., Hill, J.M., Strittmatter, P.A., Salinari, P. and Weigelt, G.: 1998, Interferometry with the large binocular telescope, *SPIE* **3350**.

Herbst, T.M., Ragazzoni, R., Andersen, D., Böhnhardt H., Bizenberger, P., Eckart, A., Gaessler, W., Rix, H.-W., Rohloff, R.-R., Salinari, P., Soci, R., Straubmeier, C. and Xu, W.: 2002, LINC-NIRVANA: A fizeau beam combiner for the large binocular telescope, *SPIE* **4838–110**.

Hill, J.M. and Salinari, P.: 2002, The large binocular telescope project, *SPIE* **4837–15**.

Hinz, P.M.: 2002, Interferometry with the large binocular telescope, *SPIE* **4838–13**.

Ragazzoni, R., Herbst, T., Andersen, D., Bizenberger, P. Rix, H., Rohloff, R.R., Arcidiacono, C., Diolaiti, E., Esposito, S., Farinato, J., Riccardi, A., Vernet-Viard, E. and Salinari, P.: 2002, NIRVANA: a visible MCAO channel for LBT, *SPIE* **4839–65**.

# VLTI INSTRUMENTS AND TOOLS

# INTRODUCING THE NEAR INFRARED VLTI INSTRUMENT AMBER TO ITS USERS

ROMAIN G. PETROV
*Universite de Nice-Sophia Antipolis, Parc Valrose, F-06108 Nice, France*
*E-mail: petrov@unice.fr*

THE AMBER CONSORTIUM†
*Universite de Nice-Sophia Antipolis, Parc Valrose, F-06108 Nice, France*
*Observatoire de la Cote d'Azur, BP 4229, F-06304 Nice, France*
*Universite Joseph Fourier, BP53, F-38041 Grenoble, France*
*Max Planck Institute fur Radiosatronomie, Auf den Huegel 69, D-53121 Bonn, Deutschland*
*Osservatorio Astrofisico di Arcetri, Largo E. Fermi 5, I-50125 Firenze, Italia*

**Abstract.** AMBER is the General User near infrared focal instrument of the Very Large Telescope Interferometer. It is a single mode, dispersed fringes, three telescope instrument. A limiting magnitude of the order of H=13 will allow tackling of a fair sample of extra galactic targets. A very high accuracy, in particular in color differential phase and closure phase modes gives good hope for very high dynamical range observations, possibly including hot extra solar planets. The relatively high maximum spectral resolution, up to 10000, will allow stellar activity observations. Between these extreme goals, AMBER has a wide range of applications including Young Stellar Objects, Evolved Stars, circumstellar material and many others.

**Keywords:** astronomy, optical interferometry, instrumentation, high angular resolution, very large telescope, differential interferometry

## 1. Introduction

AMBER is a near infrared, three beam, dispersed fringes, single mode, focal instrument for the interferometric mode of the Very Large Telescope (VLTI). The VLTI (Glindemann et al., 2002) feeds AMBER with two or three beams produced by any of the telescopes present on the Paranal Mountain top. Each beam has been partially corrected for atmospheric turbulence and is stabilized in direction, position and optical path. Besides a very wide scientific program (Malbet et al., 2002a), the near infrared offers the possi bility to use nearly diffraction limited images. The three beams yield closure phases, which have interesting calibration properties and offer the possibility to reconstruct images. Still, AMBER will mainly be an efficient 'model fitting' instrument thanks to its large spectral coverage, from 1.1 to 2.4 $\mu$m and spectral resolutions (35, 1000 and 10000). The dispersed fringes increase: the

† The AMBER Consortium has been established by the five Institutes in the affiliation list and is supported by more than 50 engineers and researchers. More information can be found on the web site: www.obs-nice.fr/amber

*Astrophysics and Space Science* **286:** 57–67, 2003.

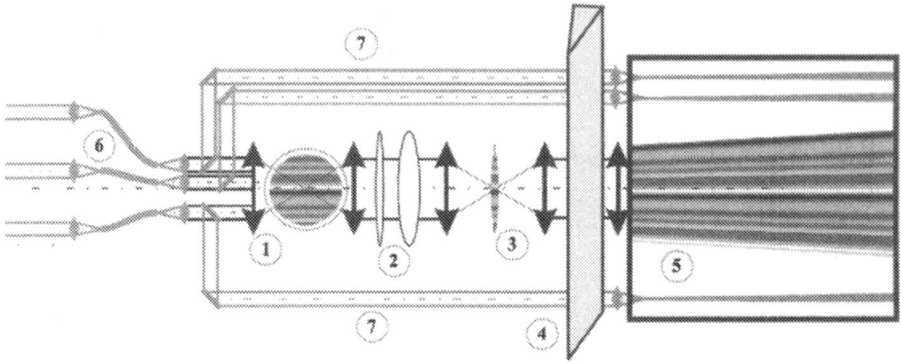


*Figure 1.* Basic concept of AMBER. 1: multi axial beam combiner; 2: cylindrical optics; 3: anamorphotic image in spectrograph slit; 4: spectrograph; 5: dispersed fringes on 2D Hawaii detector; 6: spatial filtering of incoming beams with single mode fibers; 7: channels for photometric monitoring.

instantaneous frequency plane coverage; the instrument accuracy when a (set of) spectral channel(s) can be used as a reference. They allow measures on unresolved and/or very high dynamics systems. The spatial filtering with single mode fibers is intended to achieve the highest possible accuracy on absolute as well as on differential visibility measurements.

Figure 1 summarizes the key elements of the AMBER concept. AMBER has a multi axial beam combiner (1) which produces fringes in the Airy pattern common to several non-redundantly spaced parallel beams. A cylindrical optics system (2) feeds this Airy pattern into the entrance slit (3) of a spectrograph (4), which disperses the fringes on an array detector (5). The medium (1000) and highest (10000) spectral resolutions imply to cool the spectrograph and to equip it with cold slits and pupil stops. Each incoming beam goes through a single mode optical fiber (6) which reduces all atmospheric and optic aberrations into an unknown optical path difference (OPD), which can be frozen, and an unknown coupling factor into the fibers. The coupling factors are monitored in real time by photometric beams (7) which analyze at each wavelength the transmitted flux. Since a single mode fiber is efficient only over a limited wavelength range, before feeding the beams in the fibers we split the total spectral range with dichroics and we have different spatial filters for the J, H and K bands. In addition to these basic features AMBER is equipped with many beam cleaning tools (polarization filters, refraction and dispersion correctors...) and calibration devices (artificial coherent sources, calibrated phase delays, beam commuters...). More details about the instrument can be found in (Petrov et al., 2002a; Malbet et al., 2002b; Chelli et al., 2002).

## 2. Amber Measurements and Science

By reconfiguring the VLTI ATs over the available Paranal summit positions, good use of super synthesis and of the increase in u-v sampling due to our large wavelength coverage permits reconstructing an image at the diffraction limit of the interferometer. However, such a process will be quite slow and future AMBER users should mainly think about it as an instrument designed to constrain models of astrophysical sources thanks to a limited number of angular measurements with a rich wavelength information. The following paragraphs explain what are the key AMBER measurements, together with some examples of the kind of information they can provide and with their main calibration requirements and methods.

### 2.1. AMBER MEASUREMENTS

For each elementary frame and in each AMBER spectral channel, we have an interferometric and three photometric signals processed by the same optics and the same dispersive elements. After the correction of detector cosmetics and background subtraction, a generalization of the ABCD algorithm is used (Chelli et al., 2002) to establish a linear relation between the values measured in each pixel and the complex coherence for each baseline. The matrix is calibrated using artificial sources and records of various interferograms affected by different known phase delays. It combines the effects of many parameters such as the detector gain table and bad pixels, the exact shape of the beams after the fiber output and the pupil stop in the spectrograph. For each baseline l-m, we measure the visibility $V_{lm}(\lambda)$ and the phase $\Phi_{lm}(\lambda)$.

#### 2.1.1. *Calibrated visibility as a function of wavelength:* $V_{lm}(\lambda)$

After time averaging and various bias corrections and calibration steps (Chelli et al., 2002), $V_{lm}(\lambda)$ yields the calibrated visibility in each spectral channel. For a simple object, $V_{lm}(\lambda)$ will give constraints on its angular size in the direction of the baseline $B_{lm}$. If a certain range of spatial frequencies $B_{lm}/\lambda$ is explored, thanks to baseline changes or to the spectral coverage, visibility measurements yield the different scales of the structures present in the source. Several levels of attenuation at different spatial frequencies can indicate that structures of various sizes are present. For example the super giant star Betelgeuse shows at first very fast visibility attenuation with frequency, which is the signature of an extended dust envelope. At higher frequencies, a limb darkened disk model can fit a fraction of the visibility curve. Finally, at the highest frequencies, where the disk contribution is expected to be very low, the visibility remains significantly higher than zero, which can be interpreted as the signature of a structure smaller than the stellar disk such as a stellar spot.

The use of measurements at different wavelengths increases the frequency coverage and can constrain the interpretation. For example, in a circumstellar envelope or an AGN BLR, one can compare the visibility in two narrow spectral channels in a spectral line. Each spectral channel corresponds to an area of given radial velocity. The comparison between the measures constrains the relative sizes of the equal velocity areas and the velocity field (Marconi et al., 2002).

The accuracy of the visibility measurements is critical. A few % accuracy yields gross size information (which in a limited number of cases can be very valuable). But, for example, limb darkening or different radial distribution of material in an envelope or in a dust torus will produce visibility variations of the order of 1% or less. Detecting features which are faint or small with regard to a main source (binary stars with a strong magnitude difference and ultimately extra solar planets, star spots, structures in envelopes or disks...) is particularly demanding. Even the hottest giant extra solar planets such as 51 Peg, have visibility signatures smaller than $10^{-4}$. The same will apply for 'planetary gaps' in accretion disks or for stellar structures on otherwise quite unresolved stars. In spite of all precautions: single mode spatial filtering, frozen piston, short exposures and dispersion correction, the visibility will still be sensitive to instrumental and atmospheric effects. The main tool to calibrate it is to observe as often as possible a reference star with unit or known visibility. The fundamental limit comes from the speed with which is it possible to switch between science and reference source, and to lock AOs and fringe tracker on both sources. Our best expectation for the period of such a calibration cycle is 10 minutes (closer to 30 minutes in the first months or years of VLTI operation).

#### 2.1.2. *Differential visibility and phase*

One or several spectral channels can be defined as being a reference channel (when several wavelengths are used, the information is averaged over $\lambda$). Then the ratio between the visibility in each spectral channel and the visibility in the reference channel yields the differential visibility $V_{lm}(\lambda)/V_{lm}(\lambda_0)$. The interesting point is that $V_{lm}(\lambda)$ and $V_{lm}(\lambda_0)$ are measured simultaneously which increases the accuracy of the calibration since many effects will affect all the channels in the same way (or in ways related by known equations). The absolute phase $\Phi(\lambda)$ has no meaning unless an absolute metrology (Glindemann et al., 2002) is available. However, it is immediately possible to measure the absolute phase differences between any spectral channel and the reference channel: $\Phi_{lm}(\lambda) - \Phi_{lm}(\lambda_0)$. The Differential Phase can bring specific astrophysical information. For example, if one observes in different channels of a line as described above, the Differential Phase will constrain the angular distance between the two different velocity areas. For sources with relatively simple or partially known velocity fields combining the variation of the visibility modulus and phase with wavelength is particularly interesting. The modulus variation will constrain the relative sizes and intensities of the equal velocity zones while the phase will constrain their relative position. Combining these values with some knowledge of the velocity field can allow to almost reconstruct images. A good example is likely to be circumstellar envelopes or AGN observed in emission lines (Marconi et al., 2002). Among many other applications we have multiple system with small number of stars, envelopes combining a global rotation and a global expansion, knots in a jet with a similar global direction but different velocities... Both Differential measurements are particularly interesting when the object is known to be unresolved in some spectral channels and resolved in others. Then, if the channel containing an unresolved source is used for reference, the differential complex visibility in the other channels is the exact object visibility and the image reconstruction process can be as efficient as if we had a 'phase referencing system'. A specific interest of the differential phase is that, unlike the differential modulus, it can be calibrated using an internal modulation instead of an external reference star. The measured phase difference $\Delta\Phi_{lm}(\lambda)$ will be the sum of:

$$\Delta\Phi_{lm}(\lambda) = \Delta\Phi_{lm,*}(\lambda) + \Delta\Phi_{lm,a}(\lambda) + \Delta\Phi_{lm,I}(\lambda) \quad (1)$$

where $\Delta\Phi_{lm,*}(\lambda)$ is due to the source, $\Delta\Phi_{lm,a}(\lambda)$ is the contribution of the atmosphere, which is very strongly dominated by the differential dispersion in the tunnels and $\Delta\Phi_{lm,I}(\lambda)$ is the instrumental chromatic phase difference between the beams l and m. The dominant and most difficult to eliminate term is the instrumental one $\Delta\Phi_{lm,I}(\lambda)$. AMBER plans to solve this problem by inverting the beams as close as possible to the entrance of the instrument, before any chromatic optics in AMBER (dichroics, fibers, beam splitters, cryostat windows, spectrograph chamber, detector...). Then we obtain a new phase difference $\Delta\Phi_{ml}(\lambda)$:

$$\Delta\Phi_{ml}(\lambda) = -\Delta\Phi_{lm,*}(\lambda) - \Delta\Phi_{lm,a}(\lambda) + \Delta\Phi_{lm,I}(\lambda) \quad (2)$$

The difference between equations (1) and (2) eliminates the instrumental term $\Delta\Phi_{lm,I}(\lambda)$. The atmospheric term $\Delta\Phi_{lm,a}(\lambda)$ results from the differences in average temperature, pressure and water vapor be tween the two beams. The first two terms have a globally multiplicative effect, except for saturated water vapor lines, and can be fitted in the data itself particularly if we use a large spectral coverage.

It has been shown (Vannier et al., 2002) that if we have differential phase measurements with the UTs, at low spectral resolution over the full J,H,K spectral coverage, limited only by photon, background and detector noise, the variation of phase with wavelength produced by the presence of a Jupiter mass hot extra solar planet around a solar type star at 10 pc can be detected. The SNR will be higher than 20 for 51 Peg (separation = 0.05 AU) and the SNR will be around two for a separation of about 0.2 AU. Such an observation will yield: the orbital parameters and thus the mass of the planet; the spectrum and thus the temperature and composition of the planet atmosphere. These measurements are one of the most ambitious AMBER goals, since the corresponding phase variation ranges between a few $10^{-5}$ and a few $10^{-4}$ radians. The conditions to achieve such an extreme accuracy are that the instrumental and atmospheric variation must be smaller than the photon noise over the calibration period. After a systematic study of all factors affecting this measurement (Vannier et al., 2002) it appears that the result is achievable with

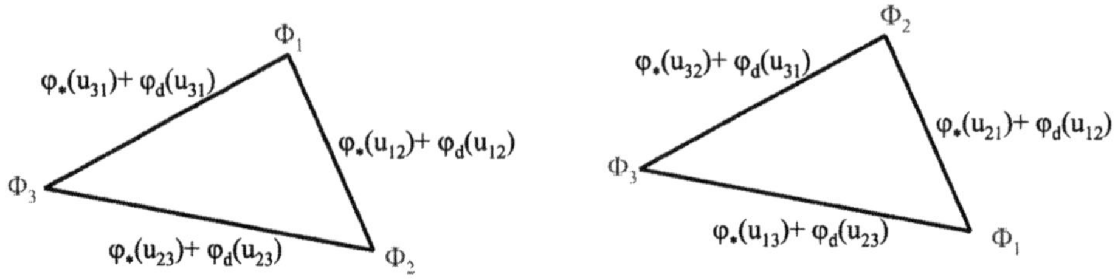


*Figure 2.* Closure phase before and after beam commutation.

a 30 s calibration cycle based on an internal beam commutation and out of reach if one has to use the 300 s cycle using a reference star. A specific attention has to be dedicated to the atmosphere term, which is not corrected by the beam commutation. As explained above, with two telescopes we expect to compute this term from the data itself. If this fails, with three telescopes this dispersion term can be eliminated using the closure phase relation (Segransan et al., 2001).

### 2.1.3. *Closure phase* $\Phi_{123}(\lambda)$

In the closure phase relation:

$$\Phi_{123}(\lambda) = \Phi_{12}(\lambda) + \Phi_{23}(\lambda) + \Phi_{31}(\lambda)$$

all atmospheric and most of instrumental terms are cancelled. The exceptions result from the fringe detection process (error in the 'pixel to visibility' matrix (PTVM) due to a change in the detector gain table, in the beams shape or overlap after the fibers, in the spectrograph geometry...). Fortunately, we recently discovered (Petrov et al., 2002b) that commuting two of the three beams calibrates the closure phase relation. Let's assume that we measure the phases for the three baselines B12, B23 and B31. Each beam i is affected by an optical path difference (OPD) which translates in a phase $\Phi_i$. The spatial frequency $u_{lm} = B_{lm}/\lambda$ yields the sum of source phase $\phi_*(u_{lm})$ with a detection phase error $\phi_d(u_{lm})$, as illustrated in Figure 2.

Since $\phi_*(u_{ij}) = -\phi_*(u_{ji})$ and $\phi_d(u_{ij}) = -\phi_d(u_{ji})$, the difference between the closure phase $\Phi_{123}(\lambda)$ without and $\Phi'_{123}(\lambda)$ with commuting beams 1 and 2 yields

$$\left(\Phi_{123}(\lambda) - \Phi'_{123}(\lambda)\right)/2 = \phi_*(u_{12}) + \phi_*(u_{23}) + \phi_*(u_{31}) = \phi_{*,123}(\lambda) \quad (3)$$

where $\phi_{*,123}(\lambda)$ is an astronomical closure phase measurement free from any atmospheric or instrumental effect, if the beam commutation has been made fast enough for the instrumental terms to remain unchanged. This way to obtain a perfectly calibrated closure phase is unique to the VLTI because of its three large telescopes and to AMBER because of its beam commuting device. In addition to this very interesting calibration 'perfection', the closure phase is the key for image reconstruction. Moreover, even a limited number of closure phases can strongly constrain a model. For example a zero closure phase at medium frequencies and a non zero one at higher frequencies may indicate (and allow to locate) a small non symmetric feature on an otherwise globally symmetric object.

#### 2.1.4. *Differential photocenter*

When the source is unresolved (i.e. smaller than $B/\lambda$) the phase tends to be proportional to the photocenter of the object at that given wavelength: $\Phi_{lm}(\lambda) = 2\pi(\vec{B}_{lm}/\lambda).\vec{\epsilon}(\lambda)$, where $\vec{B}_{lm}.\vec{\epsilon}(\lambda)$ is the scalar product of the two vectors. The photocenter is defined as the barycenter of the object brightness distribution $o(r, l)$ at the wavelength $\lambda$:

$$\epsilon(\lambda) = \frac{\int r\, o(r, l)\, dr}{\int o(r, l)\, dr} \tag{4}$$

Since phase differences can be measured even when they are extremely small compared to $2\pi$, this allows to extract, from differential phases, the variation of objects photocenter as a function of wavelength; even when the source is much smaller than the resolution limit. For example, for a magnitude 5 source observed with a resolution 1000 during 5 hours with an UT, one can measure photocenter displacement of about 0.1 $\mu$as. For a magnitude 10 star with UTs or a magnitude 7 star with ATs the accuracy on the photocenter displacement remains 1 $\mu$as for the same resolution 1000. This allows obtaining information on sources or structures much smaller than the resolution limit, such as stellar spots (Petrov, 1989).

One of the most interesting applications is likely to be the study of the Broad Line Regions (BLR) of Active Galactic Nuclei (Marconi et al., 2002). The study of the photocenter has numerous other applications, including stellar rotation, differential rotation, position of stellar rotation axes in multiple systems, stellar diameters of otherwise unresolved objects, asteroseismology, cepheids, structure and kinematics of unresolved disks and envelopes... The key fact to remember is that for objects with spatially structured spectral features, the limiting resolution of the AMBER/VLTI can be much smaller than the classical $B/\lambda$ (1 or 2 mas) limit. Photocenter measurements of unresolved objects are linked to differential phase measurements because it is easy to show that when a source of angular size $a$ is unresolved, the photocenter and the differential phase decrease proportionally to $a$, while the visibility modulus approaches 1 like $a^2$ and the closure phase decreases like $a^3$.

### 2.2. Observing and calibration cycles of AMBER

The observing modes of AMBER are set by: the number of telescopes (2 or 3); the spectrograph set up (resolutions $\approx 35$; $\approx 1000$ ; $\approx 10000$ ; spectral coverage; central wavelength); the availability of a fringe tracker. It seems that, from the very beginning, AMBER observations will benefit from an on axis fringe tracker (FINITO (Gai, 2001)).

In all cases the observing and data processing sequences remain the same and are defined by the following parameters.

- The required accuracy: for very precise measurements of the visibility modulus, one should use very short frame times, even with a fringe tracker. The minimum frame time is of the order of 10 ms for 40 spectral channels (J,H and K band in the low resolution mode). For higher sensitivity (and lower accuracy except in the differential modes) one could use longer frame times (50 to 100 ms) even with poor fringe tracking. Finally for high spectral resolution, correct fringe tracking must support frame times up to a few tens of seconds.
- The kind of calibration selected: visibility measurements will give priority to frequent calibration using one or more reference stars. Differential measurement will use relatively fast internal beam commutation and use external reference stars only from time to time to check the instrument.

In all these very similar modes, we set up the instrument and perform the corresponding PTVM calibration. Then we read frames, with frame times selected as explained above, and we combine them in exposures (successive frames recorded on the same source with the same instrument set up). Each exposure yields measurements of the complex coherence but a complete calibration needs a combination of exposures of different sources (science target, reference star, sky background, calibration lamps...) or with different beam commutations. All sequences of operations needed to perform a given measurement, i.e. a given exposure cycle, are automatically executed by the instrument. The sequences are described in standard templates, with parameters set by the User as a function of its scientific objectives. For each given exposure cycle there is a standard way to reduce the data for calibrated measurements, which will be delivered to the standard User. However, all data will be recorded to allow more sophisticated post processing.

### 2.3. Summary of AMBER top level specifications and expected performances

Table I summarizes the main specifications of AMBER after the various tradeoffs made during the design phases.

The fringe tracker sets the limiting magnitude of AMBER. The current conservative estimate for $\lambda/20$ fringe tracking is H = 13. Table II gives the SNR which can be expected in each spectral channel on a source with visibility close to 1. We assumed 4 hours observations divided in 50 to 100 s frames for a star of magnitude H=12 with the UTs or H=9 with the ATs, for various spectral resolutions. It is important to remember that: the error in the differential phase can be estimated to be $\sigma_\phi = 1/(\sqrt{2}\text{SNR})$ radians; the error in the photocenter displacement is given by $\sigma_\epsilon = \sigma_\phi(\lambda/B)/2\pi$; for lower visibility sources, the SNR decreases with the visibility modulus.

For example, for a magnitude H=12, at resolution 1000 we will get $\sigma_\phi = 3 \times 10^{-3}$ radians and $\sigma_\epsilon = 2\,\mu$arcsec (K band, 100 m baseline), which is suf-

TABLE I

Summary of the top level specifications of AMBER

| Characteristic | Specification | Goal |
|---|---|---|
| Number of beams | 3 | |
| Spectral resolution $R$ | $\approx 35$, 1000 and 10000 in K band, 2nd order in J, intermediate in H | |
| Spectral coverage | J,H,K from 1 to 2.4 $\mu$m | |
| Instantaneous spectral coverage | Simultaneous observation of the full spectral domain for $R = 35$ | |
| Instrument contrast | 0.8 | 0.9 |
| Absolute visibility accuracy | $3\sigma_V = 0.01$ | $\sigma_V = 10^{-4}$ |
| Differential visibility accuracy | $3\sigma_{V/V_0} = 10^{-3}$ | $3\sigma_{V/V_0} = 10^{-4}$ |
| Differential phase accuracy | $3\sigma_{\Delta\Phi} = 10^{-3}$ rad | $\sigma_{\Delta\Phi} = 10^{-5}$ rad |
| Instrument contrast stability | $10^{-2}$ over 5 minutes | $10^{-3}$ over 5 minutes |
| Differential phase stability | $10^{-3}$ rad over 1 minute | $10^{-4}$ rad over 1 minute |

TABLE II

SNR (or $1/\sigma_V$ or $1/\sqrt{2}\sigma_\phi$) per spectral channel for a star of magnitude H=12 (left, fringe tracking limit) with the UTs or H=7 (right) with the UTs (and 3 magnitudes less with the ATs). 4 hours of observation divided in 50 s frames

| Telescopes and star magnitude | UTs (H=12) or ATs (H=9) | | | UTs (H=7) or ATs (H=4) | | |
|---|---|---|---|---|---|---|
| Bands | J | H | K | J | H | K |
| Resolution = 35 | 508 | 843 | 1111 | 5080 | 8430 | 11110 |
| Resolution = 1000 | 95 | 158 | 208 | 950 | 1580 | 2080 |
| Resolution = 10000 | 30 | 50 | 66 | 300 | 500 | 660 |

ficient to resolve all quasars and Seyfert 1 BLRs with these kind of magnitudes. Another example, is that for a magnitude 7 star, we should have at resolution 35, a differential phase error $\sigma_\phi = 6 \times 10^{-5}$ radians, sufficient to resolve many 51 Peg like planets, if of course we manage to be limited only by these fundamental noises.

Eventually, a field separator will allow tracking of fringes and measuring wavefronts on stars up to 1 minute away from the science source. When such a reference source exists, the limiting magnitude of AMBER can approach 20 in a few hours. However, with the current performance of fringe trackers, the corresponding sky coverage will be very poor. To have good sky coverage, we need fringe sensors with detectors limited by photon noise in the near infrared. This might be achievable in a few years.

## 3. Conclusion

All subsystems of AMBER have been delivered and are now in the final global integration and test phase. The first fringes in the K band with all instrument components have been obtained in January 2003. We expect to ship the instrument to Paranal during the summer 2003, after a full laboratory commissioning of the hardware and the software. The commissioning with two large telescopes (UTs with Adaptive Optics) should take place before the end of the year and a partial access to the instrument offered to the Consortium and to the Community starting in April 2004. The full operation with three telescopes will probably be open in October 2004. The summary of the AMBER consortium Guaranteed Time Program will be released in April 2003. The initial pressure factor was of about two on the UTs and three on the ATs and we had to make a severe selection. We now have two proposals about the distance scale in the Universe, 15 extra galactic programs, mainly about Quasars and other AGNs, 22 programs about Star formation covering many topics such as parameters of PMS stars, disks, outflows, jets, young binaries and young stellar clusters. Ten programs are about low-mass companions, with a very developed program on Pegasides planets. More than 22 programs tackle various aspects of the late stages of stellar evolutions and about 15 programs are dedicated on fundamental properties of stars (age, diameters, rotation, activity, seismology, magnetism) and of stellar envelopes (structure and kinematics). The UTs are mainly asked for extra galactic sources and extra solar planets. This leaves a lot of objects and room for additional programs and ideas and we are quite confident that AMBER will permit a large harvest of new results.

## References

Chelli, A., Tatulli, E., Malbet, F. et al.: 2002, AMBER data processing, *SPIE* conference **4838** Interferometry for Optical Astronomy II, Waikoloa, Hawaii, USA, 2002, in publication.

Gai, M., Bonino, D., Corcione, L. et al.: 2001, in: J. Bergeron and G. Monnet (eds.), *FINITO: A Fringe Sensor for the VLTI in Two and Three Beam Configuration*, ESO symposia Scientific drivers for ESO future VLT/VLTI instrumentation, pp. 328–330.

Glindemann, A., Ballester, P., Bauvir B. et al.: 2002, Very large telescope interferometer: a status report, *SPIE conference* **4838** Interferometry for Optical Astronomy II, Waikoloa, Hawaii, USA, 2002, in publication.

Malbet, F., Bloecker, T., Foy, R. et al.: 2002, Astrophysical potential of the AMBER/VLTI instrument, *SPIE* conference **4838** Interferometry for Optical Astronomy II, Waikoloa, Hawaii, USA, 2002, in publication.

Malbet, F., Petrov, R.G., Tallon-Bosc, I. et al.: 2002, System analysis of the AMBER instrument on VLTI, *SPIE* conference **4838** Interferometry for Optical Astronomy II, Waikoloa, Hawaii, USA, 2002, in publication.

Marconi, A., Maiolino, R. and Petrov, R.G.: 2002, Extragalactic astronomy with the VLTI: a new window on the universe, *Proceedings of JENAM 2002 Conference*, Porto, Portugal, September 2002 (these proceedings).

Petrov, R.G.: 1989, Differential interferometry, in: D. Alloin and J.M. Mariotti (eds.), *Diffraction Limited Imaging with Very Large Telescopes*, NATO ASI Series C, **274**, Kluwer. Ac., pp. 249, 272.

Petrov, R.G., Malbet, F., Weigelt, G. et al.: 2002, Using the near infrared VLTI instrument AMBER, *SPIE* conference **4838** Interferometry for Optical Astronomy II, Waikoloa, Hawaii, USA, 2002, in publication.

Petrov, R.G., Vannier, M., Lopez, B., Bresson, Y., Robbe-Dubois, S. and Lagarde, S.: 2002, Orbits and spectra of extra solar planets with the VLTI focal instrument AMBER, in: G. Aime and R. Soumer (eds.), *Proceedings of the Workshop on Very High Dynamics Imagery and Extra Solar Planets*, Nice, France, May 2002. EDP Science Series.

Robbe-Dubois, S., Antonelli, P., Bresson, Y. et al.: 2002, VLTI focal instrument AMBER: results of the laboratory commissioning of the warm optics, *SPIE* conference **4838** Interferometry for Optical Astronomy II, Waikoloa, Hawaii, USA.

Segransan, D. et al.: 2001, How AMBER will contribute to the search of brown dwarfs and extrasolar planets, *SPIE* proceedings **4006**, Munich, March 2000, pp. 269–276.

Vannier, M., Petrov, R.G., Robbe-Dubois, S. et al.: 2002, Optimizing a Differential Interferometric Instrument, in: J. Bergeron and G. Monnet (eds.), *ESO Symposia Scientific Drivers for ESO Future VLT/VLTI Instrumentation*, pp. 340, 346.

# PREPARATION OF THE AMBER INTEGRATIONS

CARLA GIL[1,2], KARINE PERRAUT[2] and OLIVIER PREIS[2]
[1] *Centro de Astrofísica da Universidade do Porto, Rua das Estrelas, 4150-762 Porto, Portugal*
[2] *Laboratoire d'Astrophysique de l'Observatoire de Grenoble, BP 53 38041 Grenoble Cedex 9, France*

**Abstract.** The Astronomical Multi-Beam Recombiner (AMBER) is a near infrared/red focal interferometric instrument. Its integration takes place in Grenoble where each sub-system is tested, aligned and the AMBER requirements validated. In a preliminary phase the environment of the AMBER integration room was characterized. Several tests were made in order to determine, and when required to reduce, environmental constraints (temperature, turbulence and vibrations of the optical table).

## 1. The AMBER Instrument

AMBER is the near infrared/red Astronomical Multi-Beam Recombiner for the VLTI. It is a project which involves five institutes: Départment Fresnel de l'Observatoire de la Côte d'Azur, Département d'Astrophysique de Nice – Sophia Antipolis, Laboratoire d'Astrophysique de Grenoble, Max-Planck Institut für Radioastronomie and Osservatorio Astrofisico di Arcetri. It is developed to combine the light from two or three telescopes of the Very Large Telescope (VLT). It can combine the light fro m the Unit Telescopes (UTs) and/or the Auxiliary Telescopes (ATs). AMBER will operate in the J, H and K bands, from 1 to 2.5 $\mu$m, with a spectral resolution up to 10000 (Petrov et al., 2002).

## 2. Characterizing the Environment of the AMBER Integration Room

### 2.1. AMBER SPECIFICATIONS

The purpose of this study is to understand how this different parameters could influence the performances of the instrument (visibility and phase stability/accuracy requirements in 5 minutes are $10^{-3}$). The AMBER beam stability requirements have an accuracy and stability specifications of $\pm$ 3".

Before starting the integrations several specifications for the integration room were determined. It was decided that the heating sources would remain outside the room, that the integration room would be a clean room (with a fan ceiling, anteroom, clean clothes), that we would use the air conditioning to regulate the

*Astrophysics and Space Science* **286:** 69–72, 2003.

temperature and finally the use of a table pillar independent of the rest of the building to avoid vibration transmissions.

This preliminary phase consisted in a systematic study to characterize the temperature gradient and turbulence induced by the air conditioning, the turbulence generated by the fan ceiling and vibrations of the optical table inside the integration room.

### 2.2. Testbenches

In order to measure angular fluctuations a flat mirror was put at one extremity of the AMBER table and aligned with an Electronic Autocollimator which allows to accurately measure 2-axis angular displacements. This collimator was set at the other extremity of the table. The incident and reflected light propagates under the fan ceiling whose pieces are about 1-m above the beams. To try to correlate turbulence effects to temperature variations, five temperature sensors were used.

We also used an accelerometer in order to measure and record the vertical vibrations of AMBER optical table.

### 2.3. Results

#### 2.3.1. *Turbulence*

We wanted to qualify a quiet level by recording $X$ and $Y$ angular variations when the fan ceiling was off and computing the rms fluctuations of the $X$ and $Y$ positions. To determine the effects of the fan ceiling for various powers of the fan, we recorded $X$ and $Y$ angular variations when the fan ceiling was on. One computed the amplitude of these variations as well as the time required to come back to the quiet level.

We concluded that without fan ceiling or with a 30% fan power, the $X$ and $Y$ angular fluctuations remained lower than $\pm$ 0.6", which is fully acceptable as regards to the accuracy and stability specifications of $\pm$ 3". For a fan power up to 50%, the time required to come back to the quiet level remains of the order of half an hour (see Figure 1). The results obtained clearly show that air conditioning operation obviously induces an increase of the $X$ and $Y$ angular fluctuations. People who entered in the hut do not clearly disturbed the $X$ and $Y$ angular positions in a short term.

#### 2.3.2. *Temperature*

During the previous steps, the air conditioning remained on and servo-controlled at a temperature of 18°C. We carried out recordings with the air conditioning off. Temperature recordings show that our air conditioning controls the temperature within an accuracy of $\pm$ 3°C, which is significantly larger than the specifications at Paranal (better than 1°). To avoid thermal effects (as those observed in Figure1 in the right panel) we decided to stabilize the temperature by means of another device.

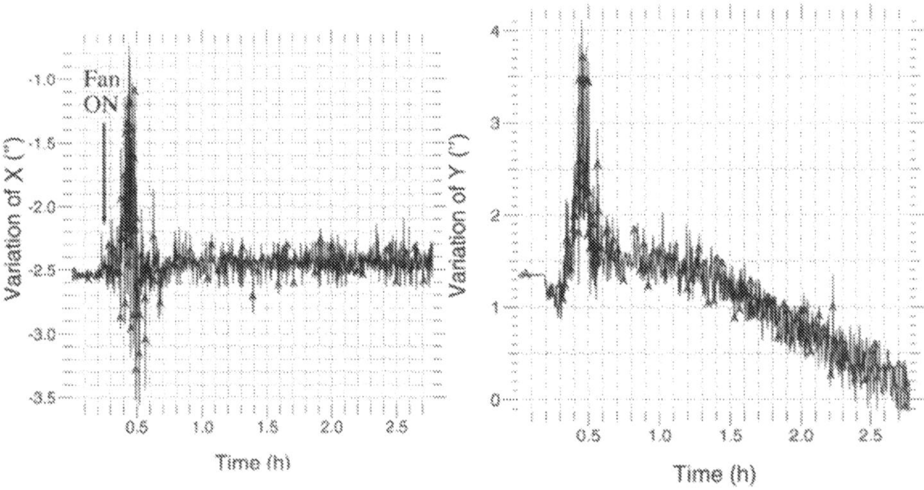


*Figure 1.* Angular variations (") in $X$ (left) and $Y$ (right) directions vs. time (hours). Fan was switched on after 1/4 h. We can see that after switching on the fan there is a period of about half an hour before coming back to a quiet level. The $Y$ variation reveals a negative drift due to temperature variations ($\pm$ 3°C). Ordinates have arbitrary units.

2.3.3. *Vibrations*

In parallel, one determined for each hour how many times the vibration of the table exceeded 100 $\mu$V, which correspond to a vibration amplification of 100 $\mu$g.

The results of the vibrations monitoring indicate that the main vibration mode is centered around 18 Hz. Due to the proximity to a road it was shown that the traffic effects can be detected but the induced vibrations on the AMBER table remained lower than 200 $\mu$g.

## 3. Integration of AMBER

During a first phase all the components of the subsystem were tested by the institutes of the AMBER consortium responsible for it. Each sub-system was fully integrated and tested before delivery for its integration on the AMBER table.

Then a second phase started in LAOG: the active integrations. This phase consist in the assembly, integration and test of AMBER in Grenoble. After all subsystems were delivered in LAOG we started aligning each sub-system with the others: the spectrograph and the optomechanics are integrated and tested together. Then, the spectrograph and the detector are integrated and tested together. The proper drive and control of the opto-mechanics by the software is checked in parallel. After these steps, AMBER is tested. The software is integrated and tested with the whole instrument and the proper processing and execution of templates are checked.

## 4. Perspectives

Integration is under progress and has been validated by preliminary fringes. Several weeks of tests are foreseen in Spring 2003.

## Acknowledgements

We would like to thank all the AMBER team and in particular all those who have helped in the preparation of this article. We also wish to thank Julien Charton for his help during the acquisition of the data. Gil was partially supported by grant POCTI/1999/FIS/34549 approved by FCT and POCTI, with funds from the European programme FEDER.

## References

Gil, C. and Perraut, K.: 2002, AMBER Memo AMB-INT-010, Turbulence and temperature measurements inside the AMBER integration room.

Malbet, F. et al.: 1999, AMB-REP-004, ESO/VLTI, Conceptual Design Review.

Perraut, K. et al.: 2000, VLT-PLA_AMB_15830-7001, Assembly, Integration and Test Plan.

Petrov, R. et al.: 2000, AMBER: the near-infrared focal instrument for the Very Large Telescope Interferometer, in: Pierre J. Lena and Andreas Quirrenbach (eds.), *Proc. SPIE, Interferometry in Optical Astronomy*.

Preis, O.: 2002, AMBER Memo AMB-INT-008, 24h Vertical vibration monitoring of the optical table for AMBER.

# MIDI – THE 10 $\mu$M INSTRUMENT ON THE VLTI

CHRISTOPH LEINERT[1], UWE GRASER[1], FRANK PRZYGODDA[1], LAURENS B.F.M. WATERS[2], GUY PERRIN[3], WALTER JAFFE[4], BRUNO LOPEZ[5], ERIC J. BAKKER[4], ARMIN BÖHM[1], OLIVIER CHESNEAU[1], WILLIAM D. COTTON[6], SIEDS DAMSTRA[7], JEROEN DE JONG[4], ANNELIE W. GLAZENBORG-KLUTTIG[7], BERNHARD GRIMM[1], HIDDO HANENBURG[7], WERNER LAUN[1], RAINER LENZEN[1], SEBASTIANO LIGORI[1], RICHARD J. MATHAR[1], JEFFREY MEISNER[4], SÉBASTIEN MOREL[8], WERNER MORR[1], UDO NEUMANN[1], JAN-WILLEM PEL[9], PETER SCHULLER[1], RALF-RAINER ROHLOFF[1], BRINGFRIED STECKLUM[10], CLEMENS STORZ[1], OSKAR VON DER LÜHE[11] and KARL WAGNER[1]

[1]*Max-Planck-Institut für Astronomie, Heidelberg, Germany*
[2]*Astronomical Institute, University of Amsterdam, Netherlands*
[3]*DESPA, Observatoire de Paris-Meudon, France*
[4]*Sterrewacht Leiden, University of Leiden, Netherlands*
[5]*Observatoire de la Côte d'Azur, Nice, France*
[6]*NRAO, Charlottesville, Virginia, USA*
[7]*ASTRON, Dwingeloo, Netherlands*
[8]*ESO, Garching, Germany*
[9]*Kapteyn Observatory, University of Groningen, Netherlands*
[10]*Thüringer Landessternwarte Tautenburg, Germany*
[11]*Kiepenheuer-Institut für Sonnenphysik, Freiburg, Germany*

**Abstract.** After more than five years of preparation, the mid-infrared interferometric instrument MIDI has been transported to Paranal where it will undergo testing and commissioning on the Very Large Telescope Interferometer VLTI from the end of 2002 through large part of this year 2003. Thereafter it will be available as a user instrument to perform interferometric observations over the 8 $\mu$m–13 $\mu$m wavelength range, with a spatial resolution of typically 20 milliarcsec, a spectral resolution of up to 250, and an anticipated point source sensitivity of N = 3–4 mag or 1–2.5 Jy for self – fringe tracking, which will be the only observing mode during the first months of operation. We describe the layout of the instrument, laboratory tests, and expected performance, both for broadband and spectrally resolved observing modes. We also briefly outline the planned guaranteed time observations.

**Keywords:** VLTI, MIDI, interferometry, mid-infrared, thermal infrared

## 1. Introduction

Up to now, interferometry in the mid-infrared around 10 $\mu$m has been performed only in narrow-band heterodyne operation. The quite successful ISI instrument (Bester et al., 1990) therefore is limited to rather bright objects of about 100 Jy or more. The mid-infrared interferometric instrument MIDI, covering the full 8

*Astrophysics and Space Science* **286:** 73–83, 2003.

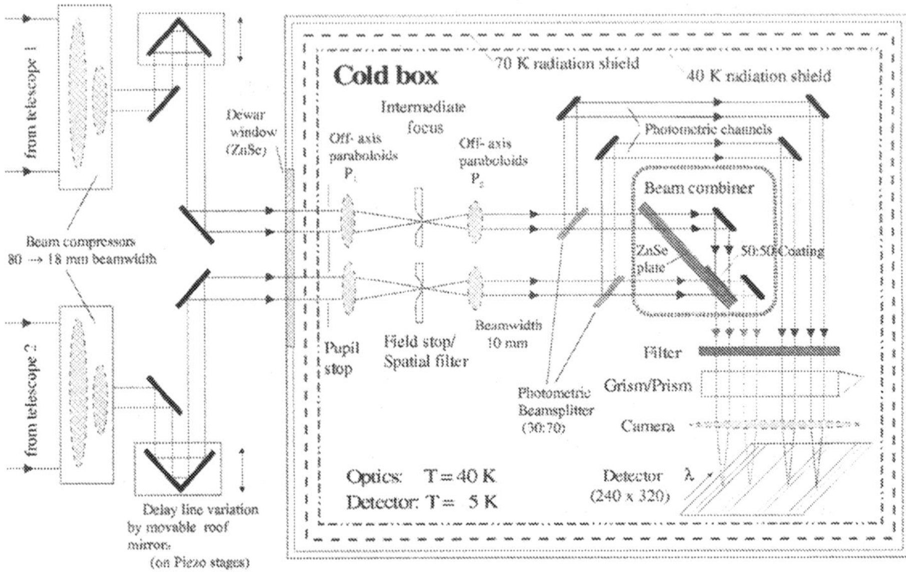


*Figure 1.* Optical concept of MIDI. Beams are called A (bottom) and B.

$\mu$m–13 $\mu$m range of the N band, is one of the two first generation instruments of the Very Large Telescope Interferometer (VLTI), the second one, AMBER, working in the near-infrared. After successful commissioning MIDI will be the first instrument doing direct combination 10 $\mu$m interferometry worldwide. Its main advantage with respect to the ISI heterodyne interferometer will be the higher sensitivity, due both to the larger diameter telescopes of the VLTI and the ability for broadband operation. Observations of sources with brightness 1Jy (N = 4 mag) appear possible initially, and a range of objects fainter by a factor of 10–100 should become accessible once external fringe stabilisation by a dedicated fringe tracker has been introduced at the VLTI.

## 2. Principle of Measurement

The design of the instrument has been driven by the necessity to suppress as well as reasonably possible the strong thermal emission from the environment (Figure 1). Most of the instrument optics have to be in a dewar and to be cooled to cryogenic temperatures 40±5 K and to be surrounded by a wall at the same temperature in order not to disturb the detector by residual thermal emission. The detector itself operates at 5-10 K. A second radiation shield at about 77 K has to be introduced just inside the dewar walls to block the thermal radiation of the dewar housing. Good suppression of the thermal radiation of external background surrounding the field-of-view requires designing for both a cold pupil and a cold field stop inside the instrument. We choose to have beam combination in (or close to) the reimaged pupil and to detect the signal in the image plane. This appeared to us the most practical and promising way to provide the baffling against the very high and varying thermal background and to reduce the effects of variable pixel responsivity.

The instrument has been built to combine the beams from two telescopes at a time, which gives the best sensitivity. In Figure 1, from the left the afocal beams from two telescopes of the VLTI are approaching the instrument. Their nominal diameter is 80 mm, and they are reduced to 18 mm diameter by a beam compressor provided as part of the VLTI infrastructure, represented here for simplicity by two lenses.

After the four folding mirrors of a small experiment-internal delay line, the compressed beams enter the cryostat ('Cold box') through the entrance window ('Dewar window'). Each telescope pupil is imaged by the VLTI delay line optics onto a cold pupil stop to provide the needed suppression of thermal emission from outside the beams. Next, an intermediate focus is formed, where different slits or diaphragms (i.e. spatial filters) can be introduced for additional suppression of unwanted radiation. If no spatial filters are used, the detector pixels, which are much smaller than the image of the telescope Airy disk, provide an alternative and flexible way to limit the spatial region admitted for the measurement. Then the beams are recollimated (again, reflective optics are represented by lenses for simplicity) and move on to combine on the surface of a 50%:50% beam splitter. This coating is indicated on the lower back side of the ZnSe plate. Here is the heart of the instrument.

From the beam combiner onwards, the two interfering beams have a common optical axis. Actually, there are two of such overlaid beams, one outgoing to each side of the beam combiner. These two outputs are modulated in flux depending on the optical path difference of the interfering beams, but with opposite sense because of energy conservation.

We measure the degree of coherence between the interfering beams (i.e. the object visibility at the actual baseline setting) by artificially stepping the optical path difference between the two input beams rapidly over at least one wavelength within the coherence time of $\approx 0.1$ s. This is done with help of the piezo-driven roof mirrors forming part of the small delay lines just outside the cryostat. The result in both channels is a signal modulated with time ('temporal fringe'), from which the fringe amplitude can be determined. The large and not precisely known thermal background forces us to determine the total flux separately by a chopped measurement, chopping between the object and an empty region of the sky, and determining the source flux by subtraction. By dividing fringe amplitude by total flux the visibility is obtained.

The measured visibility gets reduced if the flux is not the same in the two incoming beams. For good accuracy of the visibility determination, the relative fluxes of the combining beams have to be known at the time of the fringe measurement. To this end, part of the incoming light can be extracted out of both of

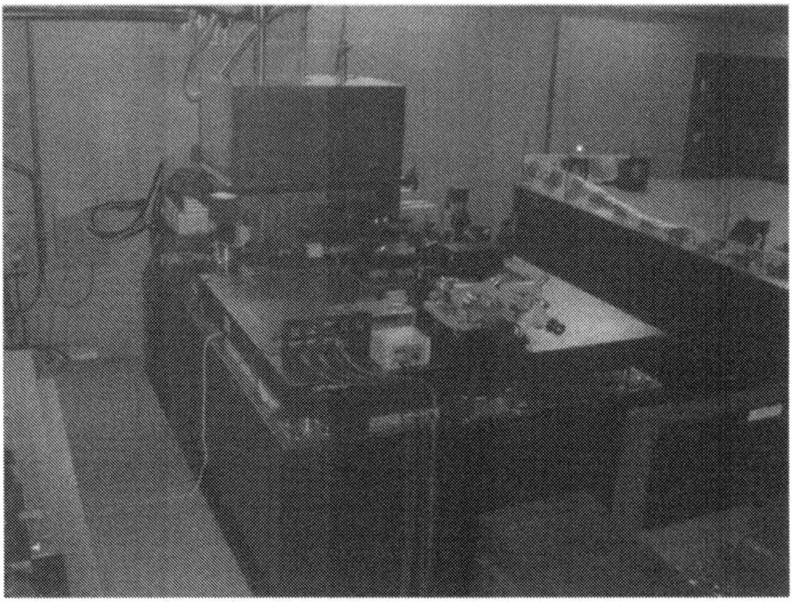

*Figure 2.* MIDI in the interferometric laboratory on Paranal. The telescope beams come in from lower right, the cold optics are hidden in the almost cubic dewar housing in the back, and the warm optics in positioned on the optical table in front of it.

the incoming beams before beam combination, and these so-called 'photometric beams' or 'photometric channels' are also imaged onto the same detector and in the same way as the interferometric signal.

Spectral resolution is obtained by filters or with a prism or grism in the parallel beam after beam combination. This provides moderate spectral resolution and also can be used to keep the detector from saturating very quickly, within the order of 1 ms for a broad-band filter.

In principle, for the measurements with MIDI the full interferometric field of $\pm 1''$ is available simultaneously. This is a small area on the sky, but about 8 times larger than the FWHM of the Airy disk of a single telescope, to which interferometric measurements are normally limited. In any case, this available field helps to identify the objects on the telescope.

## 3. Realisation

The instrument realised according to this concept can be seen in Figure 2 at its final place in the interferometric laboratory of the VLTI on Paranal. Figures 3 and 4 show how the concept evolved to a design and finally to the instrument hardware. In particular the cold optics (Figure 4) came out quite compact. This is on purpose

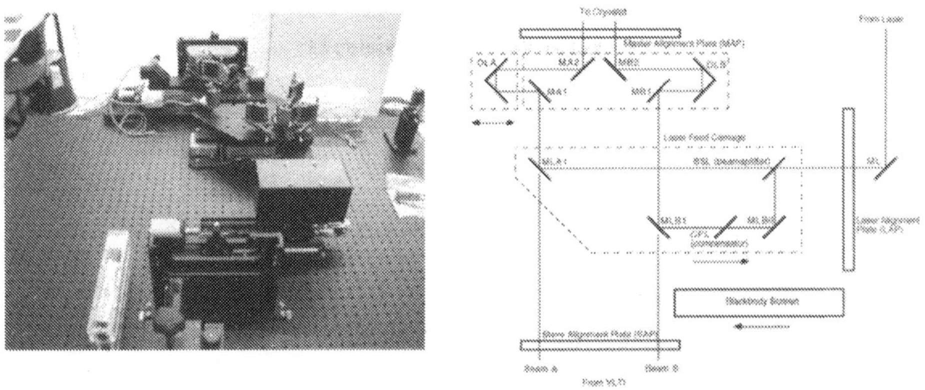

*Figure 3.* warm optics hardware and schematics

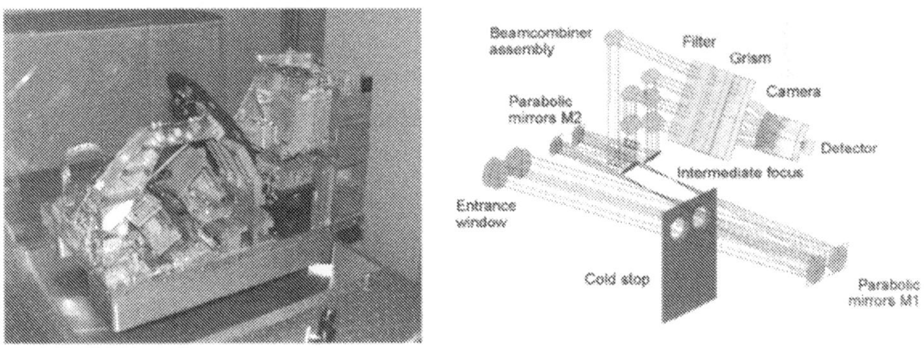


*Figure 4.* cold optics hardware and schematics

to keep internal flexures and vibrations at a low level (Glazenborg-Kluttig et al., 2003).

## 4. Options

For most of the optical elements, like field stops, cameras, filters, dispersing elements, the instrument provides flexibility by several choices, which are summarised in the following tabular presentation. Out of the many possible combinations only those most useful for actual observations will be offered, depending on the outcome of the commissioning runs. A future extension to observations at to 20 $\mu$m is in preparation.

| | | |
|---|---|---|
| **cameras** | field camera | λ/D = 3 pixels |
| | spectral camera | λ/D = 1x2 pixels (1 in spectral direction, 2 in spatial direction) |
| | pupil camera | pupil diameter = 40 pixels |
| **dispersing elements** | grism | resolution per two pixels : 220 at 10 $\mu$m |
| | prism | resolution per two pixels: 35 at 10 $\mu$m |
| | none | for observations with filters |
| **filters** | 10 filters | including full N-band, short N band (8.0–9.5 $\mu$m), long N (10.5–13.1 $\mu$m), narrow bands |
| **focal plane** | no diaphragm | 2 mm diameter (oversized) |
| | full field | 1 mm diameter (2.6 $''$ on UTs) |
| | slit width | 50 $\mu$m, 100 $\mu$m, 200 $\mu$m |
| | single pinhole | 0.4 mm diameter (1.0$''$ on UTs) |
| | triple pinholes, spacing 0.29 mm | 70 $\mu$m, 110 $\mu$m, 170 $\mu$m (diameter) |
| **beamcombiner unit** | high sensitivity | no photometric beams |
| | with photometric beam extraction | 30 % of light goes to photometric channels |
| | open | no beam combination |

## 5. Observing Modes

The observing modes of MIDI and the preparation for interferometric observations at mid-infrared wavelengths are the topic of a separate paper in these proceedings (Przygodda et al., 2003). Here, we give a short overview on how data will be taken with this instrument.

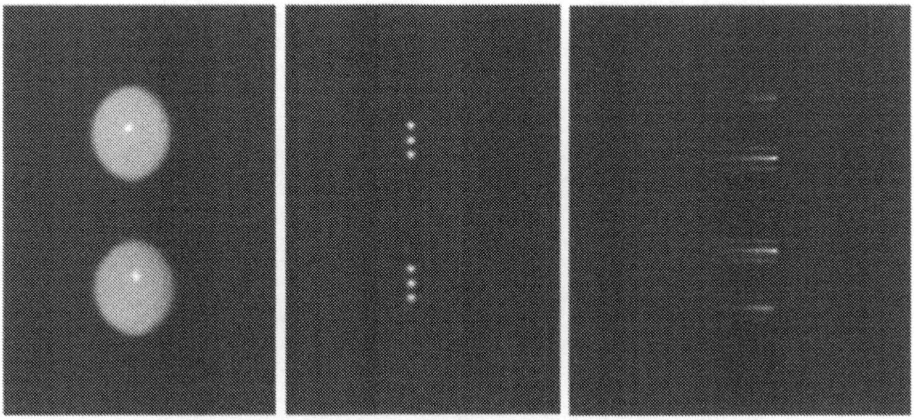

*Figure 5.* Typical examples for the light distribution on the detector. From left to right: point source in the open field which is filled by background radiation; triple pinholes illuminated by background radiation; prism with slit and triple pinholes and point source in the central pinhole (see text for details).

The detector is a Si:As impurity band conduction (IBC) array produced by Raethyon, 320×240 pixels in size. It is read out in 16 bands, each 20 pixels wide, which gives rise to the band structure weakly seen in the panels of Figure 5. The readout noise is ≈ 1000 electrons. The readout electronics, quite flexible, allow a full frame readout time of 5.6 ms, which time can be further reduced by reading only rows of relevance for the current observation (so-called 'hardware windowing'). Standard operation is first to take the exposure, then to read out ('integrate then read'). More details are given by Ligori et al. (2003).

The full long dimension of the detector is only used with the higher dispersion grism observations, its full height only when the photometric beams are extracted. For most observing modes only part of the detector array is illuminated, for which a few examples are shown in Figure 5. On the left, no diaphragm has been inserted in the focal plane. The thermal background radiation of the laboratory environment fills the field for the two incoming beams A and B (up), while a laboratory source – fed into both beams at the same time to simulate a star – produces the point-like bright spots. In the middle, a triple pinhole has been used for both beams with the laboratory source switched off. The illumination is due to the thermal background radiation, always strong at 10 $\mu$m. A slit would occupy the same position as the triple pinholes. On the right, the triple pinhole has been used again and the prism has been inserted. The dispersed spectrum has the short wavelengths on the right end. In each case, the middle spectrum is the brightest, because the laboratory source has been centered on the middle one of the triple pinholes. The separation in vertical direction is smaller than in the two first panels because the spectroscopic camera with its shorter focal length has been used here. Finally, for these measure-

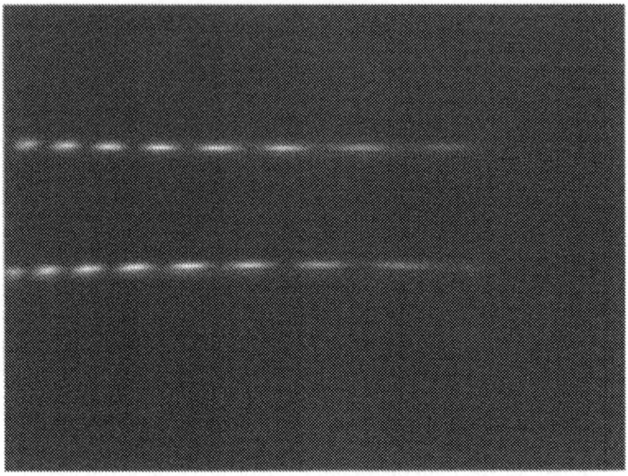

*Figure 6.* Dispersed fringes measured in the laboratory. The optical path difference between the two interfering beams was about 140 $\mu$m.

ments the photometric beams have been extracted by inserting a beam splitter. This leads to the two additional spectra on the top and bottom of the panel.

Interference patterns for determination of source visibility are produced in two ways. For undispersed data, the optical path difference (OPD) between the two input beams is varied by moving one of the roof mirrors of the small internal delay line. This results in a fringe pattern, which is damped according to the width of the filter used (see example in Przygodda et al., 2003). For dispersed data, the OPD is offset on purpose from zero by several wavelengths. Then the fringe pattern appears at any instant as a modulation of the spectrum with wavelength (Figure 6).

## 6. Anticipated Performance

While the obtainable spatial resolution is given by the geometrical arrangement of the telescopes of the VLTI and the wavlength of observation, the sensitivity of the instrument is determined mostly by the properties of the atmosphere. First, the thermal emission of the VLTI mirror train of about $2\times10^{11}$ photons per s per Airy disk results in a correspondingly high photon noise against which the signals have to be detected. Second, the integration is limited to about 0.1 s, since the fringe motion due to the turbulent atmosphere would smear the fringe pattern at substantially longer integration times. It still has to be proven that fluctuations of sky brightness are of lower importance compared to background photon noise and that the assumed integration times are available. However, the given estimates of limiting magnitude for self-fringe-tracking operation are based on this assumption.

The sensitivity of the instrument can be increased if external stabilisation of the fringes is provided, which should become available on the VLTI within the coming

TABLE I
Basic parameters of the MIDI instrument

| | | |
|---|---|---|
| Telescope diameter | UTs | 8 m |
| | ATs | 1.8 m |
| Baselines available | UTs | 47 . . . 130 m |
| – one at a time | ATs | 8 . . . 200 m |
| Resolution | λ / D for 100 m | 0.020″ at 10 μm |
| Wavelength coverage | | N band (8 μm–13 μm) |
| | expandable to | Q band (17 μ–26 μm) |
| Field of view | UTs | ± 1″ |
| | ATs | ± 5″ |
| Airy disk (FWHM | 8 m telescopes | 0.26″ |
| at 10 μm) | 1.8 m telescopes | 1.14″ |
| Input beam diameter | from UTs | compressed to 18 mm |
| | from ATs | 18 mm |
| Throughput of VLTI | to instrument | 30 %–35 % |
| Sampling time for | for fringe motion | 100 ms |
| fringe measurement | $\leq$ 1 μm rms | |
| Atmospheric stability | for chopping | 200 ms |
| Background (in Airy disk) | from sky | $2 \cdot 10^{10}$ photons/sec |
| | from VLTI | $2 \cdot 10^{11}$ photons/sec |
| Detector(320 × 240 pixels) | full well capacity | $\approx 1.0 \cdot 10^7$ electrons |
| | read noise | $\leq$ 1000 electrons |
| Limiting N-magnitude | UTs | 3–4 mag (1–2.5 Jy) |
| (self-fringe tracking) | ATs | 0–0.8 mag (20–50 Jy) |
| Limiting N-magnitude | UTs | 8-9 mag (10–25 mJy) |
| (external fringe tracking) | ATs | 5–5.8 mag (200-500 mJy) |
| Accuracy of visibility | | ± 5 % |
| Phase measurement | internally | differential within band |
| | external referencing | possible |

year. Then the fringe patterns of individual short observations can be added blindly and coherently, and theoretically a gain proportional to square root of integration time could be obtained. By adding up $10^4$ individual 100 ms integrations this could improve the limiting magnitude by up to 5 mag. The sensitivities given in Table I refer to the use of the full N band. For spectrally dispersed measurements with the grism, the limiting magnitudes will be reduced by 1–1.5 mag.

## 7. Scientific Programme

The scientific potential of the MIDI instrument has been discussed by Lopez et al. (2000). Based on the slightly optimistic sensitivity estimates of the preceding section, a guaranteed time programme has been selected to fill the 300 hours of guaranteed observing time available to the instrument team on the UTs. This listing gives an impression what may be feasible to observe with the MIDI instrument. It has to be kept in mind that the attempt for direct planet detection is untypical. It is a programme of extremely high risk for possibly high reward.

The actual observations on the VLTI will be performed in the form of so-called 'observation blocks' and with procedures described for the case of observations with MIDI by Bakker et al. (2003).

| Topic | Telescopes | |
|---|---|---|
| | UTs | ATs |
| Dust Tori in Nearby Active Galactic Nuclei | 65 h | – |
| Inner disks of low-mass young stellar objects | 65 h | 90 h |
| Inner disks around intermediate-mass young and Vega-type stars | 62.5 h | 100 h |
| Massive young stars | 52.5 h | 305 h |
| The dusty environment of hot stars | 2 h | 68 h |
| Cool Late Type Stars and related objects | 25 h | 450 h |
| Extra-solar planets and brown dwarfs | 25 h | – |

## 8. Next Steps for 10 $\mu$m Interferometry with MIDI

First observing runs on the baseline UT1-UT3, in self-fringe-tracking mode and therefore limited to brighter objects have started in the summer of 2003. After this the following development steps are foreseen:

- Fall of 2003 – additional UT baselines
  – adaptive optics, external fringe tracking, allowing to go to fainter sources
- Spring of 2004 – observations with the 1.8 m telescopes (ATs) of the auxiliary array
- Fall of 2004 – extension to 20 $\mu$m observations
  – addition of 10 $\mu$m single mode fiber in the focal plane for beam cleaning
- 2005 – dual-beam capability on the VLTI, allowing to observe faint sources

With these added possibilities, MIDI should become an instrument applicable to a wide range of astronomical studies.

## Acknowledgements

We thank the electronic and mechanical workshops in Heidelberg, Dwingeloo and Freiburg for their accurate and careful work, without which the instrument would not have become reality. And we thank the instrument responsibles at ESO, Markus Schöller, Andrea Richichi and Peter Biereichel, for positively critical partnership.

## References

Bakker, E.J., Przygodda, F., Chesneau, O. and Jaffe, W.: 2003, *MIDI* scientific and technical observing modes, *SPIE* **4838**, 905–916.

Bester, M., Danchi, W.C. and Townes, C.H.: 1990, Long baseline interferometer for the mid-infrared, *SPIE* **1237**, 40–48.

Glazenborg-Kluttig, A.W., Przygodda, F., Hanenburg, H., Morel, S., Pel, J.-W.: 2003, *Realization of the MIDI Cold Optics, SPIE* **4838**, 1171–1181.

Ligori, S., Graser, U., Grimm, B. and Klein, R.: 2002, in: Kona, W. Traub (ed.), *Experiences with the Raethyon Si:As IBC Detector Arrays for Mid-IR Interferometric Observations, SPIE* **4838**, 774–785.

Lopez, B., Leinert, Ch., Graser, U. et al.: 2000, The astrophysical potentials of the MIDI VLTI instrument, *SPIE* **4006**, 54–67.

Przygodda, F., Chesneau, O., Graser, U., Leinert, Ch. and Morel, S.: 2003, Interferometric observations at mid-infrared wavelengths with MIDI, these proceedings,

# INTERFEROMETRIC OBSERVATION AT MID-INFRARED WAVE-LENGTHS WITH MIDI

FRANK PRZYGODDA, O. CHESNEAU, U. GRASER and CH. LEINERT
*Max-Planck-Institut für Astronomie, Heidelberg, Germany*

S. MOREL
*European Southern Observatory*

**Abstract.** MIDI, the *MID*-Infrared *I*nterferometric Instrument for ESO's Very Large Telescope Interferometer (VLTI), will be the first instrument for combining mid-infrared light directly in order to obtain angular resolution up to 10 mas (assuming a 200 m baseline) in a wavelength range from 8 to 13 $\mu$m. Currently in the phase of commissioning at Paranal, the start of its scientific operation is expected for summer 2003. Direct interferometry at thermal infrared wavelengths demands special requirements on the instrument and also on the procedures of preparation of data reduction. Hereafter MIDI's different observing modes are described and an example for an interferometric observation is given.

**Keywords:** VLTI, MIDI, interferometry, interferometric observation, infrared observation, mid-infrared, thermal infrared

## 1. Introduction

### 1.1. Interferometry at Thermal-Infrared Wavelengths

Observing at mid-infrared wavelengths has its special attributes due to the characteristics of the earth's atmosphere. Compared to the optical and the near-infrared regime, the size of turbulence cells and also the time scale of their stability is larger. The Fried parameter $r_0$, which is connected to the effective size of the cells, reaches approximately 5 m at an observing wavelength of 10 micron. The atmospheric coherence time, within the atmosphere can be considered as stable, averages to about 100 ms. These numbers give the impression that observing at mid-infrared wavelengths is an easy undertaking, but one property was not mentioned yet: the extremely high and temporally changing background radiation of the sky. Compared with the signal of, for instance, a N = 0 mag object, the flux per arcsec from the sky is about 10 times larger. The usual way to overcome this problem is the application of a technique called 'chopping and nodding'. 'Chopping' means the quick tilting of the secondary (or another convenient) mirror with a frequency of several Hz while images are taken alternately at the two orientations. After the subtraction of the images, the sky background is mostly canceled and mainly the object's signal resides. To cancel also the small but non-avoidable influence of

*Astrophysics and Space Science* **286:** 85–91, 2003.

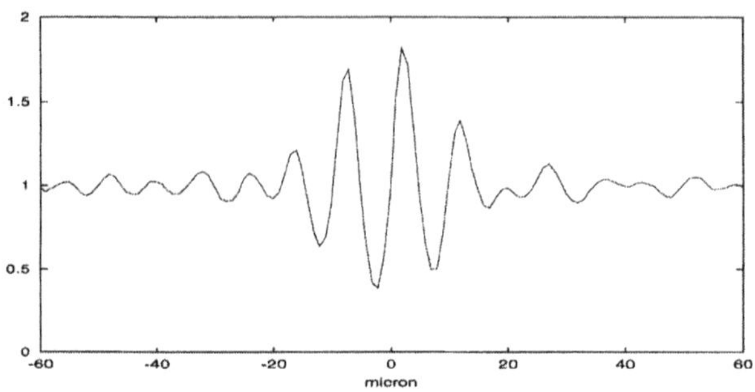


*Figure 1.* Fringe packet obtained during an OPD scan of 140 micron. Plotted is the normalized flux of the interferometric output channel (real MIDI test data).

the telescope on the background flux, which is connected to the orientation of the chopping mirror, the whole telescope can be pointed to an off-source position. This so-called 'nodding' of the telescope, together with a repetition of the chopping-procedure at the new position, allows the nearly complete cancellation of the disturbing background. Before this technique will be explained more detailed in the context of MIDI we start with a general description of the visibility measurement.

## 1.2. Fringes and visibility-measurement

The main objective of the MIDI instrument (Ch. Leinert et al., 2000) is the coherent combination of the two input beams in order to achieve object information with high angular resolution. The combination takes place in MIDI's beam combining unit where the light coming from the two telescopes interferes at a 50%/50% beam-splitter plate. In this type of interferometer, also known as *on-axis pupil plane interferometer*, the result of the interference is imaged after the combination in form of two interferometric output channels on the detector. If one of the interferometric channels shows destructive interference of the light, the other shows constructive, and the other way round. By introducing an OPD (= Optical Path Difference) between the two input beams, which is larger than the coherence length of the observed light, the interference pattern disappears. Therefore, a scan of the OPD around its zero position results in the measurement of a 'fringe packet' (see Figure 1). The strength of the modulation of the signal, the so-called 'fringe contrast', depends on the object's visibility function and the baseline used. The fringe contrast is the observable which contains the object information with high algular resolution.

Two different methods for an accurate measurement of the fringe contrast are described in the following:

- Fourier-Method: Monitoring the flux of the interferometric output channels during a long OPD-scan across the zero-position results, like described above,

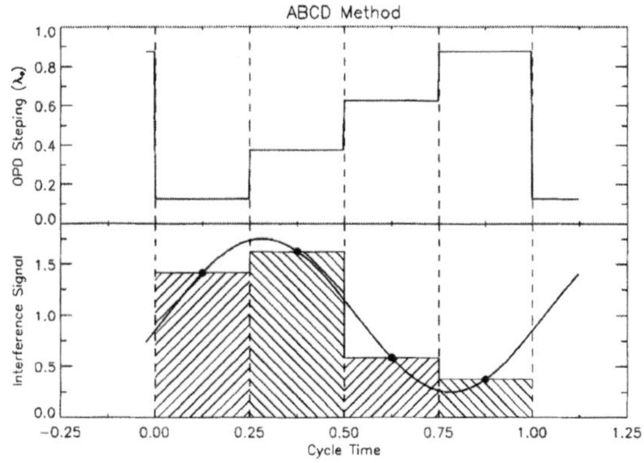


*Figure 2.* Measuring the fringe contrast by scanning the fringe in four steps. *upper graph:* OPD steps, *lower graph:* measurement of the flux at four points.

in the measurement of a fringe packet. The Fourier transform of this dataset shows a broadened peak at the frequency of the fringe modulation. The area of the peak is proportional to the object visibility. This method is insensitive to OPD shifts (introduced for instance, by the atmosphere or the hardware) as long the whole fringe packet in within the range of the scan.

– ABCD-Method: This faster method just needs the flux in the interferometric output channels at four different points. For this, MIDI's piezo-driven internal delay line performs four equidistant steps, with a distance of $\pi/2$ in terms of the measured wavelength (see Figure 2). This method needs an exact fringe tracking to avoid phase shifts within the measurement time of the four values. Also, it must be ensured that the measurement is executed at the center of the fringe packet. The object visibility $V$ is proportional to $\sqrt{(A-C)^2 + (B-D)^2}$, with the flux A,B,C,D at the four steps.

Since in both methods the non-avoidable contrast loss due to instrumental and atmospheric effects is not determinable, in *any* case the measurement of a reference object with known visibility is necessary for calibration.

## 2. MIDI's Observation Modes

### 2.1. Non-dispersed modes

As a mid-infrared instrument, MIDI also uses the chopping and nodding procedure. A special difficulty is the quick finding and tracking of the fringes during the on-source position. This process needs a complex and fast control system for the MIDI instrument and the whole VLTI infrastructure including the two telescopes (S. Hippler et al., 2000). An easier way to deal with the disturbing background is MIDI's so-called 'virtual chopping'. Here, the background flux at the objects

*Figure 3.* Detector images taken in two different observing modes: 'High Sens Mode' (left) with only the two interferometric output channels and 'Photometric Mode' (right) with two additional photometric channels. Every channel shows an image of the triple pinhole used here. Slits have the same orientation as the triple pinholes.

position on the sky is derived from the flux at two near off-source positions by using an interpolating routine. For this method, either a long slit or a special set of pinholes can be inserted in MIDI's intermediate focus. These pinholes or 'field stops' consist of three single holes. The source is put to the center one, while the other two are dedicated to the background measurement. Figure 3 shows detector images taken with this kind of field stop. Each of the interferometer output channels appears as three single spots.

To increase the accuracy of the interferometric measurement (see below), MIDI has also the option to image two photometric channels. This 'Photometric Mode' enables the monitoring of the flux in the input beams before the interference takes place. For this, 30% of the light is extracted and bypasses the beam combining unit. Since the fringe amplitude depends naturally also on the flux ratio of the two input beams, the 'true' visibility can be calculated by taking into account the information from the photometric beams. Here, in difference to the more sensitive 'High Sens Mode' without photometric beams, the precision of the fringe contrast measurement is higher. (for a detailed description see also: V. Coudé du Foresto et al., 1997)

### 2.2. Dispersed modes

An unique feature of MIDI is the option of interferometric broad-band observations in a spectral range from 8 to 13 micron. For this, two dispersive elements can be inserted in the optical path. A prism allows observations with a spectral resolution of about R=25, a higher resolution of R=230 can be obtained by inserting the grism. By using these elements the images of all output channels are dispersed along the x-axis on the detector (see Figure 4). These modes, which would be used with a long slit or a triple pinhole, allow the measurement and comparison of the fringe contrast (see next section) at different wavelengths at the same time. This technique is also known as *differential interferometry*.

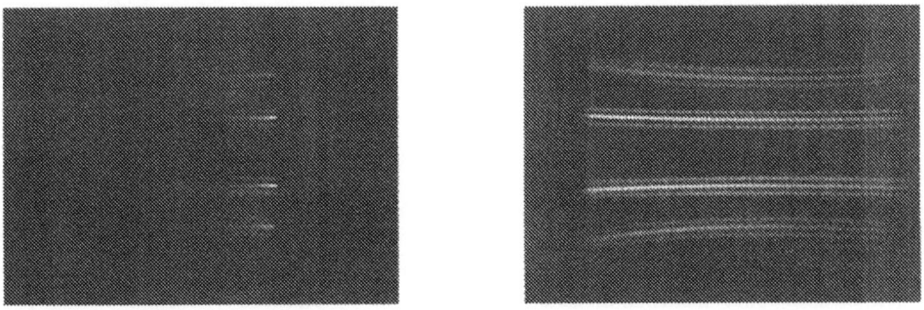

*Figure 4.* Dispersed modes with the prism (left) or the grism (right). Interferometric and photometric channels are spread along the x-axis.

### 2.3. PERFORMANCE OF MIDI

In the following, an overview of MIDI's expected performance is given. Since the instrument in not yet tested together with the VLTI infrastructure, the values may differ in reality.
Visibility accuracy: $\Delta V / V = 5\ \%$
(problem is mainly calibration, not instrument)
Limiting magnitude: (broad band operation or with prism)
– with UTs: N $\approx$ 3–4 mag (1–2.5 Jy)
– with ATs: N $\approx$ 0.8 mag (20–50 Jy)
– with dispersion by grism : 1–1.5 mag less sensitive than undispersed
– with external Fringe Tracking: up to 5 mag more sensitive

## 3. Example: Observation of Z CMa

Even more than with other instruments, the observation with an interferometer needs an accurate preparation. Useful tools for that are simulation programs like ASPRO (by Laboratoire d'Astrophysique de Grenoble), SIMVLTI (by MPIA Heidelberg) or ESO's VLTI visibility calculator. Those software packages make it possible to get an idea of the expected visibility values for given parameter setups. For a better understanding, the preparation of an observation of Z CMa is given as an example hereafter.

Z CMa is a FU Orionis type star with an IR companion at a separation of 0.1 arcsec (known from speckle observations, C. Koresko et al., 1991). An infrared excess indicates the existence of dust in the system. Observations with MIDI would clarify the location of the dust: is there a circumbinary or a circumstellar distribution of the dust? The first step is to create a model of the system with the known (and assumed) parameters (Table 3). Figure 5 shows the output of SIMVLTI for the chosen parameters. The upper right graph shows the 2-dimensional visibility function in the spatial frequency domain (also known as uv-plane) which is calculated by a Fourier

TABLE I

Model parameter for Z Cma

| | |
|---|---|
| total brightness | N = -1.28 mag |
| separation | 0.1 arcsec |
| position angle | 120 deg |
| two circumstellar Gaussian disks assumed | diameter 10 mas (FWHM) |
| flux ratio | 1:2 |
| investigated VLTI baseline setups | UT1–UT2 (56m), UT1–UT3 (103m) |
| observing time | 5 hours |

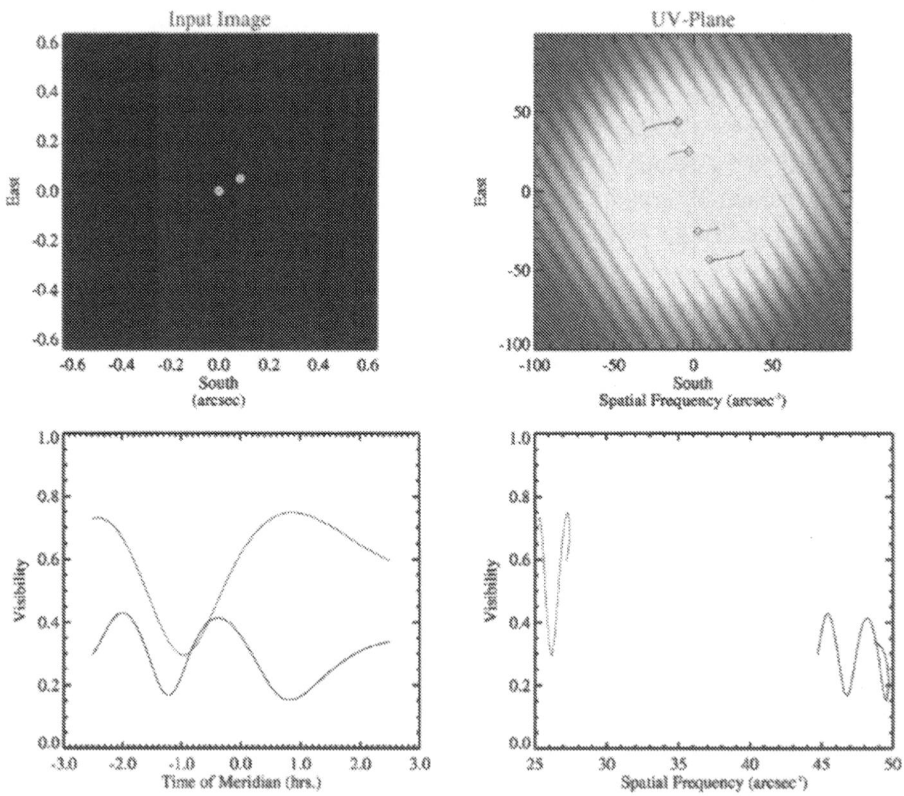


*Figure 5.* Screen-shots from SimVLTI. *Upper left:* intensity distribution on the sky, *upper right:* 2D object visibility function and superposed uv-tracks, *lower left:* expected visibility vs. observing time (for the two baselines), *lower right:* same, but visibility vs. spatial frequency

transform of the given intensity distribution. The intensity distribution of the model can be understood as a convolution of a Gaussian disk with two delta functions with a distance of the separation of the binary. Since a convolution of two functions becomes a multiplication of their Fourier transforms in the frequency domain, we find there a sinusoidal pattern (resulting from the delta peaks) multiplicated with another Gauss function (resulting from the Gaussian disk).

The program calculates also the so-called *uv-tracks*. These are tracks in the frequency domain, where the object visibility function can be measured with the selected baseline configuration during the observation time. Responsible for the track's shape is the projected length and orientation of the baseline with respect to the object which changes with time because of the earth's rotation. With a cleverly chosen configuration it is possible to detect the sinusoidal pattern as well as the Gaussian decrease of the pattern to higher frequencies.

The parameters of the sinusoidal pattern are connected to the separation, position angle and flux ratio of the binary. Note, that they will not be measurable, if the projected baseline is perpendicular to the position angle of the binary. The measurement is a verification of the binary separation but gives also the flux ratio at 10 $\mu$m.

Obtaining the parameters of the Gaussian shape by measuring the visibility function at lower and higher spartial frequencies (using UT2-UT3 and UT1-UT3) allows to determine the size of the disks. A larger (i.e. circumbinary) disk would result in a smaller FWHM of the Gaussian part in the visibility function. Therefore, a interferometric measurement with the chosen parameters can give an answer to the question of the size and location of the disks.

The basic message of this example is that the baseline configuration must be adapted to the object parameters in a way that the obtained data is able to give answers to astronomical questions. Therefore a proper preparation of the observing run is essential.

## References

V. Coudé du Foresto et al.: 1997, *A&AS* **121**, 379.
S. Hippler et al.: 2000, *SPIE* **4006**, 92.
C. Koresko et al.: 1991, *AJ* **102**, 2073.
Ch. Leinert et al.: 2000, *SPIE* **4006**, 43.

ASPRO:
`http://www-laog.obs.ujf-grenoble.fr/~jmmc/download/aspro/`
SimVLTI:
`http://www.mpia-hd.mpg.de/MIDI/SIMVLTI/SimVLTIdownload.html`
VLTI Visibility Calculator:
`http://www.eso.org/observing/etc/preview.html`

# GENIE – THE DARWIN DEMONSTRATOR

MALCOLM FRIDLUND and PHILIPPE GONDOIN
*Research and Scientific Support Department, European Space Agency, ESTEC*

**Abstract.** Darwin is a cornerstone mission of the Horizons 2000+ program of the European Space Agency. It has the express purpose of carrying out the first *direct* search for terrestrial exoplanets, and to achieve unprecedented spatial resolution in the infrared wavelength region. The detection and study of terrestrial exoplanets promises to usher in a new era in science and will affect a broad spectrum of disciplines. Further, the time line for implementation of such an instrument is now likely to be of the close order of 10 years, leading to possible answers to one of mankind's most fundamental questions in the second decade of the 21 st century. It has been found that in order to realize am interferometer in space in the next 10–15 years, it is necessary to prepare the way through a number of intermediary steps – both on the ground and in space. In this context, we here describe GENIE – a ground based nulling interferometry experiment to be implemented at the VLTI in a partnership between ESA and ESO.

**Keywords:** exo-planets, interferometry, The Darwin-mission

## 1. Introduction

The European Space Agency identified *interferometry in space* as an important topic already in the original Horizon 2000 plan. During a number of conferences/ workshops held during the 1980's, in order to define the scientific case for interferometry, the search for terrestrial exoplanets always figured prominently. It was, however, felt that the technological readiness was not then available. Advances during the 90:ties in the fields of optics, metrology and space technology lead an external survey committee to make a recommendation in 1993 for the inclusion of interferometry from space into the extension of the Horizon 2000 program (H2000+) as a cornerstone candidate. Within the interferometric context, three topics were identified for further study:

1. Astrometry;
2. The search for terrestrial exoplanets, including the characterization of their properties and atmospheres and the possible detection of biospheres through remote sensing;
3. Astrophysical imaging at a spatial resolution 2–3 orders of magnitude higher than that foreseen with the James Webb Space Telescope (JWST).

Of these the first was considered to be (relatively) 'simpler' in implementation, and resulted in the GAIA proposal and study carried out by the European Space Agency. As a result of this study, GAIA is not any longer an interferometric mission.

*Astrophysics and Space Science* **286:** 93–98, 2003.

The Darwin, however, concerns itself with the two latter concepts. Already during the definition phase of the Darwin mission, the possibility of addressing both topics (exoplanets and imaging) in one mission was given great weight. In the current paper we nevertheless focus on the search for exoplanets, and refer to the Darwin report (ESA-SCI(2000)12) for the imaging part.

Darwin was studied industrially between 1997 and 2000, and we now have a baseline design for a possible mission. This consists of 6 1.6m telescopes each on a free-flying satellite. These spacecraft fly in a hexagonal pattern with baselines between 50 and 250 meters, and transmitting the beam from each to a central hub satellite, the beams are there combined in a nulling interferometer, extinguishing the light from the central star but allowing the light from any nearby planet to be transmitted thorough and analyzed spectroscopically.

With the emergence of the new Cosmic Vision 2020 plan for European Space Research, and the introduction of the Eddington mission into that plan, as well as the inclusion of the CNES mission COROT into the ESA program, one can now say that Europe has a stepwise but ambitious approach to the study of terrestrial exoplanets and the question of life in the universe.

## 2. Formulation of the Problem

### 2.1. THE SCIENTIFIC CASE

*To detect and study Earth-type planets and characterize them as possible abodes of life* summarizes the scientific case for the Darwin mission, but nevertheless does not contain the complete picture. The simple fact is that Darwin is not only an astronomy mission, but also contains elements from geophysics (including atmospheric physics), biophysics, organic chemistry and philosophy and the humanities. Thus being totally cross disciplinary, the preparation to answer the question which put in simple terms is: *'Are we alone in the universe'* necessarily involves a host of disciplines. The data from a successful Darwin mission, is expected to impact all of those and more. It is no doubt one of mankind's longest standing questions. Although for centuries this issue has been the topic of vigorous philosophical and religious debate, we have finally arrived at a time when technology have advanced to a state that allows this question to be properly addressed.

In order to answer question like:

1. How unique is the Earth as a planet?
2. How unique is life in the Universe?

we need to basically observe other stars and **directly** determine the existence and characteristics of any accompanying bodies. This has hitherto been impossible because of the influence of the star on the attempts to observe any planet circling it.

The detection problem is essentially a matter of contrast and dynamic range. The star and planet will be very near each other on the sky, and the star will out shine the planet with up to 9 or 10 orders of magnitude. Here picture a simple old fashioned wax candle placed very near a modern coastal light house. Now place this 'model solar system' at the coast of the island of Mallorca in the Mediterranean and try to observe the wax candle from the city of Madrid (this simile adapted very freely from M. Ollivier's theses, 1999). The planet need to be characterized as what concerns mass, orbital parameters, atmospheric composition and temperature. All of this requires the repeated detection of the planet with a good *signal to noise ratio*.

The definition of an unambiguous signpost of life is another matter completely. First we need to define what is life? Then we need to determine how life affects its environment. Finally we need to define observable s that can be obtained with the level of technology foreseen for the mission. In the context of Darwin we have so far avoided the first of these questions by instead specify the requirements of the mission based on life as we have observed it on Earth. We thus disregard speculations about forms of life based on a chemistry different from that found on our planet. We then try to imagine the difference that would occur if life was nonexistent on Earth. Following Owen (1980) and Angel et al. (1986), we can then use information on how life disturbs the equilibrium in the Earth's atmosphere, as our criterion.

To fully answer the questions raised above the following appears necessary:

- Detect planets within the Habitable Zone (HZ – also referred to previously as the 'life zone' – a term coined by Frank Drake). This is the orbital radii where water would be expected to be found in a liquid state. In the Darwin study, the HZ is defined in terms of a black body temperature, and does not per se take into account atmospheric pressure etc. This of course assumes that life is based on the existence of liquid water.
- Determine the planets orbital characteristics. This means we need to repeat the observations several times.
- Observe the spectrum of the planet. Does it have an atmosphere? Determine the effective temperature and total flux. Diameter of planet (emitting area × albedo)
- Determine the composition of the atmosphere viz. the presence of water and ozone/oxygen in an Earth type planet, mainly inert gases in a Mars/Venus type planet and Hydrogen/Methane atmospheres in Jupiter type planets.

## 3. Nulling Interferometry

As mentioned before, directly detecting exoplanets is essentially a matter of contrast and dynamical range. The star around which the planet revolves is going to be $\geq 10^9$ times brighter than the planet, if we chose visual light to observe the planet in. The planet is also going to be very close. An Earth type planet in an orbit in the HZ, i.e. $\approx 1$ AU from its sun, is going to be 1 arcsec away at a distance of

1 parsec. Unfortunately, the closest potential target we have is $\alpha$ Centauri at 1.1 parsec. In order to get a suitable number of targets, we have to reach at least 10 parsec out and preferably 25 parsec, thus requiring an angular separation of 0.1 and .04 arc seconds respectively. This is for the Solar type targets. For K-type and M-type stars, which will dominate our sample, the separation distance will be much smaller, because of the demand of observing planets within the zone where we expect water to be liquid. It is clear, that using conventional telescopes, our detectors would have to handle an impossible dynamical range in order to separate out the planetary light.

The problem is alleviated by going to the infrared. As can the relative contrast between primary and planet drops several orders of magnitude by going to the thermal infrared. This was first pointed out by Bracewell (1978) and Angel et al. (1986). These authors also pointed out the advantage of utilizing lines of the $H_2O$, $O_3$ and $CO_2$ molecules as tracers of life as we know it, and that are found in the mid-IR wavelength range. Kasting et al. (1985) and Legér et al. (1993) have demonstrated that $O_3$ is a very nice tracer for $O_2$ since the former have a logarithmic dependence on the concentration of the latter. We then have suitable tracers of Earth type atmospheres in a wavelength band between 6$\mu$m and about 20$\mu$m. We still can only allow $10^{-6}$ of the stellar light to remain in the input feed to our spectrograph if we should have any hope of detecting the planetary light and register its spectrum in a reasonable time. As outlined above coronographic methods on monolithic telescopes have not been demonstrated to be able to accomplish such an extinguishing of light at the relevant spatial scales. Further, it should be stressed that an instrument would have to be of a diameter $\geq 10$ m if operated in the visual wavelength range. It has not been demonstrated that the biomarkers found in this band are as useful and unambigous as those in the mid-IR. To overcome these hurdles, the technique of 'nulling interferometry', as suggested by e.g. Bracewell and McPhie (1979) and Angel (1989) is baselined for Darwin. A nulling interferometer (a Bracewell interferometer) can best be described by considering two telescopes. By restricting the beams coming from each of the telescopes to the diameter of the point spread function, and after parallelizing the beams. We then introduce a phase shift of $\phi = \pi$ in one of the ray paths, we will achieve destructive interference on the optical axis of the system (in the combined beam). At the same time, we will have constructive interference a small angle $\theta$ away. This angle $\theta$ depends on the distance between the two input telescopes. The output of this system is a set of interference fringes (a 'map'), with a sharp 'null' in the center of the 'map'. By placing the central star under this null, and the zone where e.g. $H_2O$ will be liquid under a bright fringe, one can in principle search for planets in the 'life zone'. Now, by using more telescopes we can achieve a symmetric pattern around the star, with a deep central null on top of the star. The actual shape and transmission properties of the pattern around the central 'null' depends on the configuration, and the distance between the telescopes. Essentially, if we had an ideal case with a star and a single planet and no disturbing sources, such as exo-solar zodiacal dust, i.e. dust left over from collisions between asteroids, comets and such in the target system, the detection of a positive flux would imply that a planet is present, if the star is well and truly 'nulled out'. This means that the detector could consist of a single element. We wish, however, to also carry out spectroscopy with as high resolution as possible and a linear array is thus indicated. In the real world, there is of course a significant amount of background radiation coming from dust. The zodiacal dust in our own solar system is strong enough to be seen (in visible light) as a bright band, from the ground in dark locations. At a wavelength of $10\mu$m, this radiation is dominating (the zodiacal dust temperature within the Habitable Zone will of course also be close to 300 K and thus the peak of the emission is radiated around $10\mu$m) and a significant background signal is thus present in the inner Solar system. If we were observing the Solar system at interstellar distances, the zodiacal dust would actually be about 400 times brighter than the Earth at these wavelengths (Legér et al., 1996). In order to separate out signal of the planet from this background one need to modulate the signal from the planet and from the zodiacal dust at different frequencies. This is done either by switching between different combination schemes (certain geometrical arrangements of the telescope array), or by moving the individual telescopes around or a combination of both methods (see Legér et al. 1996 for a thorough discussion of this topic).

## 4. GENIE – The Ground Based Precursor to Darwin

Darwin is already supported by a massive technology development program. One of the key elements for Darwin is to develop and verify 'nulling' interferometry. Representative breadboards with a associated star/planet simulator providing the necessary input signals are currently in the testing stage with European industry. The next step in technology testing is obviously to design a real nulling interferometer, which can demonstrate the technology required for Darwin, as well as provide necessary experience to the organisations that will provide the final space based instrument.

An associated pre-cursor activity is the observation of the exo-zodiacal light from target systems. It was early realised both by NASA and ESA that this can be carried out from the ground, using the nulling technology demonstrator breadboards in conjunction with a large enough interferometer. The instrument of choice for the ESA is off course the Very Large Telescope Interferometer (VLTI) of the European Southern Observatory (ESO). By using the breadboard together with the VLTI, its qualification would then also provide needed and in itself interesting scientific information. An agreement of collaboration has thus been set up between ESA and ESO in order to implement such an instrument. This instrument will be the GENIE (= Groundbased European Nulling Interferometer Experiment), which will be built in cooperation between ESA and ESO. The GENIE project has under

2002 gone through a large amount of preparatory activities at both ESA and ESO, resulting in an Invitation To Tender (ITT) to industry for 2 parallell design study contracts. To be kicked-off approximately in April 2003, threse design studies will run for 12 months and result in a contract for the actual construction phase to be issued immediately afterwards. First light is planned for 2006, and a major technology evaluation and scientific programme will be carried out during some years after this. It is intended to let the instrument become a common user facility at the VLTI after that phase.

## References

Angel R.: 1989, in: P. Bely, C. Burrows and G. Illingworth (eds.), *The Next Generation Space Telescope*, STScI, Baltimore, pp. 81–88.

Angel, J.R., Cheng, A.Y. and Woolf, N.J.: 1986, *Nature* **322**, 341–343.

Bracewell R.N.: 1978, *Nature* **274**, 780–781.

Bracewell R.N. and McPhie, R.H., 1979, *Icarus* **38**, 136–147.

Kasting, J.F., Holland, H.D. and Pinto, J.P.: 1985, *J. Geophys. Res.* **90**, 10,497–10,510.

Léger, A., Pirre, M. and Marceau, F.J.: 1993, *A&A* **277**, 309–313.

Léger, A., Mariotti, J.M., Mennesson, B., Ollivier, M., Puget, J.L., Rouan, D. and Schneider, J.: 1996, *Icarus* **123**, 249–255.

# PRIMA FOR THE VLTI – SCIENCE

F. DELPLANCKE*, F. DERIE, F. PARESCE, A. GLINDEMANN, F. LÉVY, S. LÉVÊQUE and S. MÉNARDI
*European Southern Observatory, Karl Schwarzschild Strasse 2, D-85748 Garching bei Muenchen*
*(*Author for correspondence and questions regarding PRIMA; E-mail: fdelplan@eso.org)*

**Abstract.** The four main scientific objectives of PRIMA – the Phase-Referenced Imaging and Micro-arcsecond Astrometry facility for the VLTI – will be described:

- extra-solar system characterization with astrometry, to detect planets and evaluate their mass, and imaging of the dust accretion disk,
- galactic center study with astrometry(dynamics of the bulge stars) and imaging at 10 $\mu m$ (piercing the gas and dust clouds surrounding the galactic center),
- observations of AGNs and other extra-galactic objects, too faint to be observed without PRIMA, for which partial imaging is needed to constrain their structure models,
- micro-gravitational lensing event resolution (imaging and astrometry of their photo-center) in the Galactic Bulge and Magellanic Clouds, helping to determine directly the lens mass and distance.

**Keywords:** interferometry, VLTI, imaging, astrometry, PRIMA

## 1. Introduction

Performances of ground-based interferometry are limited by the atmospheric turbulence. Indeed, the differential piston between the various telescope apertures induces a random wobble of the fringe position with time. It is thus impossible to integrate the fringes during a long time, as the wobble blurs them and reduces their visibility to zero. Typical integration times are some milliseconds in K-band and some hundreds of milliseconds at 10 $\mu m$. This keeps the limiting magnitude of the VLTI to K<13 with the 8-meter Unit Telescopes (UT) equipped with adaptive optics and to K<10 with the 1.8-meter Auxiliary Telescopes (AT) with tip-tilt correction. Faint objects, and in particular most of the extra-galactic targets, are in consequence out of reach.

To solve this problem, a principle similar to adaptive optics can be applied: if a bright star can be found in the close vicinity of the faint target, fringes could be measured at high speed on the star and stabilized using an OPD (Optical Path Difference) control loop. Then, the fringes on the faint target would be partially stabilized and one could integrate longer on them.

Fringe stabilization using an off-axis guide star is the first aim of PRIMA, the Phase-Referenced Imaging and Micro-arcsecond Astrometric facility. This would increase the VLTI limiting magnitude up to K$\sim$19 with the UTs and K$\sim$15 with

*Astrophysics and Space Science* **286:** 99–104, 2003.

the ATs. At 10 $\mu m$, the increase would be from 5 to 11 with the UTs, and from 2 to 8 with the ATs.

Moreover, as two targets will be observed at the same time, the differential phase (i.e. the difference in OPD at which the fringes are observed) between the two targets can be measured. If one of the targets can be taken as a phase reference (zero or constant phase), the angular distance between the targets as well as the intrinsic phase of the object can be extracted from the differential phase. If the differential OPD is measured with an accuracy of 10 nm, the angular distance between the targets can be evaluated with a resolution of 10 $\mu$arcsec, hence the micro-arcsecond astrometric aim of PRIMA. With the same differential OPD accuracy, the object intrinsic phase can be measured with an accuracy of 2° in K-band, essential for high-dynamic range synthetic aperture imaging.

## 2. On the Importance of the Phase in Interferometry

When reconstructing an image from interferometric data in synthetic aperture mode, two parameters are fundamental for the quality of the reconstruction: the total number of independent baselines and the accuracy of the measurements. A good UV-coverage (i.e. coverage of the plane corresponding to the image spatial frequencies distribution) obtained with a large number of independent baselines from 8 meters to several hundreds of meters is essential. But the accuracy on the measurement of the fringe phase (position) for each baseline is also critical. An image reconstructed from pure amplitude measurements without information on the phase cannot be recognized, while in the reverse case, the shape of the object can be partially recovered. Typically, an accuracy of 3.6° on the phase is required for good quality imaging.

Thus, in order to do interferometric imaging, the phase is absolutely needed. Unfortunately, the phase, i.e. the position of the fringe packet relative to a reference point, is varying strongly due to the atmospheric turbulence. Two ways of retrieving the phase information are possible: phase-closure and phase-referenced imaging.

In phase-closure, the simultaneous interferometric use of $N$ telescopes gives a maximum of $(N-1) \cdot (N-2)/2$ independent closure phases, independent of the atmospheric turbulence. In phase-referenced imaging, each observation in dual-feed (reference star and object observed simultaneously), with a facility like PRIMA, gives the object relative phase for that particular baseline.

The VLTI, with its very large number of independent baselines (potentially 259 with the ATs and 6 with the UTs), with its particular geometry, with the delay lines already foreseen for dual-feed, and with its current instrumentation, is particularly suited to use the phase-referenced imaging technique. Compared to phase-closure with 3 telescopes, PRIMA does not require more observation nights but reaches fainter objects. However, it requires a very good long term time stability of the full VLTI system.

## 3. PRIMA Principle

The PRIMA principle scheme is described in detail by Delplancke et al. (2000) and Derie et al. It is made of 4 sub-systems: the Star Separator, the Fringe Sensor Unit, the Differential Delay Lines, and the incremental Metrology. The Star Separator is the dual-feed facility of the telescopes. Installed at their coudé focus, it picks up two sub-field of 2" each in a 2' field-of-view and sends both collimated beams toward the rest of the VLTI optical train. The Fringe Sensor is an instrument in charge of acquiring fringes on the bright object, measuring their position accurately for closed loop stabilization, and determining their group delay (position of the fringe packet) which is the information related to the OPD. The Differential Delay Lines are used to compensate for the differential delay due to the angular separation between both objects, and to be the high speed actuator in the fringe stabilization closed loop. Finally, the incremental Metrology (Lévêque et al., 2002) is in charge of measuring very accurately (accuracy of 5 nm $rms$), the internal differential OPD, which will give ultimately the astrometric and phase information. The zero of the incremental Metrology is set by using the Star Separator such that the same star is sent down both of the feeds and that the fringes are centered both on the Fringe Sensor and on the instrument. In the astrometric case, the instrument would be a second fringe sensor. Any instrument (including yet-to-be-defined second generation instruments) can be used in the imaging mode.

The differential phase between the reference star and the object measured by PRIMA is given by:

$$OPD_{ref.star} - OPD_{object} = \overline{\Delta S} \cdot \overline{B} + \frac{\lambda}{2 \cdot \pi} \cdot \phi_{object} + OPD_{atm} + OPD_{int}$$

where $OPD_{ref.star}$ is the position of the fringes for the reference star, $OPD_{object}$ the position of the fringes for the studied object, $\overline{\Delta S}$ the angular separation vector on the sky between the star and the object (i.e. the astrometric signal), $\overline{B}$ the baseline vector, $\phi_{object}$ the object intrinsic phase (i.e. the imaging information), $OPD_{atm}$ the random differential OPD due to the atmospheric turbulence, and $OPD_{int}$ the internal differential OPD introduced by the VLTI.

## 4. Extra-Solar Systems

Interferometry with PRIMA, both in the imaging and astrometric modes, will bring very important information on extra-solar systems, complementary to what is brought by other techniques like classical or adaptive optics imaging and radial-velocity methods.

The astrometric mode will contribute to the detection and characterization of planets around nearby stars. Indeed the presence of a planet in orbit around a star gives rise to a wobble of the star around the center of mass of the system. This slight astrometric displacement is of the order of some tens to some hundreds of micro-arcsecond for Jupiter and Uranus-like planets around M stars at a distance of up to some tens of parsec. This is within reach of interferometric differential astrometry.

The problem is to find two astrometric reference stars within 30" from the star, in order to get the star wobble in two dimensions on the sky. Happily, the stars, which are within reach of PRIMA astrometric accuracy, are relatively close by and, in consequence relatively bright. In most of the cases, they can be used as fringe tracking star. Hence we have to find two faint stars ($K<15$) close by. The resulting sky coverage is more than 50% when looking around the galactic plane ($\pm$ 20° galactic latitude). Moreover, these stars are generally far enough to be unresolved by the interferometer and thus are suitable as phase reference stars.

Compared to radial-velocity techniques, astrometry can lift the indetermination on the planet mass (the famous $\sin i$ problem), is applicable to pre-main sequence stars, where the radial-velocity technique cannot be applied due to their photospheric and chromospheric activity, and is more sensitive to large planets far from the star than close to it. Both techniques are thus complementary.

The imaging mode, used with AMBER and MIDI will give access to star forming regions and accretion disks. The resolution of these instruments goes down to a 1 AU resolution for stars located up to 1 kpc with AMBER and 100 pc with MIDI. In most of the cases, again, the star itself will be bright enough to fringe track and only one faint and unresolved star must be found as phase reference in a radius of 10 to 60" around the star.

Access to the inner core of the accretion disk or planet formation nebula will complement coronographic images of the same object obtained with or without adaptive optics. It will help understand planetary and stellar formation.

## 5. Galactic Center

Recent studies (Schödel et al., 2002) using speckle and adaptive optics techniques have analyzed the movement of stars around our Galactic Center. They have shown that one star orbits around an invisible object of a mass of $(3.7 \pm 1.5) \cdot 10^6$ solar masses. This mass confirms the hypothesis of the presence of a black hole at the center of our galaxy.

To refine the evaluation and description of the central mass, continued measurements of star orbits with better accuracy are needed. Astrometry of these central stars at $< 0.1$ milli-arcsecond level would offer exciting prospects for the exploration of relativistic motions at 10–100 Schwarzschild radii from the central black hole.

In addition, imaging at 10 $\mu m$ of the Galactic Center would pierce through the thick absorbing clouds surrounding the central black hole and provide unprecedented resolution down to 10 AU or ~1000 Schwarzschild radii of the central object.

## 6. Active Galactic Nuclei and Extra-Galactic Objects

The vast majority of extra-galactic objects, like active galactic nuclei, cannot be reach by any interferometer without dual-feed and fringe stabilization. Thus PRIMA, with the UTs, is essential to this kind of science.

Obviously, AGNs are the highest priority targets for observations with PRIMA. Their light distribution is very well suited for interferometric observations, with a high flux small source surrounded by a not too faint environment. The resolution of the VLTI with PRIMA is sufficient to observe, in the closest AGNs, dust tori (inner and outer radii, density, temperature), accretion disks and jets.

Central regions of nearby galaxies are also suitable targets for PRIMA.

## 7. Gravitational Micro-Lensing

Photometric studies of gravitational micro-lensing of stars in the Galactic Center or in the Magellanic clouds do not allow determining the mass of the lens (Paczyński, 1996). Indeed, from the amplitude and duration of the light increase only, the distance to the lens, its transversal velocity and its mass cannot be disentangled. However, micro-lensing events does not only increase the total flux received from the background star, also split the star in several images separated by some milli-arcseconds. If the separation between the images can be determined and if the distance to the background star is known, the mass of the lens can be extracted (Delplancke et al., 2001). Micro-lensing also causes a wobble of the image photometric center measurable with PRIMA.

If the lens has a companion (e.g. a planet), the photometric light curve shows a secondary peak. But this could also be due to a single lens combined with a parallactic effect. If the image can be resolved, both cases can be distinguished: a multiple lens splits the image into 3 or 5 spots instead of 2. From the image, the mass of the companion can be evaluated.

The VLTI and PRIMA with an imaging resolution of 2 milli-arcsecond and an astrometric resolution of 10 micro-arcsecond have the capability of resolving micro-lensing events and would give a direct access to the mass of the lenses, helping to constrain our galaxy mass distribution model. PRIMA is required due to the limiting magnitude: micro-lensing events are usually faint ($V > 13$). But, with PRIMA, the VLTI should be able to observe at least some events per year.

## References

Delplancke, F., Lévêque, S., Glindemann, A., Kervella, P. and d'Arcio, L.: 2000, Phase-referenced imaging and micro-arcsecond astrometry with the VLTI, in: P. Léna and A. Quirrenbach (eds.), *Proceedings of the SPIE Conference – Astronomical Telescope and Instrumentation*, Munich, Germany, March 2000.

Delplancke, F., Górski, K.M. and Richichi, A.: 2001, Resolving gravitational microlensing events with long-baseline optical interferometry – Prospects for the ESO Very Large Telescope Interfereometer, *A&A* **375**, 701–710.

Derie, F., Delplancke, F., Glindemann, A., Lévêque, S., Ménardi, S., Paresce, F., Wilhelm, R. and Wirenstrand, K., 2002, PRIMA technical description and implementation, To be published in *Proceedings of the Genie Workshop – Hunting for Planets*, Leiden, The Netherlands, June 2002.

Lévêque, S., Salvadé, Y., Scherler, O. and Daendliker, R.: 2002, High accuracy laser metrology enhances the VLTI, *Laser Focus World*, April 2002.

Paczyński, B.: 1996, Gravitational microlensing in the local group, *ARA&A* **34**, 419–460.

Schödel, R. et al.: 2002, A Star Orbit in a 15.2-year orbit around the supermassive black hole at the centre of the Milky Way, *Nature* **419**, 694–696.

# THE VLTI CALIBRATORS PROGRAM: GETTING READY FOR HIGH ACCURACY MEASUREMENTS

ISABELLE PERCHERON[1,2], ANDREA RICHICHI[2] and MARKUS WITTKOWSKI[2]
[1]*NEVEC, Leiden Observatory, Niels Bohr Web, NL-2300 Leiden, The Netherlands*
[2]*European Southern Observatory, Karl-Schwarzschildstr. 2, D-85748 Garching bei München, Germany*

**Abstract.** The need for high angular resolution, high sensitivity measurements has increased since a few decades. More large, long baseline interferometers (VLTI, Keck Interferometer KI, CHARA) are coming on line to fulfill these needs. To exploit fully the scientific potential of an interferometer, astronomical calibrators have to be observed in close sequence with the scientific targets to provide a measurement of the Instrumental Transfer Function which will be used to remove the instrumental and atmospheric effects in the scientific data. This requires establishing a network of astronomical calibrators, either unresolved or with a diameter known with an accuracy of only a few tenths of percent to be comparable to the internal accuracy of the instruments. The aim of the ESO/NEVEC VLTI Calibrators Program is to build such a database for the VLT Interferometer. This database will include sets of high accuracy calibrator objects suitable for the different configurations of the VLTI with the different instruments (VINCI, MIDI, AMBER) and for a wide range of baselines (up to 200 m). After an introduction on the VLTI instrument and a short description of some of the problems of calibration in Interferometry, statistics on the preliminary list of VLTI calibrators will be shown. Since the first fringes in the K band with the commissioning instrument VINCI, data on calibrators have been accumulated with several baselines. We will explain the strategy used to observe and process the data. Some of the preliminary measurements will be shown and some of the limitations will be discussed.

**Keywords:** high accuracy, sensitivity, high angular resolution, diameter, astronomical calibrators, instrumental transfer function

Long baseline interferometry is used since the pioneering experiments to achieve high angular resolution measurements of astronomical objects. Facilities in France (GI2T) or in the United States (PTI, NPOI, IOTA) have been producing scientific results, but the new generation of long baseline, large aperture ground based interferometers will provide unprecedent results concerning angular resolution (long baseline), sensitivity (large aperture) and wavelength coverage (from visual to infrared). Like the other ground based instruments, interferometers are affected by the atmospheric turbulence which limits the quality, the precision and the sensitivity of the measurements. While it is possible to reduce the problems by selecting observational sites of good quality and by adopting techniques such as adaptive optics, the process of calibration of the instrumental visibility remains fundamental.

*Astrophysics and Space Science* **286:** 105–112, 2003.

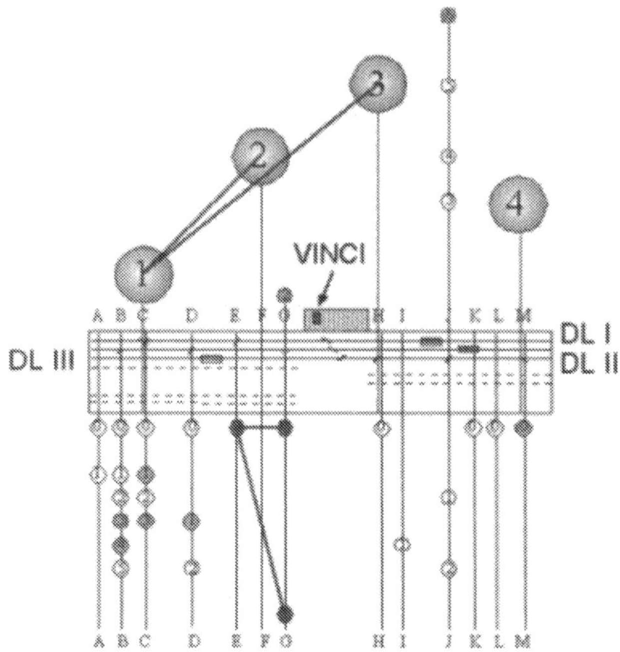


*Figure 1.* Configuration of the VLTI.

TABLE I
Characteristics of the VLTI scientific instruments

| instrument | wavelength ($\mu$m) | spectral resolution ($\lambda$ /$\Delta$ $\lambda$) | accuracy ($\Delta$ V/V) |
|---|---|---|---|
| VINCI | K : 1.9-2.4 | 5 | $10^{-3}$ |
| MIDI | N (Q) : 8-22 | 30-260 | $10^{-2}$ |
| AMBER | J,K,L : 1.0-2.4 | 35-14000 | $10^{-4}$ |

## 1. The VLTI and Its Instruments

The ESO Very Large Telescope Interferometer (VLTI) includes four 8.4m Unit Telescopes (UTs) and a number of 1.8m auxiliary telescopes (ATs) (Glindemann, 2002). Since the beginning of the operation in March 2001, two 0.4m siderostat telescopes (SIDs) have been used with the VLTI commissioning instrument VINCI. The SIDs and ATs can be moved over an array of 30 stations. The 254 independent baselines ranges from 8 to 205m with various orientations (see Figure 1).

The VLTI consists or will consist in a near future of several subsystems and instruments. Table I shows only the characteristics of the instruments which are related directly to the calibration problem. More informations on the VLTI instruments can be found in these Proceedings.

## 2. Instrument Calibration and Calibrator Objects

### 2.1. THE PROBLEM OF INSTRUMENT CALIBRATION

The problem of instrumental calibration is well-known in ground based interferometry and has been addressed by different authors, in the early days by Hanbury Brown (Hanbury Brown, 1974) and then later by different authors (a set of bibliographical references from authors related to interferometers such as Mark III, NPOI, PTI, IOTA, SUSI, GI2T, I2T, ISI can be found in (Richichi, 2002)). Instrumental calibration is getting more complex with the range of high accuracy and sensitivity obtained with the new generation interferometers (KI, VLTI). The astronomical calibrators must be known with a precision similar or better than the internal accuracy of the instrument (a few tenths of percent).

In the case of the VLTI, the problem is further complicated by the location in the Southern hemisphere, where only few potential calibrators have been measured. For example with the first set of VLTI calibrators and the 66m baseline, only 42% of the objects are known with an accuracy better than 1%. In the case of the 103m UTs baseline, only 13% of the calibrators are known give a visibility accuracy better than 1%.

The aim of the ESO/NEVEC Calibrators Program is to provide the different VLTI instruments with a set of calibrators, known with a good accuracy. We choose to select a preliminary set of potential VLTI calibrators which are observed routinely to be re-measured with the VLTI.

### 2.2. CHOOSING AN ASTRONOMICAL CALIBRATOR

The main requirements to choose an object to calibrate scientific interferometric data are :

- a single source, unresolved or with a well-known diameter
- no photometric variability
- no Infrared excess and compact atmosphere
- The spectral type and Luminosity Class should be comparable with those of the scientific target. The spectral type should also be restricted to early type.
- the object must be bright enough in the observation band
- the angular distance of the pair calibrator-science object is as small as possible.

Different approaches to choose calibrators are possible. To select calibrators with a known diameter, the CHARM catalog (Richichi, 2002) can be used. This catalog includes objects measured with direct methods (Long baseline interferometry,

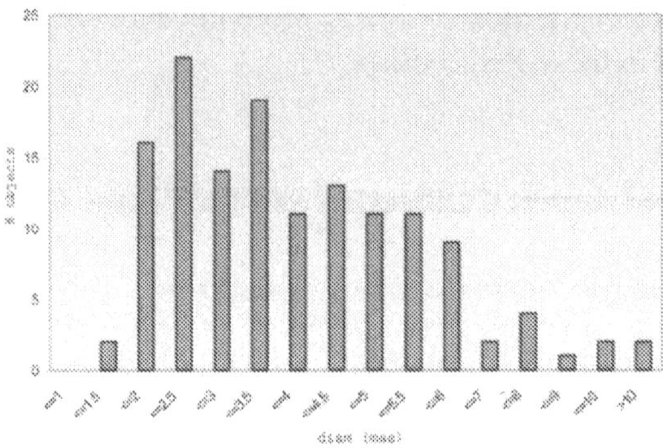


*Figure 2.* Distribution of the angular diameters for the VLTI potential calibrators

Lunar Occultations) as well as indirect measurements (Cohen (1999), Blackwell (1994)). More recently Bordé (Borde, 2002) refined the list of Cohen calibrators by imposing selection criteria for interferometry. A second approach is to carefully select objects from sky surveys (Hipparcos, Tycho) and estimate their diameters from spectro-photometric methods (Van Belle, 1999) .

## 3. The VLTI Calibrators

Because of the configuration of the VLTI Commissioning instrument VINCI associated with the siderostats, the first list of around 150 potential calibrators is mostly restricted to bright objects ($K <= 2$) with an angular size between 1.5 to 10 mas (Figure 2). Some fainter objects were included to be observed during the interferometric runs with the UTs. Because of the brightness requierement due to the size of the siderostats telescopes, most of these objects are semi resolved with the VLTI baselines and are extracted from the CHARM Catalog (Richichi, 2002). We choose to select the bright Cohen objects (Cohen, 1999) and some bright objects measured several times by the different interferometers (Nordgren, 2001).

To obtain a better sky coverage (Figure 3), some bright objects not fullfilling all the requirements (for example on variability or spectral type) are included. In this case, a quality flag is set.

### 3.1. Strategy

Together with routine observations to calibrate the scientific objects or to monitor the instrument, the calibrators are also observed during dedicated nights. These dedicated nights are aimed towards the following goals:

- the diameter of the calibrators is re-measured with the VLTI to obtain a better accuracy, some of the calibrators data are processed like scientific data and their diameters are measured.

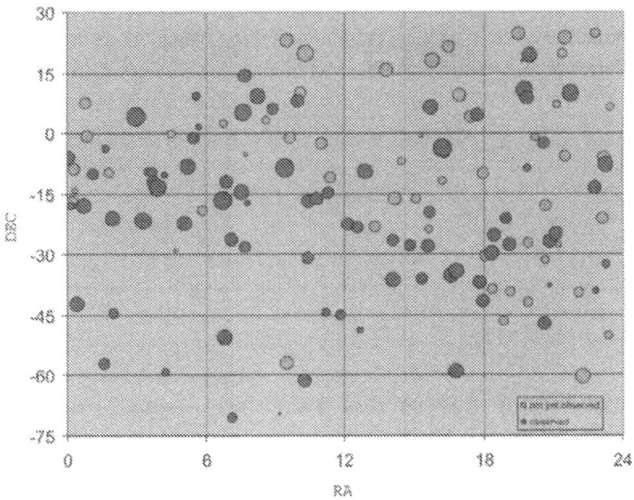


*Figure 3.* sky distribution. The size of the dot is proportional to the brightness of the object (from bigger dot : K=-2 to smaller dot : K=2.87). The light blue dots show the objects not yet observed in spetember 2002.

- some objects are followed during several hours to better understand the fluctuations of the transfer function due for example to zenith angle or turbulence parameters,
- pairs of objects with different angular separations are observed to determine the maximum separation between a scientific target and its associated calibrator.

Suitable calibrators are also observed at different baselines to further constraint their diameter values.

### 3.2. VLTI DATA

The observations with VLTI are done with a system of templates and Observing Blocks (OB) equivalent to those used during VLT observations. Typically, one OB includes between 300 and 500 interferograms. The data produced by the instrument are transferred to different workstations for on-line processing, archiving, off-line analysis and Quality Control. The pipeline software runs the on-line analysis, displays the data and creates different sets of files (FITS files and Quality Control operational files). Data quantities characterizing the instrument and the observation are measured by the on-line processing and logged in a Quality Control operational files (QC log files). A more complete description of the Observations and the VLTI Data Flow System can be found in (Ballester, 2001).

Table II shows the number of observations and objects observed from March 2001 to September 2002. During this period more than 70% of the OBs acquired are taken on 100 potential calibrators.

TABLE II

VINCI observations from March 2001 to September 2002. This table gives only the 3 baselines where the most observations have been done.

| Baseline | number of nights | number of Objects | number of calibrators | number of OBs |
|---|---|---|---|---|
| E0-G0 | 171 | 99 | 38 | 6388 |
| E0-G1 | 144 | 116 | 75 | 6463 |
| U1-U3 | 13 | 42 | 17 | 612 |

To select the calibrator observations and to have a quick look at the quality of the data, we use the Quality Control Operational files. For each OB taken on a calibrator, an instrumental Transfer function is calculated using

$$\mathrm{TF} = V^{ref}/V^{ref}_{theo}$$

where $V^{ref}$ is the measured visibility and $V^{ref}_{theo}$ is the theoretical visibility calculated from the calibrator diameter.

Figure 4 shows one of the night dedicated to the calibrators observation. It illustrates several problems:

- The VLTI measurements of the calibrator 1 obtained during the night are more stable than the accuracy obtained on each Transfer Function. In this case the precision on the diameter of the calibrator is known only to a few percents, while the internal stability is lower than 1% . In our first list the accuracy is not enough for most of the calibrators. The errors on the diameter go from .5% to 17% with the majority between 1 and 5%. This error on the diameter translates directly in an error on the predicted Visibility for the calibrator. In the case of a 16m baseline the value of the visibility of the calibrators object is known to less than .1%, but in the case of a 140m baseline, this translates to an error on the Visibility between 1 and 5% for the majority of the objects.
- The calibrators 1, 2 and 3 have been observed alternatively. The instrumental Transfer functions obtained on these 3 objects vary from 65% to 71%. The differences could be explained by the use of an underestimate or overestimate diameter. These diameters will be re-calculated using VLTI data.

### 3.3. Measurement of the VLTI calibrators

As seen in the previous section, the diameters of the calibrators have to be re-measured with VLTI. The precision obtained on these measurements will be better than the initial precision.

The measurements of the VLTI calibrators is done by calibrating the different calibrators against each other. As a first iteration, we chose a calibrator and re-

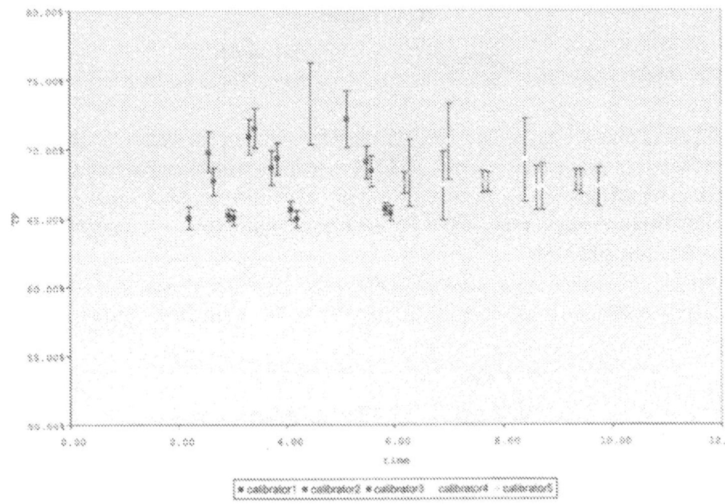

*Figure 4.* Night dedicated to calibrator observations (22nd January 2002).

measure all the astronomical calibrators observed the same night with this primary calibrator. We then re-iterate using another object as primary calibrator, until all the solutions converge.

### 3.4. FUTURE

After the processing of all the calibrators data, an updated list of VLTI calibrators will be released with new accurate VINCI measurements on several objects. To accommodate the needs of larger baselines (140 m and 189 m for the VLTI period 70 and 71) and to get a better sky coverage, the VLTI list will be extended to more than one thousand potential VLTI calibrators. A tool for the selection of the VLTI calibrators will be distributed by ESO to the community. The calibrator parameters will be calculated on line depending on the instrument mode (wavelength, dispersion). Different filters will allow the expert user to further constraint the criteria of the calibrator selection. For a non expert user, default parameters will be proposed.

## 4. Conclusion

The processing of the observations of the calibrator objects showed the limitations of the preliminary list of VLTI calibrators. Because of the sensitivity and the high accuracy of the VLTI, it is necessary to re-measure carefully most of the objects following a well define strategy to reach a precision comparable with the internal accuracy of the instrument. The problem of high accurate instrumental calibration is a common problem to the community and needs to be studied in collaboration with the European and international groups.

## References


Ballester, P. et al.: 2001, Data Flow System for the Very Large Telescope Interferometer, *SPIE Proc.* **4477**.

Blackwell, D.E. and Lynas-Gray, A.E.: 1994, Stellar effective temperatures and angular diameters determined by the infrared flux method (IRFM): revisions using improved Kurucz LTE stellar atmospheres, *A&A* **282**, 899.

Bordé, P., Coudé du Foresto, V. et al.: 2002, A catalog of calibrators stars for long baseline stellar interferometry, *A&A* **393**, 183.

Cohen, M., Walker, R. et al.: 1999, Spectral Irradiance Calibration in the Infrared. X. A Self-Consistent Radiometric All-Sky Network of Absolutely Calibrated Stellar Spectra, *AJ* **117**, 1864.

Glindemann, A. et al.: 2002, The Very Large Telescope Interferometer – A Status Report, *These Proceedings*.

Hanbury Brown, R. et al.: 1974, The angular diameter of 32 stars, *MNRAS* **167**, 121.

Nordgren, T.E. et al.: 2001, Comparison of stellar angular diameters from the NPOI, the Mark III optical interferometer, and the Infrared Flux Method, *Astron. J.* **122**.

Richichi, A. and Percheron, I.: 2002, CHARM: A Catalog of High Angular Resolution Measurements. *A&A* **386**, 492.

Van Belle, G.T.: 1999, Predicting Stellar Angular Sizes, *PASP* **111**, 1515–1523.

# 2D/3D DUST CONTINUUM RADIATIVE TRANSFER CODES TO ANALYZE AND PREDICT VLTI OBSERVATIONS

ILARIA PASCUCCI[1], THOMAS HENNING[1], JÜRGEN STEINACKER[2] and SEBASTIAN WOLF[3]

[1]*Max-Planck-Institut für Astronomie, Königstuhl 17, D-69117 Heidelberg, Germany*
[2]*Astrophysikalisches Institut und Univ.-Sternwarte, Schillergässchen 2-3, D-07745 Jena, Germany*
[3]*California Institute of Technology, 1200 E California Blvd, 220-6 Pasadena, CA 91106, USA*

**Abstract.** Radiative Transfer (RT) codes with image capability are a fundamental tool for preparing interferometric observations and for interpreting visibility data. In view of the upcoming VLTI facilities, we present the first comparison of images/visibilities coming from two 3D codes that use completely different techniques to solve the problem of self-consistent continuum RT. In addition, we focus on the astrophysical case of a disk distorted by tidal interaction with by-passing stars or internal planets and investigate for which parameters the distortion can be best detected in the mid-infrared using the mid-infrared interferometric device MIDI.

**Keywords:** accretion disks, interferometry (MIDI), radiative transfer

## 1. Introduction

The high resolution and sensitivity of the Very Large Telescope Interferometer (VLTI) will soon improve our knowledge in many fields of astrophysics. However, the first VLTI data will be in form of visibilities that require further analysis to be correctly interpreted. Hence, there is an ample need for multidimensional Radiative Transfer (RT) codes, treating complex configurations with high resolution and accuracy. Some 2D and a few 3D dust continuum RT codes have been developed and recently improved (Lopez et al., 1995; Dullemond and Turolla, 2000; Wolf, 2003; Steinacker et al., 2003). As results most of them provide not only spectral energy distributions (SEDs) but also wavelength-dependent images. These images have key importance for the interpretation of observed visibilities.

We lately demonstrated the reliability of RT codes by the first 2D dust continuum RT benchmark (Pascucci et al., 2003). Five RT codes implementing different numerical schemes have been compared in a 2D disk configuration: three of the codes apply the Monte-Carlo (MC) technique while the other two are grid-based codes. Considering the general astrophysical case of a star embedded in a circumstellar disk with an inner cavity free of dust, solutions for a set of well-defined cases that differ in optical depth and viewing angle have been provided.

The two 3D RT codes participating in the benchmark, namely MC3D (Wolf, 2003) and STEINRAY (Steinacker et al., 2002; Steinacker et al., 2003), are able to

*Astrophysics and Space Science* **286:** 113–118, 2003.

handle complex configurations without severe restriction on optical dust properties or dust distribution. Here, we demonstrate the capabilities of the codes in producing mid-infrared (MIR) images for a distorted disk. We also investigate for which parameters the distortion can be best detected by the VLTI MIR interferometric instrument MIDI.

## 2. First Image/Visibility Comparison from 2 RT Codes

The 2D RT benchmark aimed to compare both emerging temperatures and SEDs from defined disk configurations. As a natural step, we extend our comparison to the images. For this purpose we make use of two of the tested codes that apply completely different methods to solve the dust continuum RT equation: MC3D is a Monte Carlo code while STEINRAY is a grid-based code. Simulated images, and corresponding visibilities, are extremely useful for both the preparation of interferometric observations and for a correct interpretation of interferometric data.

Distorted disks are relevant to understand the formation and evolution of circumstellar disks and for their connection with the formation of low-mass stars and possibly planets. Numerical simulations show that star-disk and disk-disk encounters influence the initial disk density distribution by inducing two-armed spiral structures (Pfalzner et al., 2000). Similar arms are predicted by high-resolution numerical simulations in circumbinary disks surrounding classical T Tauri stars (Günther and Kley, 2002). D'Angelo et al. (2002) performed hydrodynamical calculations that clearly show a wave pattern induced by planet-disk interactions.

The interesting question is about the detectability of these disk features. The MIR appears to be a promising spectral region to investigate because it probes those warm regions of the disk that are not contaminated by the stellar atmospheres. Sub-arcsec resolution is mandatory to resolve those inner arms that have higher density contrast. Wolf et al. (2002) explored the possibility of detecting gaps in circumstellar disks caused by planets. Here, we focus on the detectability of spiral density waves.

As a simplified version of an azimuthally extended distortion, we consider here a ring at 25 AU with a density five times larger than the local density taken from one of the the disk models tested in the benchmark. The assumed density contrast is in agreement with the numerical simulations previously cited. These numerical simulations also show that the spiral width varies strongly with the distance from the central object and with the distance and nature of the disturbing object: typical widths range from a few tenths of an AU to several AU. For the presented calculations, we have assumed a width of 3.5 AU. Star and disk parameters are summarized in Table I for clarity.

We compute images at 20 $\mu$m for both a face-on and an edge-on case. Visibilities are calculated with the SIMVLTI tool* for the longest baseline (UT1-UT4)

* http://www.mpia-hd.mpg.de/MIDI/SIMVLTI/

TABLE I
Model parameters

| Parameter | value |
|---|---|
| Stellar mass, radius, temperature | 1 $M_\odot$, 1 $R_\odot$, 5800 K |
| Inner-outer disk radius | 1-1000 AU |
| Grain type, radius | Silicate[a], 0.12 $\mu$m |
| Optical depth at 550 nm ($\tau_V$) | 1[b] |
| Ring location, width | 25 AU, 3.5 AU |

[a] Draine and Lee (1984).
[b] $\tau_V$ is calculated in the disk midplane from the inner to the outer disk radius.

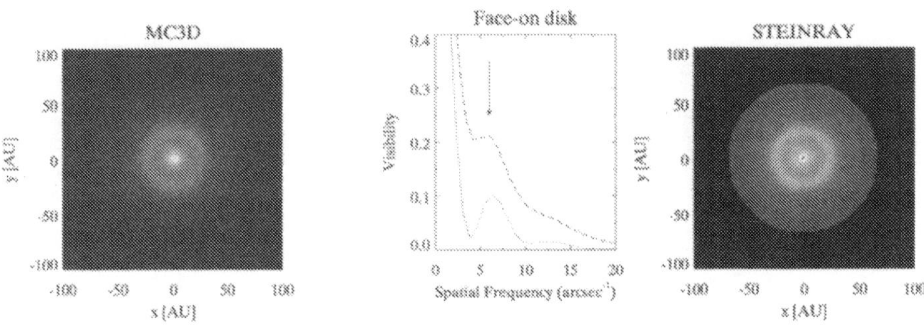


*Figure 1.* Images and visibilities for the face-on disk (model given in Table I). Left panel: 20$\mu$m image obtained with MC3D. Right panel: 20$\mu$m image simulated with STEINRAY. Middle panel: visibility curve computed for the baseline UT1-UT4: dashed line for MC3D, dotted line for STEINRAY.

available at the VLTI, a source distance of 150 pc (typical of T Tauri stars in nearby star-forming regions) and declination of -30$^o$. In Figure 1 we present the resulting images only for the face-on configuration. Images are restricted to the inner 200$\times$200 AU and intensities are scaled to the power of 0.25 to enhance the faint parts of the disk. The middle panel of Figure 1 provides the corresponding visibilities: dashed line for MC3D and dotted line for STEINRAY. A vertical arrow is drawn to visualize the spatial frequency at which the ring shows up. The ring appears with similar extent and flux in both simulations. The difference in flux in the entire image is smaller than 20%. The main finding of these simulations is that both RT codes predict a detectable ring. The discrepancy in the overall shape of the visibility curve originates from a different intensity distribution that may be due to the fact that STEINRAY predicts a slightly warmer inner disk in comparison to MC3D. Another source of deviations may be the different temperature grid used by the codes to store the results: an appropriate 2D grid is used in MC3D for these particular simulations while a 3D grid is used in STEINRAY.

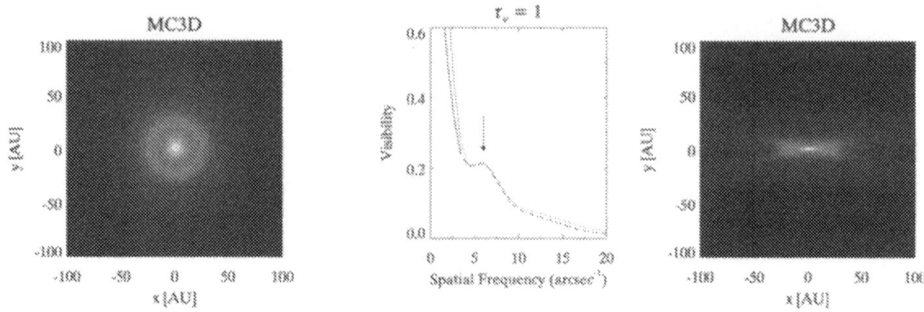


*Figure 2.* Images and visibilities for the thin disk ($\tau_V$=1). Left panel: 20$\mu$m face-on disk. Right panel: 20$\mu$m edge-on disk. Middle panel: visibility curve computed for the baseline UT1-UT4. Dashed line: face-on disk, dotted line: edge-on disk.

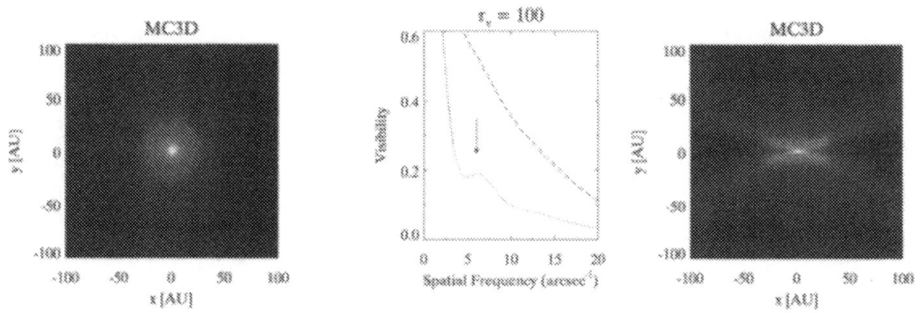


*Figure 3.* Images and visibilities for the thin disk ($\tau_V$=100). Left panel: 20$\mu$m face-on disk. Right panel: 20$\mu$m edge-on disk. Middle panel: visibility curve computed for the baseline UT1-UT4. Dashed line: face-on disk, dotted line: edge-on disk.

## 3. Investigating the Disk Parameters

In this section we investigate for which parameters the ring can be best detected by MIDI. The following RT simulations are performed using MC3D.

At first we inspect the effect of the disk inclination. Among the various computations, the two extreme cases of a face-on and an edge-on disk are presented in Figure 2. The computed visibilities clearly show that the ring detectability is only marginally influenced by the disk inclination. This could be due to the fact that the ring is almost completely optically thin for the chosen disk configuration and the observing wavelength. If we compute the same models for a disk that has a 100 times higher optical depth, we find that the ring is still detectable in the edge-on case but not anymore in the face-on configuration (see Figure 3).

Another important parameter to investigate is the ring location. We consider the original face-on disk having $\tau_V$=1 and we compute models for rings that are located in an interval from 10 to 50 AU from the star. Since the disk is resolved, the visibility is small for large spatial frequencies (dashed line of Figure 4). This

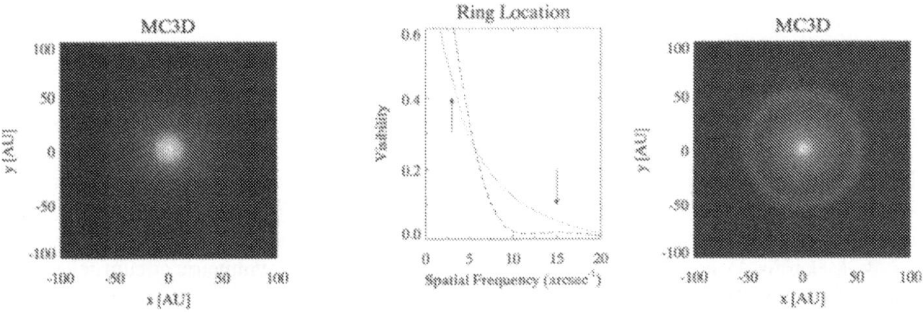


*Figure 4.* 20$\mu$m images and visibilities for a disk with a ring located at 10 AU (left panel) and at 50 AU (right panel). Middle panel: visibility curve computed for the baseline UT1-UT4: dashed line for the ring at 10 AU, dotted line for the ring at 50 AU. At these distances rings become hardly detectable in the visibility curve.

implies that is easier to detect rings that are farther away than 10 AU from the star, i.e. for spatial frequencies $< 15$ arcsec$^{-1}$. On the other hand, rings become undetectable when located too far from the star. At about 50 AU, we are in the region of exponential decrease of the visibility curve and a density increase of a factor of 5 over the local density cannot be seen in the visibility (dotted line of Figure 4). We conclude that rings can be detected by MIDI when located between 10-40 AU.

Finally, we inspect the minimum density contrast necessary to detect the ring. We cover density contrasts ranging from 2 to 8 and find that for a ring at 25 AU a minimum density contrast of approximately 4 is required. As suggested by the previous results, the minimum density contrast for the ring detectability also depends on its location.

## 4. Conclusions and Future Work

We have presented the first comparison of images/visibilities produced by two 3D dust continuum RT codes based on very different methods. Both codes predict that features like rings in a disk can be detected in the MIR with MIDI. Further analysis will be performed to study in detail how many visibilities are required to detect such a ring in a real source with realistic noise, and where the differences in the overall shape of the visibility come from.

In addition, we have used the Monte Carlo code to explore for which parameters the ring can best detected with MIDI. As a next step, we intend to use one of the density distributions predicted from numerical simulations of star/disk encounters, and study which parameters of the spiral-like pattern can be determined with interferometric observations.

## References

D'Angelo, G., Henning, T.H. and Kley, W.: 2002, Nested-grid calculations of disk-planet interaction, *A&A* **385**, 647–670.

Draine, B.T. and Lee, H.M.: 1984, Optical properties of interstellar graphite and silicate grains, *ApJ* **285**, 89–108.

Dullemond, C.P. and Turolla, R.: 2000, An efficient algorithm for two-dimensional radiative transfer in axisymmetric circumstellar envelopes and disks, *A&A* **360**, 1187–1202.

Günther, R. and Kley, W.: 2002, Circumbinary disk evolution, *A&A* **387**, 550–559.

Lopez, B., Mekarnia, D. and Lefevre, J.: 1995, Radiative transfer in axisymmetric circumstellar dust shells, *A&A* **296**, 752–760.

Pascucci, I., Wolf, S., Steinacker, J., Dullemond, C.P., Henning, Th., Niccolini, G., Woitke, P. and Lopez, B.: 2003, The 2D Continuum Radiative Transfer Problem. Benchmark Results for Disk Configurations, *A&A*, under submission.

Pfalzner, S., Henning, Th. and Kley, W.: 2000, Star-disc and disc-disc encounters, *IAUS* **200**, 193–195.

Steinacker, J., Bacmann, A. and Henning, Th.: 2002, Adaptive multi-frequency grids for three-dimensional radiative transfer, *JQSRT* **75**, 765–786.

Steinacker, J., Henning, Th., Bacmann, A. and Semenov, D.: 2003, 3D Continuum radiative transfer in complex dust configurations around young stellar objects and active nuclei. I. Calculational method and capabilities, *A&A*, accepted.

Wolf, S., Gueth, F., Henning, Th. and Kley, W.: 2002, Detecting planets in protoplanetary disks: A prospective study, *ApJ* **566**, L97–L99.

Wolf, S.: 2003, MC3D-3D continuum radiative transfer, Version 2, *Comp. Phys. Comm.* **150**(2), 99–115.

# SCIENTIFIC AND TECHNICAL RESULTS FROM VINCI USING COHERENT ESTIMATION OF FRINGE VISIBILITY *

J.A. MEISNER
*Leiden Observatory, The Netherlands*

**Abstract.** Although primarily intended as a test and alignment instrument in order to commission the VLTI, VINCI has taken useful scientific data in its first year and a half of operation. Our results employ coherent integration of fringe visibility in which the actual amplitudes of the raw scans are combined linearly after correcting for the position of the fringe within each scan. In addition to reducing the effect of noise compared to incoherent integration, the result contains a broader range of information, including an estimate of the complex visibility spectrum. Such an estimator is thus sensitive to instrumental phase and spectral characteristics, including the variable component of dispersion introduced by the excess air paths in the delay lines. Calibration of such instrumental effects demonstrates the ability to detect source phase at a fine level as will be required for direct interferometric detection of extrasolar planets. We present diameters for five stars obtained by observing the visibility null in their correlated spectra. Using coherent integration we have also observed the peculiar correlated spectra seen in many Mira variables, possibly due to changes in the apparent diameter with wavelength. Calibration of the zero-baseline power from *o* Ceti is used with other interferometric observations of this star over a period of 90 days to plot diameter variations associated with its pulsation cycle.

**Keywords:** stellar interferometry, coherent integration, phase, VINCI, VLTI

## 1. Introduction

VINCI (Kervella et al., 2000) is the first interferometric instrument installed on the VLTI and has been in operation since March 2001. Primarily intended as a test and alignment instrument in order to commission the VLTI, its design was based on the FLUOR instrument (Coude Du Foresto et al., 1998). Like FLUOR, VINCI operates at 2 microns (K band) using single spatial mode interference employing optical fibers for spatial filtering and beam combination. The interferometric signal is temporally modulated using delay-scanning.

In a year and a half of service, VINCI has made over 7000 interferometric observations (raw visibilities), from which significant astronomical results have been derived. We report here on results employing *coherent* methods of fringe visibility estimation. Among other advantages, such techniques are capable of obtaining the (optical) spectrum of correlated flux including the *phase* of interference over that spectrum. However the opportunity to garner such information requires

* VINCI data included herein is based on observations made with the European Southern Observatory telescopes obtained from the ESO/ST-ECF Science Archive Facility.

*Astrophysics and Space Science* **286:** 119–127, 2003.

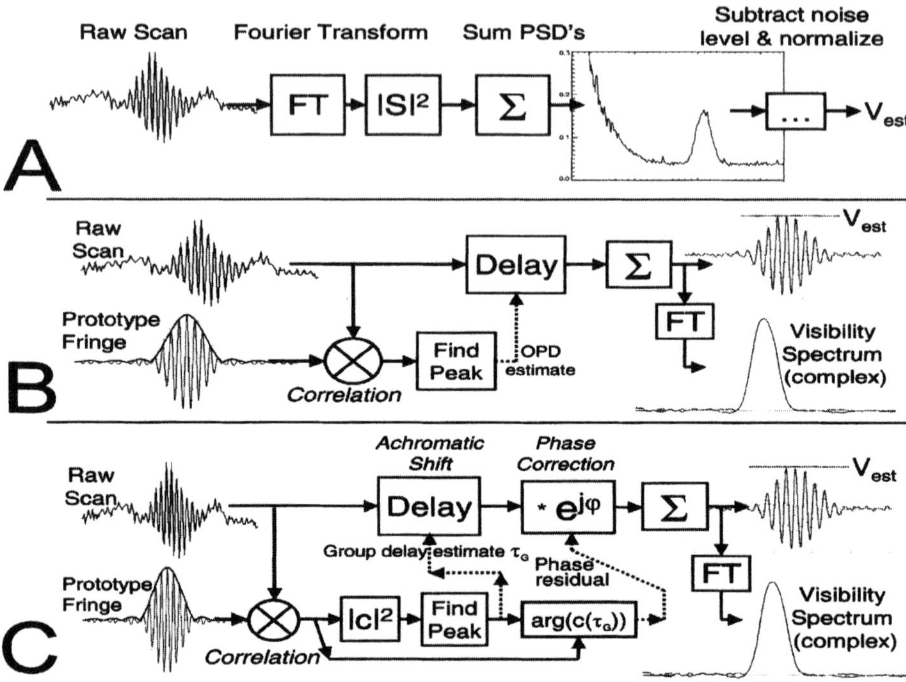


*Figure 1.* **A.** *Incoherent* integration of fringe visibility sums the spectrum of each individual scan. The net power spectrum contains a noise pedestal, above which the integrated amplitude due to interference yields the squared magnitude of visibility. **B.** *Coherent* integration centers each scan by correlating it with a 'prototype fringe.' The fringe amplitudes are then summed, to yield an actual estimate of the underlying fringe (right) from which a complex spectrum can be obtained. **C.** *Quasi-coherent* integration is similar, but each scan is centered using the *group delay* (envelope) estimator. An additional phase correction is then required to force cophasing, as the error of the group delay estimator is generally greater than a radian. The first order term of dispersion (spectral phase variation around center frequency) is destroyed by this processing which forces the phase delay at the center frequency to match the group delay prior to integration.

increased levels of instrumental calibration and characterization. In this paper we shall briefly describe the data reduction methods involved, the instrumental calibration procedures applied, and a few preliminary scientific results obtained using such methods.

## 2. Data Reduction Methods

The success of FLUOR, from which the design of VINCI was derived, was in large part due to its innovative use of single mode optical fibers. The consequent spatial filtering of the input beams enabled a high precision and repeatability of interferometric visibilities, at the 1% level, without being highly sensitivity to changes in atmospheric seeing. Differential dispersion due to unequal fiber lengths was tolerated by employing long scans which sampled the thereby broadened fringe packet, and a visibility estimator (Coude Du Foresto et al., 1997) which is utterly insensitive to dispersion. This is done by accumulating the power spectra but discarding the phase of interference on a scan-by-scan basis (Figure 1A).

Results in this paper, however, employ *coherent* integration methods, in which the actual *amplitudes* of the fringe packets are co-added (Figure 1B). This produces a linear estimate of the visibility spectrum without introducing the noise pedestal. With a semi-narrowband instrument such as VINCI, there is an advantage in the *quasi-coherent* method (Figure 1C), which does not presuppose unambiguous identification of the central fringe in each scan. This entails loss of the linear term of the visibility phase, which is difficult to estimate anyway with a narrowband instrument (except with a very high SNR). Random atmospheric dispersion fluctuations (Meisner and Le Poole, 2002) which are largely first order in phase, are conveniently ignored by the estimator. All results presented in this paper employ quasi-coherent integration of fringe visibility; one may compare a (complex) correlated spectrum so obtained with the square root of the incoherently integrated power spectrum plotted in Figure 3A (dotted line).

## 3. Calibration of Visibility Phase

Although the 'zero-delay' reference point of an interferometric observation is arbitrary, one can observe the *relative* phase delays of spectral components in the visibility curve. The phase we then measure includes both the true source phase plus instrumental components. Thus retrieval of the source phase is dependent on calibration of the instrumental phase function.

The instrumental phase has a fixed component due to the fiber beam combiner (and other optical elements in the beams' paths) plus a component due to unequal path lengths through dispersive air in the delay lines and light paths. Therefore if we can determine the fixed instrumental phase function of the optical components and beam combiner $\Phi_B(\nu)$, and we know the phase function per meter of delay line air $d\Phi_A/dl$, then knowing the excess air path $D$ for a particular observation, we can find the true source phase $\Phi_{CORR}(\nu)$ from the raw visibility phase $\Phi_{RAW}$ as follows.

$$\Phi_{CORR}(\nu) = \Phi_{RAW}(\nu) - \Phi_B(\nu) - \frac{d\Phi_A(\nu)}{dl} \cdot D \tag{1}$$

The coefficient of the delay line component of dispersion, $d\Phi_A/dl$, may be based on a physical model (Meisner and Le Poole, 2002) and can be verified by comparison of observations using different air paths $D$ during a single night during which the beam combiner phase function $\Phi_B(\nu)$ was stable. More careful monitoring of the humidity in the VLTI tunnels, along with barometric and temperature measurements already recorded, will be required to fully compensate for the variable

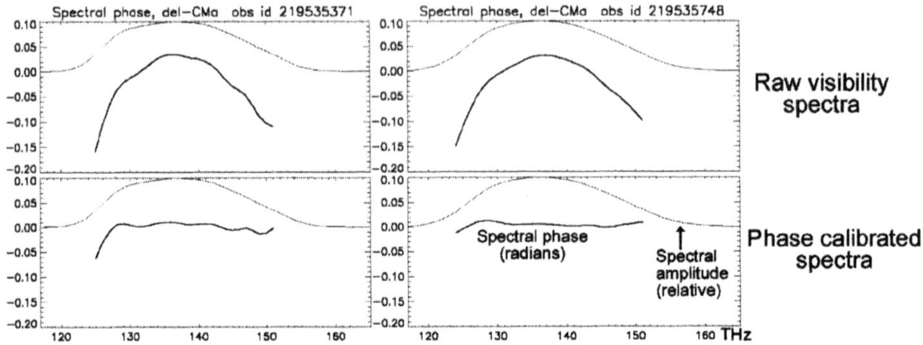


*Figure 2.* **Top:** Raw visibility phase as a function of optical frequency for two high SNR observations of $\delta$ CMa. Phase in radians plotted in bold versus optical frequency. Thin plot is the relative spectral amplitude of correlated flux, depicting energy in the K band from 125 THz (2.4 $\mu m$) to 150 THz (2.0 $\mu m$).
**Bottom:** Residual phase functions following calibration of beam combiner phase and excess air path dispersion as described in the text.

component of dispersion. Given that calibration of the dispersive air, the fixed instrumental component of spectral phase $\Phi_B(\nu)$ can be obtained from one or more calibrator observations, using stars expected to have symmetric structure (and thus zero phase). That instrumental phase function has been observed to change radically when adjustments to the fiber beam combiner have been performed, but appears to usually remain stable during a given night.

In Figure 2 the visibility phase over the K band is shown for two observations of $\delta$ CMa on 13 October 2001. Since these reductions used the quasi-coherent technique, first order components of the phase function are absent. The remaining phase function is largely instrumental in origin, and is greatly reduced after phase correction using (1), as seen in the bottom graphs of Figure 2. The residual phase has thereby been reduced to about .01 radians rms which is at the level of the measurement noise of these observations. Programs underway to interferometrically detect the light of extrasolar planets will require such phase calibration an order of magnitude better still.

## 4. Stellar Diameters from Visibility Nulls in the Spectra

One interesting feature found in several VINCI observations, occurs when the effective baseline samples the spatial frequency $f_{null} = 1.22/D$, at which the visibility of a star with a uniform disk diameter $D$ crosses zero. This will occur, for a geometric baseline $B$, at the optical frequency $\nu_{null} = c\, f_{null}/B$. If that optical frequency lies within the instrument's passband, then the visibility will cross zero at that point. Such a determination of a stellar diameter is not dependent on the instrument's visibility calibration (transfer function).

Figure 3A shows the spectrum of $\alpha$ Sco (obtained through quasi- coherent integration) observed on 2 April 2001. Ignoring the non-zero imaginary part of the spectrum (surely due to an asymmetric feature in the source), we can see that the visibility crosses zero at an optical frequency of 143.3 THz (2.09 $\mu m$). At the baseline length of 14.96 meters, this implies a spatial frequency of $f_{null} = 34.7\,arcsec^{-1}$ corresponding to a diameter (uniform disk model) of $1.22/f_{null} = 35.2\,mas$.

In Figure 3B we have used an observation of $\alpha$ Ori performed just past the first visibility null. An approximate 'calibration' of the spectrum has been performed by dividing the coherently integrated spectrum obtained by that of an arbitrary calibration star. The resulting (negative) visibility has a shape which is reasonably well matched to the expected curve shown in a dashed line; the extrapolated position at which the observed visibility crosses zero again supplies an estimate of the uniform disk diameter. Further results obtained in this manner from several VINCI observations are detailed in the following table.

| Object Name | Observation JD-2450000 | Baseline (meters) | P.A. | Null Frequency (THz) | (arcsec$^{-1}$) | UD Diam (mas) |
|---|---|---|---|---|---|---|
| $\alpha$ Sco | 2001.41747 | 14.96 | 9.5° | 143.3 | 34.67 | 35.19 |
| | 2001.42195 | 14.81 | 8.8° | 144.7 | 34.65 | 35.20 |
| | 1991.37136 | 15.99 | 20.4° | 129.2 | 33.41 | 36.52 |
| $\alpha$ Her | 1998.40935 | 15.91 | 16.2° | 150.8 | 38.80 | 31.44 |
| | 1998.41386 | 15.95 | 16.6° | 149.8 | 38.64 | 31.57 |
| | 2001.35130 | 14.73 | 12.9° | 162.0* | 38.59 | 31.61 |
| $\alpha$ Ori | 2195.34149 | 14.66 | 15.6° | 117.5* | 27.86 | 43.80 |
| W Hya | 2297.27323 | 12.53 | 59.0° | 120.5* | 24.42 | 49.97 |
| $o$ Ceti | 2295.05506 | 14.34 | 17.7° | 149.3 | 34.62 | 35.24 |
| | 2295.05907 | 14.17 | 17.9° | 150.1 | 34.39 | 35.47 |

* Frequency of a *projected* null, outside of the instrument's passband.

The diameters found here for $o$ Ceti are also plotted as triangles in Figure 5C. The 11% larger diameter found using this method, compared to the fit based on broadband visibilities, could possibly be attributed to a wavelength-dependence in the effective diameter. The larger diameter listed in the above table reflects the UD diameter at 150 THz (2.0 $\mu$m) whereas the broadband visibility is dominated by power centered around 135 THz (2.2 $\mu$m).

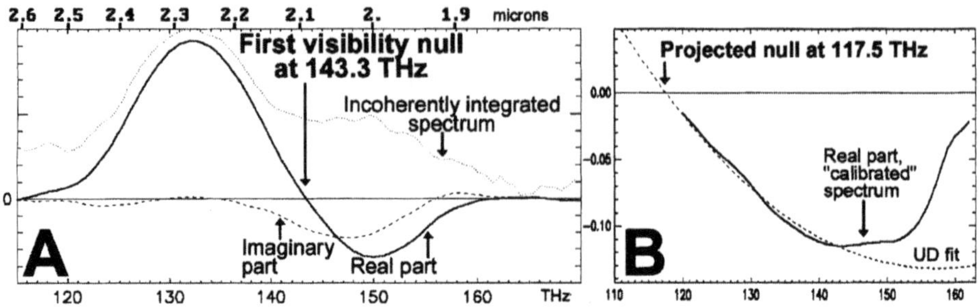


*Figure 3.* **A.** Zero-crossing of real part of coherently integrated visibility spectrum of $\alpha$ Sco allows for diameter determination. **B.** Visibility curve of $\alpha$ Ori 'calibrated' using instrumental spectrum of an average star. Fit to theoretical UD visibility function shown, with projected null just outside of instrumental passband.

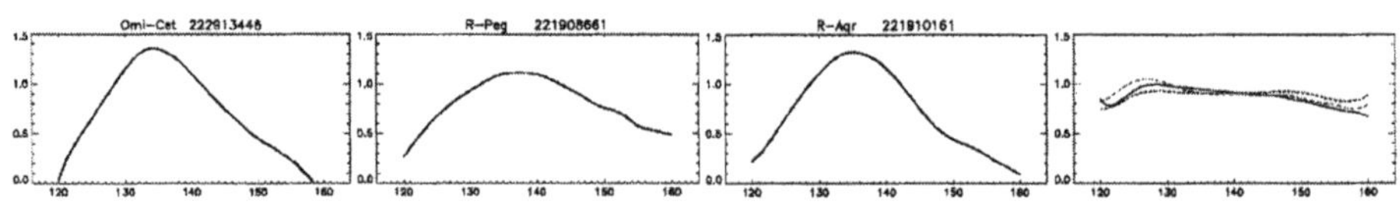

*Figure 4.* Spectra of correlated flux relative to 'calibration' star for 3 Mira variables. **Right:** Calibrated spectra of 3 random stars for comparison.

## 5. Visibility Spectra of Mira's and Diameter Determinations

One remarkable effect noticed in the spectrally resolved visibility curves is found in the case of Mira variables. Most of the Mira's observed with VINCI produce spectra which, when compared to typical stars (to account for the instrumental spectral response) are peaked around 135 THz (2.2 $\mu m$), but which fall off at both lower and higher frequencies. Three examples are shown in Figure 4, in which the relative visibility spectra of $o$ Ceti, R Peg, and R Aqr are shown after dividing by the visibility spectrum of $\psi$ Phe. Similarly obtained spectra of three random stars are plotted in the rightmost panel for comparison purposes.

It is suspected that the reduction in correlation toward the edges of the K band is evidence of an effective diameter increase at these wavelengths due to an extended stellar atmosphere, as has previously been observed interferometrically (Mennesson et al., 2002) in such stars. Such an overresolved region would have little contribution to the broadband visibility, leaving only the contribution from a more compact photosphere responsible for the correlated flux at intermediate wavelengths.

Further evidence of the existence of infrared emission in an extended region is provided by fitting of visibilities observed on 7 nights between October 2001 and January 2002 using a 16 meter physical baseline. 25 visibility points observed on 23 and 24 October 2001 are plotted in Figure 5A. A 'normal' 30 mas diameter uniform disk fit to these visibilities shown with the dashed line clearly fails.

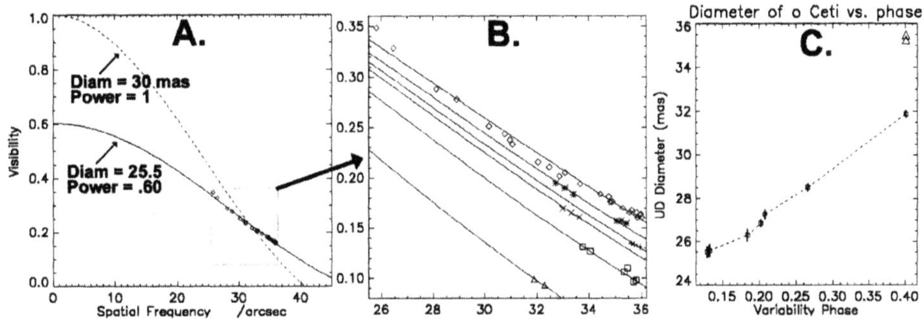


*Figure 5.* **A.** Visibility vs. baseline using a 'normal' UD model is rejected in favor of a model with a zero-baseline power reduced to .60. **B.** Detail of the nightly fits for each of the diameter solutions modelled using this zero-baseline power. **C.** Solutions of the (UD) diameter variations over 90 days of this star's variability cycle.

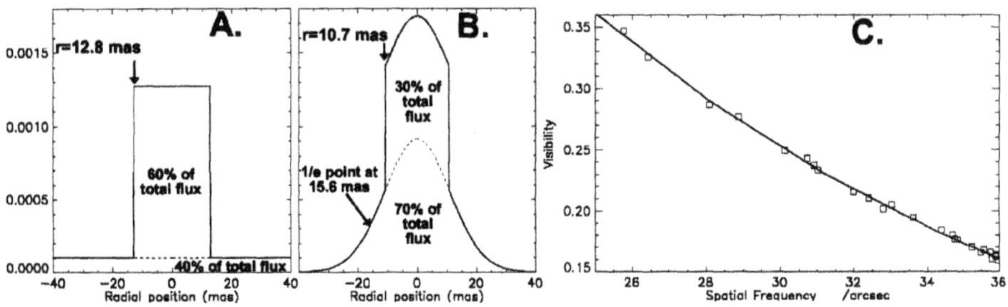


*Figure 6.* **A.** Two-parameter solution, with 60% of flux in a uniform disk of diameter 25.55 mas. Region containing the remaining flux is considered to be highly overresolved (level shown for illustrative purposes only). **B.** Three-parameter solution as described in text. **C.** Fit of visibility predicted by three-parameter solution to data.

A reasonably good fit, however, is obtained with the visibility curve for a 25.5 mas uniform disk *when reduced* by the factor 0.60. This would be consistent, for instance, with a model in which 60% of the light emanated from a compact photosphere with this diameter, and the remaining light was uncorrelated (therefore having zero contribution to the net visibility), perhaps because it emanated from a very extended region which was overresolved at these baselines (see Figure 6A).

We attempted to fit the data from each of the 7 nights using this two-parameter model as detailed in the following table. During four of these nights there was sufficient projected baseline coverage to solve for the proportion of 'correlatable power,' that is, the position on the visibility curve that would be extrapolated at zero spatial frequency (as depicted in Figure 5A). The resulting proportion, 0.60, was then forced to apply to the model for each of the seven nights, to deduce the variation in diameter over time as plotted in Figure 5C. Note that this assumption, that the zero-baseline power proportion was constant, is not well founded. However independent estimations on the four nights in which this was possible are consistent with a constant value. Figure 5B shows the fits of visibility versus projected baseline on each night using the final diameter solutions listed in the following table.

| Datapoints | | | Zero-baseline power | | Diameter (mas) | |
|---|---|---|---|---|---|---|
| JD | Phase | Num | Estimated | rms | Estimated | rms |
| Independent nightly solutions used to estimate zero-baseline power | | | | | | |
| 2205 | .13 | 13 | .622 | .019 | 25.80 | .28 |
| 2206 | .13 | 12 | .609 | .016 | 25.68 | .22 |
| 2223 | .18 | 6 | .568 | .021 | 25.84 | .22 |
| 2250 | .27 | 6 | .599 | .048 | 28.48 | .44 |
| **Average** | | | .6024 | .010 | | |
| Solutions assuming estimated zero-baseline power determined above | | | | | | |
| 2205 | .13 | 13 | (.6024) | .010 | 25.52 | .23 |
| 2206 | .13 | 12 | (.6024) | .010 | 25.58 | .26 |
| 2223 | .18 | 6 | (.6024) | .010 | 26.30 | .26 |
| 2229 | .20 | 3 | (.6024) | .010 | 26.84 | .13 |
| 2231 | .21 | 3 | (.6024) | .010 | 27.26 | .20 |
| 2250 | .27 | 6 | (.6024) | .010 | 28.52 | .15 |
| 2295 | .40 | 2 | (.6024) | .010 | 31.88 | .13 |

The visibility determinations employed the quasi-coherent estimator, from which only the real parts of the visibility at each optical frequency were summed to compute the net visibility. Thus additional noise from the imaginary parts of these visibilities was rejected; imaginary components of the visibility are not consistent with a symmetric source structure as required by this model. In fact, it was found that the rms of the imaginary part of these visibilities was usually less than .005, thereby placing a tight upper limit on the magnitude of source asymmetries which are resolvable using these baselines.

Although the two-parameter fit to the 25 visibilities obtained on 23 and 24 October 2001 (Figure 5B, top curve) is not bad, systematics in the residual errors can be observed which are inconsistent with uncorrelated noise. Applying a three-parameter fit to these data, it was possible to reduce the rms amplitude of the residual error by a factor of two, thus justifying such an approach. The stellar image is assumed to be radially symmetric, including a wide gaussian circumstellar envelope whose $1/e$ radius is to be solved for. Added to that is the flux from a compact photosphere, modeled as a uniform disk with a diameter which is also solved for. The third model parameter, again, is the proportion of flux received from each component. The net solution is depicted in Figure 6B with the solution parameters labelled.

It must be emphasized that this solution is certainly not unique among three-parameter models, and the specific model invoked is unrealistic if considered in detail. However the remarkable reduction in the $\chi^2$ of the fit thus obtained, by adding only one degree of freedom, is sufficient to conclude that the underlying image at these wavelengths is better described by the solution shown in Figure 6B rather than the simpler two-parameter model (Figure 6A). The fit to the visibility data shown in Figure 6C may be compared to the poorer fit using the simple two-parameter model as seen in the top curve of Figure 5B. Such a 'parametric imaging' capability represents the best that can be hoped for using high precision data points (SNR $>$ 100) obtained over a very limited range of baselines before the first visibility null.

## References

Coude Du Foresto, V., Perrin, G., Ruilier, C., Mennesson, B.P., Traub, W.A. and Lacasse, M.G.: 1998, FLUOR fibered instrument at the IOTA interferometer, in: Robert D. Reasenberg (ed.), *Astronomical Interferometry,Proc. SPIE* **3350**, pp. 856–863.

Coude Du Foresto, V., Ridgway, S. and Mariotti, J.-M.: 1997, Deriving object visibilities from interferograms obtained with a fiber stellar interferometer, *Astronomy and Astrophysics Supplement series* **121**, 379–392.

Kervella, P., Coude du Foresto, V., Glindemann, A. and Hofmann, R.: 2000, VINCI: the VLT Interferometer commissioning instrument, in: P.J. Lena and A. Quirrenbach (eds.), *Interferometry in Optical Astronomy, Proc. SPIE* **4006**, pp. 31–42.

Meisner, J. and Le Poole, R.: 2002, Dispersion affecting the VLTI and 10 micron interferometry using MIDI, in: *Interferometry for Optical Astronomy II, Proc. SPIE* **4838**.

Mennesson, B., Perrin, G., Chagnon, G., Foresto, V.C.d., Ridgway, S., Merand, A., Salome, P., Bordc, P., Cotton, W., Morel, S., Kervella, P., Traub, W. and Lacasse, M.: 2002, Evidence for very extended gaseous layers around O-rich Mira variables and M giants, *AJ* **579**, 446–454.

# INTERFEROMETRIC SCIENCE

# PROBING THE CLOSE ENVIRONMENT OF YOUNG STELLAR OBJECTS WITH INTERFEROMETRY

FABIEN MALBET
*Laboratoire d'Astrophysique, Observatoire de Grenoble, UMR 5571 CNRS/UJF, Grenoble, France*
*E-mail: Fabien.Malbet@obs.ujf-grenoble.fr*

**Abstract.** The study of Young Stellar Objects (YSOs) is one of the most exciting topics that can be undertaken by long baseline optical interferometry. The magnitudes of these objects are at the edge of capabilities of current optical interferometers, limiting the studies to a few dozen, but are well within the capability of coming large aperture interferometers like the VLT Interferometer, the Keck Interferometer, the Large Binocular Telescope or 'OHANA. The milli-arcsecond spatial resolution reached by interferometry probes the very close environment of young stars, down to a tenth of an astronomical unit. In this paper, I review the different aspects of star formation that can be tackled by interferometry: circumstellar disks, multiplicity, jets. I present recent observations performed with operational infrared interferometers, IOTA, PTI and ISI, and I show why in the next future one will extend these studies with large aperture interferometers.

**Keywords:** interferometry, optical, infrared, star formation, young stellar objects, pre-main sequence stars

## 1. Introduction

When trying to understand the origin of our planetary system, one has basically two approaches: (i) looking for other existing planetary systems in the universe to characterize them or (ii) investigate how stellar systems have been forming. By studying young stellar objects (hereafter YSOs) in our Universe, i.e. stars in their early stages of evolution, one focuses our attention to the second approach.

So far, these objects have been extensively observed by spectrophotometry with seeing-limited resolution that corresponds at best to a hundred of astronomical units for typical star formation regions, thus many questions are still *unresolved** because our incapability to disentangle the various phenomena at smaller scales. The advent of millimeter-wave interferometry, adaptive optics imaging and space telescope observations in the last decade allowed us to probe these systems closer to the star (down to 10 AU) but not close enough to probe inner solar system scales. Therefore long baseline optical interferometry with milli-arcsecond resolution corresponding to a tenth of a Earth orbit radius is an essential tool to unravel the nature of the physical phenomena occuring in the early stage of star and planet formation.

* Pun intended as pointed out by JDM.

*Astrophysics and Space Science* **286:** 131–144, 2003.

In this paper, Sect. 2 briefly reviews the current knowledge about YSOs, then Sect. 3 describes the first results obtained with presently operational interferometers and finally Sect. 4 introduces some expected science objectives that will result from interferometric facilities with large apertures that are being built, such as the VLT Interferometer, the Keck Interferometer, the Large Binocular Telescope or 'OHANA.

## 2. Stellar Formation at High Spatial Resolution

The star formation process of solar-type stars lasts about 10 Myr. The scenario of star formation has been categorized in classes of objects corresponding to different stages of evolution (André, 1994) from class 0 to class III. During the first 10,000 yr, the about 10 pc-wide clouds at the origin of stellar systems collapse as a result of reaching their critical mass. Since the clouds are initially rotating, they fragment in flattened rotating structures called pre-stellar class 0 cores and observed at the sub-millimeter wavelengths (Ward-Thompson et al., 1994) since due to the large opacity the energy released during the collapse cannot escape in the visible domain. After about 100,000 yr, the collapsing envelopes around class I protostars dissipate and reveal the inner part of class II systems composed of a protostar usually surrounded by an active accretion disk and a strong bipolar jet. The systems are now shining at shorter wavelengths in the infrared domain, since the main sources of radiation are the disk and the young stars still contracting. At that stage, the extent of the protoplanetary system reaches up to 1500 AU. Later on, at about 1 Myr, the gas reservoir has dissipated and the accretion stops. The class III stars, with some planetesimals and forming planets orbiting around it, are now shining mainly in the visible, a tenuous disk remains when the star arrives on the main sequence, resulting from the collisions of the planetesimals. In this paper we focus our attention on class II systems, targets of interests for optical interferometry.

The first evidence for the presence of circumstellar disks around young stars came from the advent of infrared detectors and subsequent surveys (Mendoza, 1966; Mendoza, 1968) which culminated with the IRAS survey in the early 1980's, as well as from the first UV space observations (IUE). Indeed young stars exhibit an important UV and IR excess that can even exceed the stellar luminosity. Using disk models originally developed by Shakura and Sunyaev (1973) and Lynden-Bell and Pringle (1974), several teams (e.g. Bertout, Basri and Bouvier, 1988) showed that these excesses came respectively from the radiation of the circumstellar matter orbiting at lower effective temperature in the disk and from the interaction between the disk and the slowly rotating star. These models work well for many young systems, but for some, the spectral energy distributions (SEDs) display even larger excesses, explained by passive flaring disk models (Adams, Shu and Lada, 1988).

Millimeter-wave interferometry succeeded in resolving some of the systems with sub-arcsecond resolution (see for example UY Aur rotating disk revealed by Dutrey et al., 1998). At about the same time, adaptive optics on 4-m class telescopes and the Hubble Space Telescope (HST) directly brought to view such disks at visible and near-infrared wavelengths (see e.g. HK Tau B by Stapelfeldt et al., 1998 and HV Tau by Monin and Bouvier, 2000). In fact most of these disks are observed edge-on, because the contrast between the star and the disk is decreased by the absorption of the stellar light by the central part of the disk. The disk is seen in stellar light, scattered in the upper part of the disk atmosphere. The extent of these systems is about 100–200 AU with typical spatial resolution of 10 to 20 AU.

The Herbig-Haro 30 object observed by the HST in forbidden lines and in the continuum (Burrows et al., 1996) brought a lot of information. The relevant region of investigation lies within the 10 AU from the center where the accretion disk interacts with the star and the base of the bipolar jet is forming. One does not know precisely how these different phenomena work and interact. Is the accreting matter falling onto the star through the equatorial plane or along magnetosphere field lines? Is the jet engine the rotating star or the upper part of the disk atmosphere? Where do planets form: in the outer part of the disk or at the locations known for the solar system? What is the physical phenomenon at the origin of the viscous dissipation: pure hydrodynamical turbulence or manifestations of the magnetic field? What is the disk chemistry? What is the spatial distribution of the matter? All these questions can be investigated by visible and infrared long-baseline interferometry.

The current capabilities of optical* long baseline interferometers (with [0.5–10$\mu$m] spectral coverage, baselines up to several hundreds of meters) gives us the characteristic range of spatial scales: between 0.1 and 10 AU, and a temperature range between few 100 K to 5000 K. With these numbers in mind, one can list in more detail what can be investigated by optical interferometry:

- Star surfaces: apparent diameters of young stars are typically 0.1 mas, i.e. they will be barely resolved. In addition the circumstellar matter around them might pollute diameter measurements. However weak-line T Tauri stars, which are class III stars, are believed to have lost their accretion disk and therefore some determination of stellar parameters like the radius, effective temperature or even limb-darkening might be possible.
- Multiplicity: about 80% of young stellar systems are expected to be multiple (Reipert and Zinnecker, 1993). The binary frequency has so far been investigated in the domain of spectroscopic binaries; and visual and adaptive optics binaries, and interferometric observations will fill the gap between the two. Using interferometric data to get the orbit of known spectroscopic binaries will allow to derive their mass accurately.
- Circumstellar disks: optical interferometry allows to study the thermal emission of the disks rather than the stellar light scattered by the disk. Therefore

* When using the term *optical*, I refer to both the *visible* and *infrared* wavelength domains.

one has access to quantities like the temperature and eventually their surface density distributions.

- Accretion: in the current paradigm, accretion occurs in a Keplerian disk. With some spectral resolution, one will verify this motion and even measure departure from the Kepler law (see Sect. 4). The terminal accretion is not yet understood and neither is the link with the star magnetosphere. With spectral and spatial resolution, one can solve the kinematics of the inner region.
- Ejection: one does not know yet if the origin of the jet comes from the stellar dipole/quadrupole or from the disk. Interferometry will allow to probe the base of the jets and determine their opening angle, which is a critical parameter (see Sect. 4.2.2). One can also investigate the link between the jets at small scales and larger outflows.
- Formation of planets: currently giant planets are believed to be born in the outer region of active accretion disk and to then migrate inward closer to the star. Such phenomenon perturbs the disk and may leave detectable signatures like the presence of gaps or gravitational waves in the disk.
- Star formation scenarios: YSOs correspond to a collection of different objects at different times of evolution ranging from classical T Tauri stars to weak-line T Tauri stars , and also include FU Orionis objects which are believed to be T Tauri stars for which the disk undergoes an accretion outburst, and, intermediate mass counterparts called Herbig Ae/Be stars. Observing these different classes of objects will give us clues on the evolution of young stars.

Most of these phenomena takes place in a sub-AU region where the dynamic time scale is less than a year for solar mass stars and even shorter for more massive stars. Therefore one should also consider that the investigated physics will no longer be stationary but for a large part dynamical.

Finally, I would emphasize that observing YSOs will bring us a wealth of information also useful for other types of astrophysical objects such as active galactic nuclei, cataclysmic variables, radio-jets,... since it will shed lights on universal physical processes like the nature of viscosity, the role of magnetic field, how dust grains form and grow, and the link between accretion and ejection processes.

## 3. First Results from Operational Small Aperture Optical Interferometers

In this section, I summarize the results obtained so far in optical interferometry on the brightest objects and I show that interferometry has already modified our understanding of disks.

These results come from three infrared interferometers: the *Palomar Testbed Interferometer (PTI)* located on Palomar Mountain (California, USA) and operating two pairs of siderostats in the *H* and *K* bands; the *Infrared and Optical Telescope Array (IOTA)* located on Mount Hopkins (Arizona, USA) operating in the *H* and *K* bands with two relocatable siderostats; and the *Infrared Spatial Interferometer*

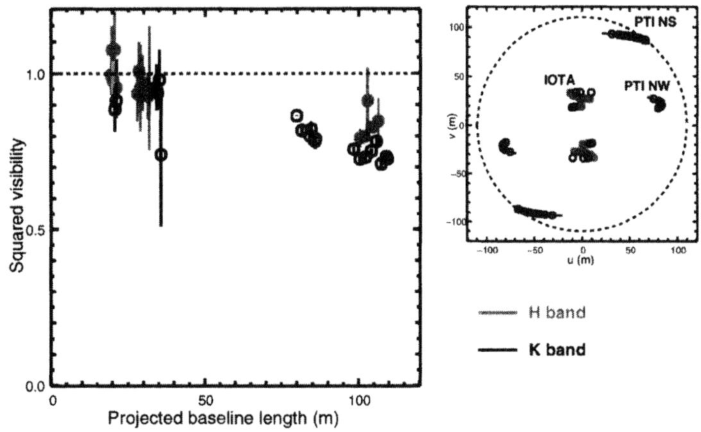


*Figure 1.* Interferometric observations (Malbet et al., 1998; Malbet and Berger, 2001) of FU Ori collected at IOTA and PTI between 1997 and 2000. Gray circles correspond to *H* band, and black circles to *K* band. Left panel: squared visibilities as a function of the projected baseline. Right panel: $(u, v)$ coverage.

*(ISI)* located on Mount Wilson (California, USA) and operating at $10\mu$m with a heterodyne detection.

The list of refereed papers that describe all YSO observations carried out todate is: FU Ori at PTI (Malbet et al., 1998); AB Aur at IOTA and PTI (Millan-Gabet et al., 1999); T Tau, MWC 147, SU Aur and AB Aur at PTI (Akeson et al., 2000); about 15 Herbig Ae/Be stars at IOTA (Millan-Gabet, Schloerb and Traub, 2001); and the massive star LkH$\alpha$ 101 at ISI (Tuthill et al., 2002).

Most of the objects observed so far are either intermediate mass objects or low-mass stars with an active accretion disk, because their magnitudes are close to the sensitivity limit of the instruments. The measurements are limited to visibility amplitudes and therefore are mainly focused on the disk physics.

### 3.1. Constraint on T Tauri disks

One interesting example is the case of FU Ori. This system is known to be composed of a T Tauri star with a disk undergoing an accretion outburst (Hartmann and Kenyon, 1996). Therefore this system is brighter by more than 2–4 magnitudes than other classical T Tauri systems and the disk emission dominates the spectral energy distribution. This object is therefore an ideal target to study disk accretion, since at near-infrared wavelengths the observations are not contaminated by other physical processes.

Figure 1 shows a typical example of data collected with IR interferometers. The visibility reduction of 15-30% at long baselines corresponds to the resolution of a structure of size of the order of 1 mas, corresponding to 0.5 AU at the distance of the Orion cloud. Moreover, the visibility drop in *H* is smaller than in *K*, meaning that the size of this structure decreases with the wavelength faster than the resolution $\lambda/B$.

The data is consistent with the standard disk model used to fit the SED. In fact the data points are lying exactly where they were expected to be from disk model predictions (Malbet and Bertout, 1995).

However the data collected in the following years in the $H$ and $K$ bands show square visibilities which differ more than expected by the standard accretion disk model. A way to investigate this departure from the standard model is to derive a direct measure of the temperature law. Adams, Shu and Lada (1988) have introduced an *ad-hoc* prescription to describe the variation of the effective temperature through the disk:

$$T = T_0 \left(\frac{r}{r_0}\right)^{-q} \quad \text{with} \quad 0.5 \leq q \leq 0.75 \tag{1}$$

A value of $q = 3/4$ corresponds to the standard geometrically thin disks and $q = 0.5$ describes some type of flaring passive disks. Beckwith et al. (1990) surveyed a large number of sources at millimeter wavelength and could derive values of $q$ which indeed range between 0.5 and 0.75.

In the case of FU Ori, Malbet (2002) obtained $q \simeq 0.6 \pm 0.2$, compatible with the SED derived value of $q \simeq 0.75$. However, in the future more precise observations might no longer be interpreted with this simple disk model. The reason for such behavior can be explained by the fact that interferometry probes the inner region of the disk whereas the SED is more sensitive to the outer regions. One speculates that the change of temperature law is due to a change of the accretion rate in the inner part of the disk due to the presence of a jet that removes material from the disk.

The case of FU Ori is simple since the dominant physical process is the accretion. In low-mass stars YSOs, some other effects can play a substantial role, like the heating of the disk photosphere by the central star (Calvet et al., 1991; Malbet and Bertout, 1991; Chiang and Goldreich, 1997). This creates an upper layer which can have higher temperature than the layer located at lower altitudes, impacting both the SED and the visibilities.

To describe this stellar reprocessing in the disk, there exists no simple models. Chiang and Goldreich (1997) presented a two-layer model that is successful in interpreting SEDs. More recently Lachaume, Malbet and Monin (2003) generalized the Chiang and Goldreich model to include the effect of viscous dissipation. They have applied their model to interferometric observations made by Akeson et al. (2000) at PTI and succeeded in fitting reasonably well both SED and visibilities. It therefore shows that interferometric data can provide valuable information on the physics of the disk.

### 3.2. Disks around Herbig Ae/Be stars are not as expected

Herbig Ae/Be stars (HAEBEs) are young stars of intermediate mass (a few solar masses). They exhibit an infrared excess like the lower mass T Tauri stars, but there has been a controversy on the origin of this excess (Hillenbrand et al., 1992; Boehm and Catala, 1994). Since they are brighter than T Tauri stars, they have been excellent candidates to be observed by infrared interferometers.

Millan-Gabet et al. (1999) observed the first Herbig Ae star AB Aurigae in 1997–1998 and discovered that the standard disk model was not compatible with the data. In brief, the object was over-resolved compared to the prediction of the model. This result was confirmed by observations of about 15 targets of the same type by Millan-Gabet, Schloerb and Traub (2001).

The best interpretation of the data is to consider that the star is surrounded by a ring of material located at sub-AU distances, namely 0.3 AU for AB Aur. Tuthill, Monnier and Danchi (2001) obtained similar results on the more massive star LkH$\alpha$ 101, and proposed that this ring is the result of the sublimation of the dust orbiting the star, such that the ring radius gives an estimate of the distance where dust sublimation occurs.

Monnier and Millan-Gabet (2002) compiled measurements made with the various interferometers and put them in a diagram with the ring inner radius as a function of the central star luminosity. The correlation seems to follow the law that can be expected by a simple model of dust sublimation in $r^{-2}$. In parallel, Natta et al. (2001) who worked on the SEDs measured by ISO were not able to reproduce the gap occuring in the near-infrared with the standard model. By truncating the inner part of the disk and heating the inner part of the disk by the central star, they were able to fit both the ISO data and the interferometric observations. The ring proposed by Millan-Gabet and collaborators would then be the inner part of a disk which is puffed up by the stellar radiation and may also shadow part of the disk (see also the model proposed by Dullemond, Dominik and Natta, 2001).

It is less clear whether T Tauri stars follow the same behavior. As a matter of fact, the regions for dust sublimation models and accretion disk are very close for these objects. More detailed studies are therefore required for these low-mass objects and should be possible with large aperture interferometers.

### 3.3. Conclusions and perspectives

One starts constraining the physical conditions in disks with different techniques (millimeter-wave interferometers, adaptive optics, HST, optical interferometry), but optical interferometry is the only current technique which constrains the inner 10 AU of the disk. The first results on low-mass T Tauri stars are consistent with standard accretion disk models, but more complex environments appear to be required for the more massive HAEBE stars with passive disks.

One needs to improve the wavelength coverage since it is a good way to constrain temperature laws. Ideally $J$, $H$, $K$, $L$ and $N$ data sets and better $(u, v)$ coverage are desirable.

The coming facilities with large apertures like the VLTI, KI, LBT and 'OHANA will certainly boost these studies by bringing higher sensitivity, better $(u, v)$ coverage, access to phases and eventually to images. Finally, it is also essential to continue the theoretical modelling effort, particularly relevant physics of the inner disks.

## 4. Expected Science from Coming Large Aperture Interferometers

Some very interesting science has been carried out with current interferometers with small apertures. In this section, I present in the YSO field the impact of anticipated improvements that will be accessible with interferometers with large apertures like the VLTI, the Keck interferometer, the LBT and 'OHANA.

### 4.1. Improvements Expected from Large Interferometers

The angular resolution and the spectral coverage offered by presently operating interferometers match almost perfectly the needs in the star formation domain (see Figure 2). YSOs are usually at the limit of detection of present small aperture telescopes with a typical $K$ limiting magnitude of 6. Only very few objects can be observed and are mainly the more massive ones like Herbig Ae/Be stars (see results presented in Sect. 3).

Usually in interferometry, the magnitudes of the objects are not the unique detection criterion but the expected visibility is also important: interference fringes must be detected and the amplitude of these fringes are proportional to the correlated magnitude, i.e. both to the flux and to the visibility amplitude at the given spatial frequency. However in young stars, since most structures are barely resolved, one does not expect the correlated magnitude to be smaller by 1 or 2 magnitudes than the object magnitude and the object magnitude remains a good criterion. The left panel of Figure 3 displays the spectral energy distribution for different types of low-mass stars located respectively at 150 pc and 450 pc: classical T Tauri stars, T Tauri stars with a flat infrared excess and FU Orionis objects. HAEBE stars and more massive stars usually lie far above these curves. On the same figure, I indicate the sensitivity limits of PTI and IOTA demonstrating that these interferometers are at the very edge of the domain. Expected performances from interferometers using 2m (ATs) and 8-10m (UTs) class telescopes like the VLTI, KI, LBT or 'OHANA show that those interferometers are sensitive enough to fully investigate these objects at least in the near-infrared. In fact the right panel of Figure 3, which plots the $K$ and $V$ histograms of the young stars listed in the Herbig and Bell catalog (Herbig and Bell, 1988) shows that with a $K$ limiting

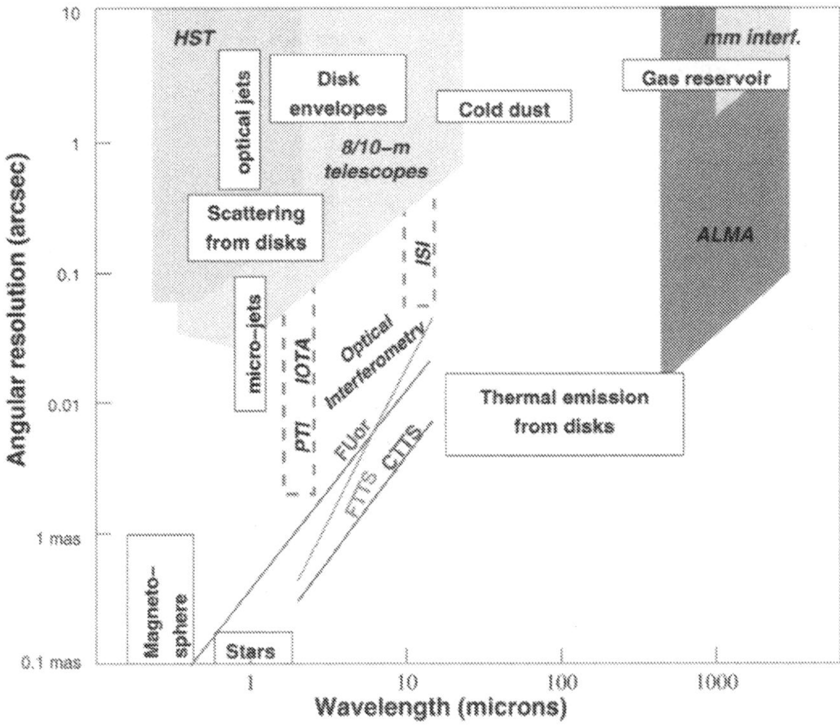


*Figure 2.* Resolution diagram showing angular resolution as a function of wavelength. Competitive instruments from optical to millimeter wavelengths are displayed together with the main issues in the star formation field. Also present are the operational infrared interferometers that have already produced some astrophysical results: IOTA, PTI and ISI.

magnitude of 10, a conservative value, several hundreds of objects are observable. With such a high number of potential targets, one can now consider carrying out surveys in order to perform statistical studies where one would look for correlations between visibilities and flux excesses, binary frequency, or veiling.

Another limitation of presently operating telescopes is the number of operational baselines. IOTA is limited to a maximum baseline length of 38m, PTI can only use long baselines of 80m and 100m (see for example the $(u, v)$ coverage of FU Orionis shown in Figure 1). This is not sufficient to obtain a complete characterization of the resolved material. In the near future, the VLTI will be able to provide a wide coverage of a 200-m large synthetic aperture thanks to 4 relocatable auxiliary telescopes. CHARA with a maximum baseline of 350m even if with only 1-m apertures will provide more angular resolution. 'OHANA with a limited number of fixed telescopes remains interesting because of its very long maximum baseline of 800 m. A high number of measurements at different spatial frequencies are also expected to lead to aperture synthesis imaging. In fact even a small number of spatial frequencies should permit to achieve parametric images

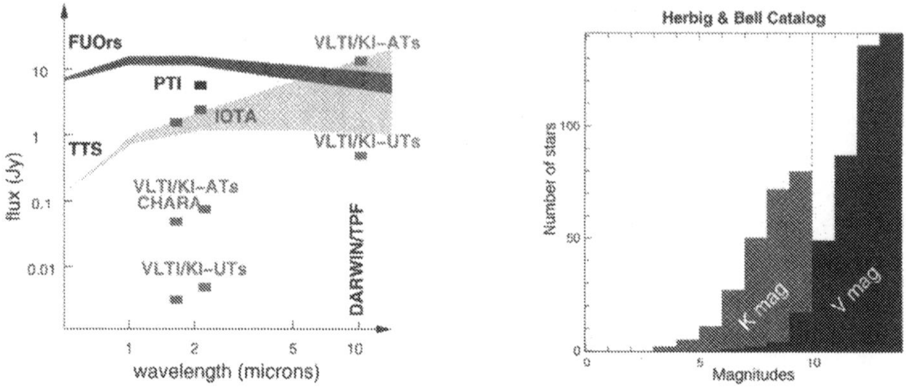


*Figure 3.* Left panel: Sensitivity diagram showing the flux sensitivity as a function of the wavelength in the optical range. Sensitivities from IOTA and PTI are represented as well as the typical flux for T Tauri stars (TTS, both classical ones and those with a flat IR spectrum) and FU Orionis stars (FUOrs). Right panel: $K$ and $V$ histograms of objects listed in the Herbig & Bell catalog (Herbig and Bell, 1988). A limiting magnitude of $K = 10$ (dashed line) is a conservative value for an interferometer with large apertures.

using regularization methods that takes into account the astrophysical nature of the sources.

I would like to emphasize also that interferometers up to now provided only broad-band visibility amplitudes. IOTA has been upgraded to 3 telescopes and will soon provide closure phases on YSOs, an observable which gives information on the symmetry of the obscured structure. A centro-symmetric object has a zero closure phase, whereas a non centro-symmetric object has a non-zero closure phase. Lachaume, Malbet et al. (2003, 2000) simulated circumstellar disks irradiated by young stars and showed that the expected closure phase with typical 100-m baselines can reach 50 degrees. AMBER (Petrov et al., 2002) on the VLTI will also provide closure phases whenever used with a triplet of auxiliary or unit telescopes. AMBER and MIDI on the VLTI will provide for the first time spectral information on YSOs in the IR domain. With such spectral information, relative visibility amplitudes and phases through the spectrum and therefore through spectral lines can be measured. This feature should bring out a radical change in the observational constraints because it includes both morphological and kinematics information.

### 4.2. Some examples of new YSO investigations

In this part, I present some non-exhaustive examples of new types of observations than can be carried out with instruments providing the type of improvements listed in the previous paragraph.

4.2.1. *Planet-disk interaction*

Wolf, Henning and d'Angelo (2001) have investigated the possibility to detect the signature of a protoplanet inside a protoplanetary disk. Some exoplanets discovered by radial velocity techniques are massive giant planets located close to the parent star. This location was not predicted by theory, but since then, several models were able to explain the presence of such planets thanks to migration mechanisms in a protoplanetary disk (Nelson et al., 1998). The idea of Wolf, Henning and d'Angelo, was to see if it would be possible to detect such a gap in accretion disks that would be the signature of the presence of a giant planet. They made the prediction for mid-IR observations (MIDI on the VLTI) and demonstrated that the difference between a perfectly smooth disk and a disk with a gap leads to a change in visibility amplitudes of the order of 5%. This number is at the limit of detection for the instrument MIDI, but within realistic expectations for current near-infrared instruments or future instruments.

4.2.2. *Opening angle of microjets*

Eislöffel and Dougados (1997) and Garcia, Foy and Thiébaut (2001) proposed to investigate the morphology of jets close to the stars. Jets are an important feature of young stellar objects, almost always present whenever there is a disk. For the moment very little is known on the interaction between the disk, the stars and the jets. In fact several magneto-hydrodynamical models have been developed and optical interferometry could be the tool to test them on very high angular resolution observations.

However, even before directly testing the physics of jets, the simple observation of the opening angle of jets would bring significant constrain on models. Up to now the resolution of a single dish telecope was not sufficient to see the base of the jets. Garcia, Foy and Thiébaut (2001) used a toy model to simulate a hollow jet and computed the expected visibility of a star-jet system. The expected visibility amplitude is of the order of 50%. Garcia et al. demonstrated that RU Lupus can be observed with sufficient signal-to-noise ratio with AMBER and medium spectral resolution.

4.2.3. *Probing the velocity field of disks at sub-AU scale*

Spectral information is indeed the feature that could bring one the most interesting information in YSOs. Lachaume, Malbet and Monin (priv. comm.) proposed to observe circumstellar disks with enough spectral resolution to be able to detect the Keplerian motion of the matter in the disk. Kenyon, Hartmann and Hewett (1988) have correlated several absorption lines of the spectrum of FU Orionis in order to detect the Doppler shift signal coming from the matter orbiting around the star. They found a double peak line that could be interpreted as the evidence of Keplerian motion in the disk. However they were not able to completely rule out the presence of a spectroscopic binary in the system. With AMBER and the spectral information, one can follow the position of the photocenter by measuring

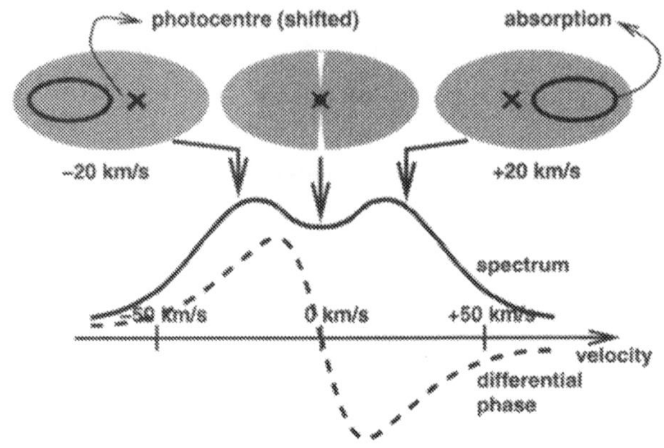


*Figure 4.* Principle to measure the Keplerian motion of matter inside a circumstellar disk (Lachaume et al., priv. comm.). See the text for details.

the wavelength-dependent phase. Figure 4 shows the principle of the measure: in the blue part of the absorption line the matter which absorb the light is located on one part of the disk and therefore the photocenter of remaining light is shifted to the opposite side of the disk. For the red part of the spectrum, the signal is shifted in the opposite direction. In the middle of the line the disk is symmetrical and the photocenter is aligned with the star. Therefore the phase will increase positively in the blue part of the line goes through zero at the systemic velocity and increase negatively in the red part. The shape of this curve can be compared to the one expected from a pure Keplerian motion. Any departure would be an interesting test of the physics of the disk. The signal will be faint, but the plan is to correlate the signals from various spectral lines as is done for radial velocity measurements.

## 5. Conclusion

I reviewed the science that has been performed so far with existing interferometers. Some important information has been obtained mainly for the most massive objects already observable. A wider range of issues can be addressed by large interferometers since a large fraction of YSOs will be observable with a large number of different instrumental configurations. This astrophysical domain is therefore undoubtedly progressing thanks to this new high angular resolution technique and the results will be of importance not only for star formation but also for general physical processes like the origin of turbulence, the interraction between accretion and ejection.

However, the wavelength coverage of current and planned instruments does not cover yet the visible part of the spectrum, the field is limited to compact objects and imaging is still a goal. For all these reasons, star formation remains a major scientific drivers for next generation interferometric instruments.

## Acknowledgements

I would like to thank J.-P. Berger, R. Lachaume and R. Millan-Gabet for very interesting discussions and inputs for this review that was initially prepared for the SPIE'2002 meeting. I am also grateful to R. Millan-Gabet and and J.D. Monnier who reread the manuscript.

## References

Adams, F.C., Shu, F.H. and Lada, C.J.: 1988, The disks of T Tauri stars with flat infrared spectra, *AJ* **326**, 865–883.

Akeson, R.L., Ciardi, D.R., van Belle, G.T., Creech-Eakman, M.J. and Lada, E.A.: 2000, Infrared Interferometric Observations of Young Stellar Objects, *AJ* **543**, 313–317.

Akeson, R.L., Ciardi, D.R., van Belle, G.T. and Creech-Eakman, M.J.: 2002, Constraints on Circumstellar Disk Parameters from Multiwavelength Observations: T Tauri and SU Aurigae, *AJ* **566**, 1124–1131.

André, P.: 1994, Observations of protostars and protostellar stages, in: T. Montmerle, C.J. Lada, I.F. Mirabel and J. Tran Thanh Van (eds.), *The Cold Universe; Proceedings of the XXVIIIth Rencontres de Moriond,* pp. 179. Frontières, Paris.

Beckwith, S.V.W., Sargent, A.I., Chini, R.S. and Guesten, R.: 1990, A survey for circumstellar disks around young stellar objects, *Astron. J.* **99**, 924–945.

Berger, J.P.: 1998, Interférométrie et formation stellaire. Perspectives pour une instrumentation en optique intégrée, *Ph.D. Thesis*, University of Grenoble.

Bertout, C., Basri, G. and Bouvier, J.: 1988, Accretion disks around T Tauri stars, *AJ* **330**, 350–373.

Boehm, T. and Catala, C.: 1994, Forbidden lines in Herbig Ae/Be stars: the [OI](1F) 6300.31A and 6363.79A lines. I. Observations and qualitative analysis, *A&A* **290**, 167–175.

Burrows, C.J. et al.: 1996, Hubble Space Telescope Observations of the Disk and Jet of HH 30, *AJ* **473**, 437.

Calvet, N., Patino, A., Magris, G.C. and D'Alessio, P.: 1991, Irradiation of accretion disks around young objects. I – Near-infrared CO bands, *AJ* **380**, 617–630.

Chiang, E.I. and Goldreich, P.: 1997, Spectral Energy Distributions of T Tauri Stars with Passive Circumstellar Disks, *AJ* **490**, 368.

Dullemond, C.P., Dominik, C. and Natta, A.: 2001, Passive Irradiated Circumstellar Disks with an Inner Hole, *AJ* **560**, 957–969.

Dutrey, A., Guilloteau, S., Prato, L., Simon, M., Duvert, G., Schuster, K. and Menard, F.: 1998, CO study of the GM Aurigae Keplerian disk, *A&A* **338**, L63–L66.

Eislöffel, J. and Dougados, C.: 1997, Outflows from Young Stars, *Science with the VLT Interferometer* **240**.

Garcia, P.J.V., Foy, R. and Thiébaut, E.: 2001, Into the twilight zone: reaching the jet engine with AMBER/VLTI, in: J.F. Alves and M.J. McCaughrean (eds.), *ESO conference*, **267**.

Hartmann, L. and Kenyon, S.J.: 1996, The FU Orionis Phenomenon, *Annual Review of Astronomy and Astrophysics* **34**, 207–240.

Herbig, G.H. and Bell, K.R.: 1988, Catalog of emission line stars of the Orion population: 3: 1988, *Lick Observatory Bulletin, Santa Cruz: Lick Observatory.*

Hillenbrand, L.A., Strom, S.E., Vrba, F.J. and Keene, J.: 1992, Herbig Ae/Be stars – Intermediate-mass stars surrounded by massive circumstellar accretion disks, *AJ* **397**, 613–643.

Kenyon, S.J., Hartmann, L. and Hewett, R.: 1988, Accretion disk models for FU Orionis and V1057 Cygni – Detailed comparisons between observations and theory, *AJ* **325**, 231–251.

Lachaume, R.: 2003, Structure des disques d'accrétion autour des étoiles jeunes de faible masse, *Ph.D. Thesis*, University of Grenoble.

Lachaume, R., Malbet, F. and Monin, J.-L.: 2003, The vertical structure of T Tauri accretion disks. III. Consistent interpretation of spectra and visibilities with a two-layer model, *A&A*, in press.

Lynden-Bell, D. and Pringle, J.E.: 1974, The evolution of viscous discs and the origin of the nebular variables, *MNRAS* **168**, 603–637.

Malbet, F. et al.: 1998, FU Orionis resolved by infrared long-baseline interferometry at a 2 AU scale, *AJ* **507**, L149–L152.

Malbet, F., Lachaume, R., Monin, J.-L. and Berger, J.-P.: 2000, Probing the inner part of protoplanetary disks with the VLTI, *SPIE* **4006**, 243–251.

Malbet, F. and Berger, J.-P.: 2001, The nature of the circumstellar matter around FU Orionis, *SF2A-2001: Semaine de l'Astrophysique Francaise*, p. 457.

Malbet, F.: 2002, Young stellar objects science with interferometry: a review, *SPIE* **4838**, in press.

Malbet, F. and Bertout, C.: 1991, The vertical structure of T Tauri accretion disks. I – Heating by the central star, *AJ* **383**, 814–819.

Malbet, F. and Bertout, C.: Detecting T Tauri disks with optical long-baseline interferometry, *Astronomy and Astrophysics Supplement Series* **113**, 369.

Millan-Gabet, R., Schloerb, F.P., Traub, W.A., Malbet, F., Berger, J.P. and Bregman, J.D.: 1999, Sub-Astronomical unit structure of the near-infrared emission from AB Aurigae, *AJ* **513**, L131–L134.

Millan-Gabet, R., Schloerb, F.P. and Traub, W.A.: 2001, Spatially resolved circumstellar structure of Herbig AE/BE stars in the near-infrared, *AJ* **546**, 358–381.

Mendoza, V., E.E.: 1966, Infrared photometry of T Tauri stars and related objects, *AJ* **143**, 1010.

Mendoza, V., E.E.: 1968, Infrared excesses in T Tauri stars and related objects, *AJ* **151**, 977.

Monin, J.-L. and Bouvier, J.: 2000, Disks in multiple systems: direct imaging of a nearly edge-on circumstellar disk in the young triple system HV Tau, *A&A* **356**, L75–L78.

Monnier, J.D. and Millan-Gabet, R.: 2002, On the interferometric sizes of young stellar objects, *AJ* **579**, 694–698.

Natta, A., Prusti, T., Neri, R., Wooden, D., Grinin, V.P. and Mannings, V.: 2001, A reconsideration of disk properties in Herbig Ae stars, *A&A* **371**, 186–197.

Nelson, R.P., Papaloizou, J.C.B., Masset, F. and Kley, W.: 1998, The migration and growth of protoplanets in protostellar discs, *MNRAS* **318**, 18–36.

Petrov, R. et al.: 2002, Using AMBER, the near-infrared VLTI focal instrument, *SPIE* **4838**, in press.

Reipurth, B. and Zinnecker, H.: 1993, Visual binaries among pre-main sequence stars, *A&A* **278**, 81–108.

Shakura, N.I. and Sunyaev, R.A.: 1973, Black holes in binary systems. Observational appearance, *A&A* **24**, 337–355.

Stapelfeldt, K.R., Krist, J.E., Menard, F., Bouvier, J., Padgett, D.L. and Burrows, C.J.: 1998, An edge-on circumstellar disk in the young binary system HK Tauri, *AJ* **502**, L65.

Tuthill, P.G., Monnier, J.D. and Danchi, W.C.: 2001, A dusty torus around the luminous young star LkH$\alpha$ 101, *Nature* **409**, 1012–1014.

Tuthill, P.G., Monnier, J.D., Danchi, W.C., Hale, D.D.S. and Townes, C.H.: 2002, Imaging the disk around the luminous young star LkH$\alpha$ 101 with infrared interferometry, *AJ* **577**, 826–838.

Ward-Thompson, D., Scott, P.F., Hills, R.E. and André, P.: 1994, A submillimetre continuum survey of pre protostellar cores, *MNRAS* **268**, 276.

Wolf, S., Henning, T. and d'Angelo, G.: 2001, Detecting gaps in protoplanetary disks with MIDI at the VLTI, in: J.F. Alves and M.J. McCaughrean (eds.), *ESO conference* **325**.

# PALOMAR TESTBED INTERFEROMETER OBSERVATIONS OF YOUNG STELLAR OBJECTS *

F.P. WILKIN[1,*] and R.L. AKESON[2]

[1] *Instituto de Astronomía, Universidad Nacional Autónoma de México, Unidad Morelia, Apdo.Postal 3-72 (Xangari), 58090 Morelia, Michoacán, México*
*E-mail: f.wilkin@astrosmo.unam.mx*

[2] *Michelson Science Center, California Institute of Technology, MS 100-22, Pasadena, CA 91125, USA*

**Abstract.** We present observations of a sample of Herbig AeBe stars, as well as the FU Orionis object V1057 Cygni. Our K-band (2.2 $\mu m$) observations from the Palomar Testbed Interferometer (PTI) used baselines of 110 m and 85 m, resulting in fringe spacings of $\sim 4\,mas$ and $5\,mas$, respectively. Fringes were obtained for the first time on V1057 Cygni as well as V594 Cas. Additional measurements were made of MWC147, while upper limits to visibility-squared are obtained for MWC297, HD190073, and MWC614. These measurements are sensitive to the distribution of warm, circumstellar dust in these sources. If the circumstellar infrared emission comes from warm dust in a disk, the inclination of the disk to the line of sight implies that the observed interferometric visibilities should depend upon hour angle. Surprisingly, the observations of Millan-Gabet, Schloerb and Traub (2001) (hereafter MST) did not show significant variation with hour angle. However, limited sampling of angular frequencies on the sky was possible with the IOTA interferometer, motivating us to study a subset of their objects to further constrain these systems.

**Keywords:** optical interferometry, young stars, circumstellar matter

## 1. Introduction

Near-infrared, long baseline interferometry is sensitive to the distribution of dust around the nearest young stars on scales of the order of 1 AU, and provides a powerful probe of models of disks and envelopes of such stars. The Herbig Ae-Be stars are pre-main sequence, emission line objects that are the intermediate mass ($1.5 - 10\,M_{\odot}$) counterparts of T Tauri stars (Hillenbrand et al., 1992). We also observed the FU Orionis object V1057 Cyg, expected to have a strong disk signature due to the high accretion rate of such objects. While the evolutionary status of the FU Orionis objects remains unclear, they are believed to be T Tauri stars undergoing an episode of greatly increased disk accretion, involving a brightening of $\sim$ 5 magnitudes. V1057 Cyg, whose outburst began in 1969–70, is the only FU Orionis object for which a pre-outburst spectrum is available, confirming its pre-main sequence nature (Grasdalen, 1973). Until now, only one FU Orionis object, FU Orionis itself, has been resolved by long baseline optical interferometry

* Partially supported by NSF International Researchers Fellows Program.

TABLE I
Observing sample

| **Name** | **Alternate** | **RA** (2000.0) | **Dec** (2000.0) | $m_V$ | $m_K$ | **Spec** | d, pc |
|---|---|---|---|---|---|---|---|
| HBC 330 | V594 Cas | 00 43 18.260 | +61 54 40.100 | 9.9 | 5.7 | B8e | 650 |
| HD 259431 | MWC 147 | 06 33 05.190 | +10 19 19.984 | 8.7 | 5.7 | B6pe | 800 |
| MWC 297 | NZ Ser | 18 27 39.6 | −03 49 52 | 9.5 | 3.1 | O9e | 450 |
| HD 179218 | MWC 614 | 19 11 11.254 | +15 47 15.630 | 7.4 | 5.9 | B9 | 240 |
| HD 190073 | V1295 Aql | 20 03 02.510 | +05 44 16.676 | 7.8 | 5.8 | A2pe | 280 |
| HBC 300 | V1057 Cyg | 20 58 53.73 | +44 15 28.4 | 11.6 | 5.7 | — | 575 |

(Malbet et al., 1998), and V1057 Cyg was chosen for study as the next-brightest such object accessible to PTI.

## 2. Source Selection and Observations

We selected a sample of 5 sources from the thesis of Millan-Gabet, chosen to satisfy the observing limitations of PTI, and to avoid known binaries (with the exception of MWC 147, whose companion is too faint to affect the current measurements). Details of the instrument are described in Colavita et al. (1999). Table I describes our final sample.

Observations of each source were interweaved with nearby calibrator stars, chosen to exclude known binaries and variable stars. System visibility was determined based upon observations of the calibrators and models of the calibrator (e.g. size based upon multiwavelength photometry). The measured raw source visibilities were then divided by the system visibility. The resulting calibrated visibilities $V^2_{obs}$ are presented in Table II. Our reported visibilities are a wideband average produced synthetically from five narrowband channels. As a consistency check, sources were calibrated first relative to one calibrator, then relative to another, and the results compared to avoid problems with unknown binarity. The stellar contribution to $V^2_{obs}$ is subsequently removed, assuming the observed spatial distribution of emission on the sky is the sum of an unresolved point source of known flux, and an extended circumstellar contribution. For the Herbig stars, MST estimated the fractions of the infrared emission due to the star and due to circumstellar emission at K. In Table II we list the fraction $f_{cs,K}$ of emission due to circumstellar matter, while that of the star is $f_{*,K} = 1 - f_{cs,K}$. For V1057 Cyg, we will assume all the infrared emission is circumstellar. Table II also gives $V^2_{cs,K}$ for the circumstellar contribution, where $V_{obs} = f_{*,K} + f_{cs,K} V_{cs}$. Because our program stars all have large infrared excesses, the corrections for stellar light are generally small. Upper limits to the visibility squared were determined for sources lacking fringes, based

TABLE II
Calibrated visibilities and stellar contribution

| **Source** | **Baseline** | $V_{obs}^2$ | $f_{cs,K}$ | $V_{cs}^2$ |
|---|---|---|---|---|
| V594 Cas | NW | $0.40 \pm 0.04$ | $0.92 \pm 0.03$ | $0.36 \pm 0.04$ |
| MWC 147 | NW | $0.58 \pm 0.03$ | $0.84 \pm 0.04$ | $0.51 \pm 0.04$ |
| MWC 147 | NS | $0.54 \pm 0.04$ | $0.84 \pm 0.04$ | $0.47 \pm 0.05$ |
| V1057 Cyg | NW | $0.63 \pm 0.03$ | 1.000 | $0.63 \pm 0.03$ |
| MWC 297 | NW,NS | $< 0.2$ | $0.84 \pm 0.03$ | $< 0.12$ |
| MWC 614 | NW,NS | $< 0.2$ | $0.71 \pm 0.10$ | $< 0.12$ |
| V1295 Aql | NS | $< 0.2$ | $0.72 \pm 0.03$ | $< 0.12$ |

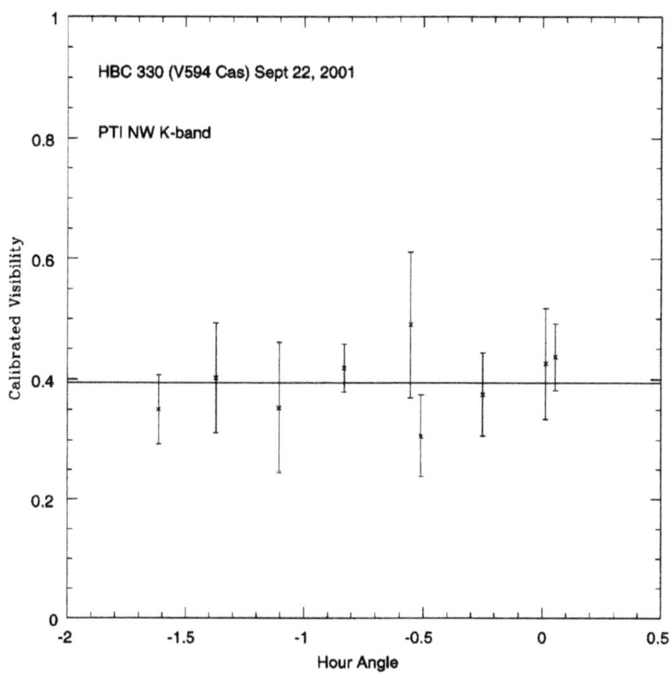


*Figure 1.* Calibrated Visibility of HBC 330 (V594 Cas) as a function of hour angle. The limited range of hour angles does not permit strong constaints on the source geometry.

upon the sensitivity of the detection algorithm and measuring the system visibility with a nearby calibrator. Figures 1–2 show some of the measured individual visibilities $V_{obs}^2$ for our resolved sources.

Fringes were obtained for a total of four sources, although for one of these, MWC 297, there are insufficient data to produce a calibrated measurement. Thus, we treat MWC 297 as an upper limit. Based upon the observed circumstellar visib-

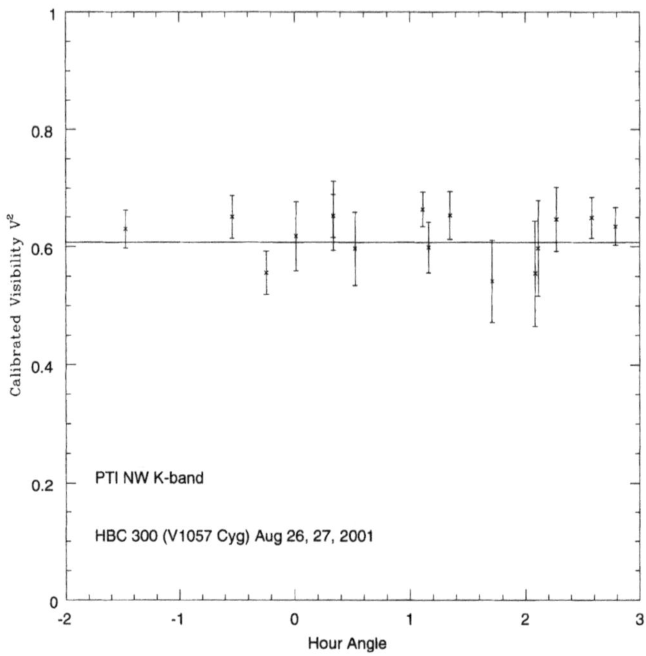


*Figure 2.* Calibrated visibility versus hour angle for the FU Orionis star V1057 Cyg. The lack of a significant trend suggests a circularly symmetric geometry on the sky.

ilities $V_{cs}^2$, Table III gives approximate source sizes based upon a circular Gaussian and a uniform disk model:

$$V_{Gauss}^2 = \left(\exp\left\{-\frac{\pi^2}{\ln 2}\left(\frac{\theta_{FWHM}}{2}\right)^2 \frac{B_p^2}{\lambda^2}\right\}\right)^2,$$

$$V_{UD}^2 = \left(\frac{2J_1(\pi\,\theta_{UD}\,B_p/\lambda)}{\pi\,\theta_{UD}\,B_p/\lambda}\right)^2.$$

Here $\lambda = 2.235\,\mu m$, $B_p$ the projected baseline, $\theta_{FWHM}$ is the FWHM in radians, $\theta_{UD}$ is the uniform disk diameter in radians, and $J_1$ is a Bessel function. The baseline lengths are 110 m in NS, and 85 m in NW. Error bars include uncertainties in our measurements and in the stellar and circumstellar fluxes, but not in the distance.

## 3. Discussion

For our observations with the largest range of hour angles and projected baseline orientation, V1057 Cygni is consistent with a circularly symmetric source. As

TABLE III
Source diameters for the Gaussian and uniform disk models

| Source | Baseline | Gaussian (mas) | Gaussian (AU) | Uniform Disk (mas) | Uniform Disk (AU) |
|---|---|---|---|---|---|
| V594 Cas | NW | $2.07 \pm 0.11$ | $1.35 \pm 0.07$ | $3.34 \pm 0.16$ | $2.17 \pm 0.11$ |
| MWC 147 | NW | $1.63 \pm 0.10$ | $1.30 \pm 0.06$ | $2.75 \pm 0.15$ | $2.20 \pm 0.12$ |
| MWC 147 | NS | $1.34 \pm 0.09$ | $1.07 \pm 0.07$ | $2.24 \pm 0.15$ | $1.80 \pm 0.12$ |
| V1057 Cyg | NW | $1.36 \pm 0.07$ | $0.78 \pm 0.04$ | $2.30 \pm 0.11$ | $1.32 \pm 0.07$ |
| MWC 297 | NW | $> 2.9$ | $> 1.3$ | $> 4.6$ | $> 2.1$ |
| MWC 614 | NW | $> 3.5$ | $> 0.84$ | $> 5.2$ | $> 1.2$ |
| V1295 Aql | NS | $> 2.7$ | $> 0.76$ | $> 4.0$ | $> 1.1$ |

an FU Ori type object, there is little doubt that its infrared excess comes from a circumstellar disk and not a spherical distribution of dust. More modeling is necessary to put limits on the possible orientation of the disk. Our measurements of MWC 147 in the NS baseline are consistent with those of Akeson et al. (2000). However, the new measurement in the NW baseline is inconsisitent with that of the NS baseline if the source is indeed a circularly-symmetric distribution on the sky. Because the baselines have differring orientations, the difference can be accounted for by an asymmetric source distribution, such as a tilted disk. We wish to confirm the new NW measurement and perform further modeling of this source.

## 4. Conclusions

We have resolved three young stellar objects at milli-arc second scales, two for the first time (the Herbig Be star V594 Cas, and the FU Orionis star V1057 Cyg). Presumably we are sampling the distribution of warm dust close to the stars. However, these data alone are insufficient to fully constrain the sources, and other explanations besides circumstellar disks (e.g. a binary companion) are possible. No significant variation of the visibility is seen as a function of hour angle on the sky, suggesting a symmetric distribution on the sky. However, for MWC 147, the measurements in two different baselines suggest an asymmetric distribution, such as a tilted disk. This is consistent with recent measurements by Eisner et al. (2003) for a similar sample of Herbig stars, three of which appear to have disks with significant inclinations.

This work has made use of software produced by the Michelson Science Center at the California Institute of Technology. F.P.W. is grateful to the Observatoire de la Côte d'Azur for a Poincaré fellowship, and to the NSF International Researchers Fellows Program for financial support.

## References

Akeson, R.L., Ciardi, D.R., van Belle, G.T., Creech-Eakman, M.J. and Lada, E.A.: 2000, *ApJ* **543**, 313.

Colavita, M.M., Wallace, J.K., Hines, B.E. et al.: 1999, *ApJ* **510**, 505.

Eisner, J.A., Lane, B.F., Akeson, R.L., Hillenbrand, L.A. and Sargent, A.I.: 2003, *ApJ*, **588**, 360.

Grasdalen, G.L.: 1973, *ApJ* **182**, 781.

Hillenbrand, L.A., Strom, S.E., Vrba, F.J. and Keene, J.: 1992, *ApJ* **397**, 613.

Malbet, F., Berger, J.-P., Colavita, M.M. et al.: 2000, *ApJ* **543**, 313.

Millan-Gabet, R., Schloerb, F.P. and Traub, W.A.: 2001, *ApJ* **546**, 358.

# PROBING THE MAGNETOSPHERE IN YOUNG STARS WITH AMBER/VLTI

C. DOUGADOS, J. BOUVIER and G. DUVERT
*Laboratoire d'Astrophysique de l'Observatoire de Grenoble, BP 53, 38041 Grenoble Cédex 9, France; E-mail: Catherine.Dougados@obs.ujf-grenoble.fr*

P.J.V. GARCIA and D.F.M. FOLHA
*Departamento de Física da Faculdade de Engenharia, Centro de Astrofísica da Universidade do Porto, Portugal*

**Abstract.** The exact geometry of the interface region between the accretion disk and the stellar surface in young stars has crucial implications for both the origin of mass-loss and the regulation of stellar angular momentum. We discuss proposed AMBER/VLTI observations of the Pa$\beta$ line emitting region that will put the first direct constraints on the magnetospheric accretion flow scenario in active young stellar systems.

**Keywords:** star formation, VLTI, accretion, magnetosphere

## 1. The Star/Disk Interface

How exactly matter is transferred from the inner accretion disk to the stellar surface is a long standing unsolved problem in star formation. The exact geometry of this interaction region has crucial implication on: 1) the regulation of stellar angular momentum, 2) wind formation processes and 3) the source of variability, three major key issues in star formation. Magnetospheric models in which matter is funneled along dipolar stellar magnetic field lines anchored at the disk inner radius are now being favored (Hartmann et al., 1994; Muzerolle et al., 1998ab, 2001, Figure 1). The location of the magnetic truncation radius (hence the size of the magnetosphere) is a critical parameter of the models. In particular, both the regulation of stellar angular momentum and the origin of mass-loss require that the disk truncation radius lies at or close to corotation with the stellar surface. With typical T Tauri star parameters, this corotation radius is estimated to lie in the range 3-7 stellar radii (i.e. $\leq 0.35$ mas at the Taurus distance of 140 pc).

Active solar-type young stars are characterized by strong optical and near-infrared hydrogen recombination lines. Magnetospheric funnel flow models have recently successfully reproduced the characteristic shape of these broad, centrally peaked emission profiles (Muzerolle et al., 1998, 2001, Figure 1). In these models, line emission originates from hot gas flowing along the dipolar magnetic field lines linking the inner disk boundary and stellar surface. This interpretation is further supported by the strong correlation observed in T Tauri stars between hy-

*Astrophysics and Space Science* **286:** 151–156, 2003.

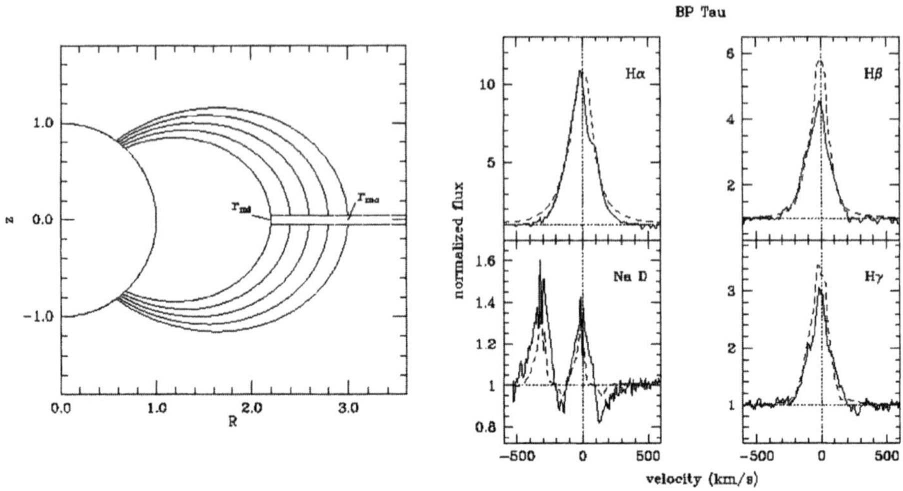


*Figure 1.* **Magnetospheric accretion flow models** *Left*: Schematic of the dipole magnetic field geometry used in the accretion flow models. *Right*: Comparison of observed and model line profiles (solid and dashed lines, respectively) of the Na D and Balmer hydrogen lines for the active T Tauri star BP Tau using the magnetospheric accretion flow models (adapted from Muzerolle et al., 1998a, 2001). The magnetosphere extends out to 3 $R_*$ in this case.

drogen emission line fluxes (especially for Pa$\beta$ and Br$\gamma$) and accretion luminosities (Muzerolle et al., 1998b).

The combination of spatial and spectral resolutions provided by AMBER/VLTI offers the first opportunity to put critical constraints on the size and kinematics of the hydrogen line emitting region and therefore directly test the magnetospheric accretion flow scenario.

## 2. Measuring the Size of the Magnetosphere with the ATs

We first show that medium resolution JH band AMBER observations with the ATs and minimum UV coverage (3 baselines) will bring critical constraints to the size of the magnetosphere in actively accreting young stars. We compare in Figure 2 expected visibilities in the Pa$\beta$ and Br$\gamma$ lines. The magnetosphere emits in the line only while both the stellar photosphere and thermal+scattered light from the circumstellar disk contribute to the continuum emission. We assume a 2D elliptical gaussian distribution for the line emission with FWHM ranging from 5 to 7 stellar diameters (0.1 mas at d = 140 pc). Although the sensitivity of AMBER/VLTI is lower in the J band than in the K band, observations in the Pa$\beta$ line (1.282 $\mu$m) are clearly favored over Br$\gamma$ (2.166 $\mu$m) to maximize both the line to continuum contrast and the achieved angular resolution. In addition, at shorter wavelengths, the contribution of thermal+scattered emission from the circumstellar disk to the

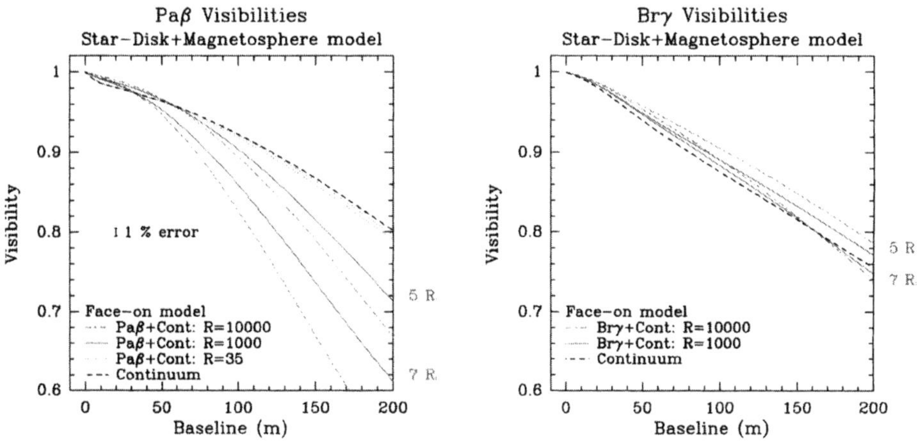


*Figure 2.* **Visibilities of the Star-Disk+Magnetosphere model at** *Left* **Pa$\beta$ (1.28 $\mu$m) and** *Right* **Br$\gamma$ (2.17 $\mu$m).** Predicted visibilities are shown for a spectral channel in the continuum (*dashed line*) and in the line at different spectral resolution: R = $10^4$ (*dot-dashed lines*), R = 1000 (*solid lines*), R = 35 (*dotted lines*). Line equivalent widths are 10 and 5 Åin Pa$\beta$ and Br$\gamma$ respectively, and assumed line velocity FWHM = 100 km/s. An elliptical gaussian distribution with FWHM = 5 and 7 stellar diameters is assumed for the line emitting region. Continuum emission includes the unresolved stellar disk (FWHM = 0.1 mas) and thermal+scattered light contribution from a fiducial accretion disk model.

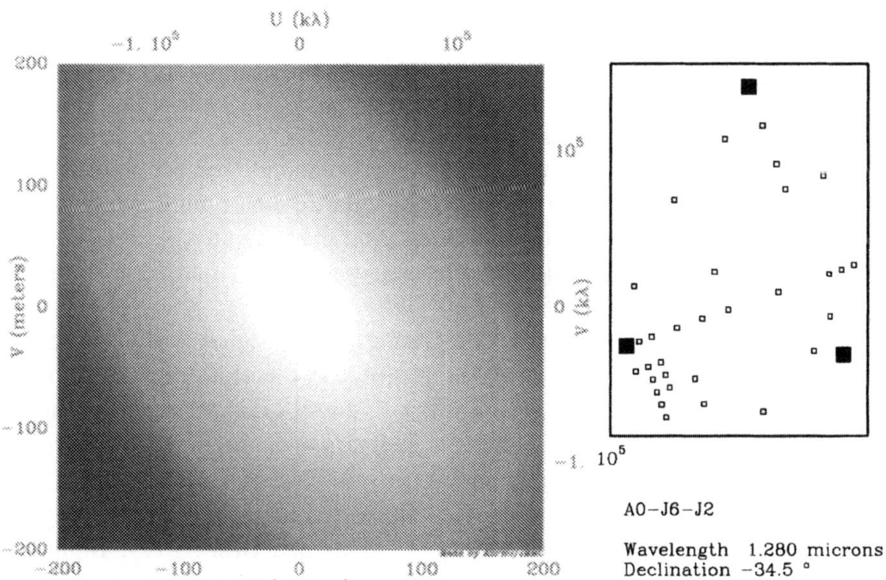


*Figure 3.* **Simulated UV tracks** for 3 ATs (A0-J6-J2) superposed on model Pa$\beta$ visibility for an extended magnetosphere with size 7 stellar radii (FWHM = 0.7 mas) viewed at an inclination to the line of sight of 40 degrees. The UV tracks span a range of 4 hours in HA. Computed for TW Hya observed on 15-03-2003.

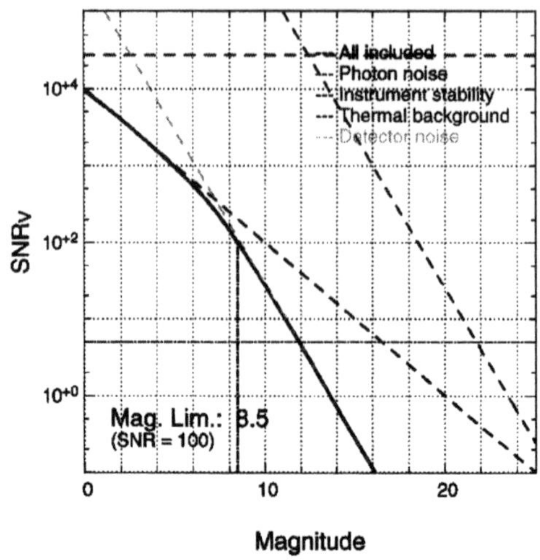


*Figure 4.* Simulated SNR achieved with the ATs at R = 1000 and tip-tilt correction, in 4 hours ($144\times100$s) on source integration. Additional parameters are: average seeing ($0.8''$), Vobj=0.8. SNR $\geq$ 100 is reached for J < 8.5.

continuum is minimized. The medium spectral resolution mode of AMBER (R = 1000) appears as the best compromise between sensitivity and line to continuum ratio. At this resolution the Pa$\beta$ line will remain spectrally unresolved. We predict drop of visibilities in the line with respect to the continuum, of 5 to 20 % at 1.28 $\mu$m for baselines B in the range 100–200 m (see Figure 2), well within the range of detectability of AMBER/VLTI. Measuring visibilities on 3 large baselines at different orientations with the ATs will allow to disentangle projection effects and constrain unambiguously the physical size of the Pa$\beta$ line emitting region (Figure 3). We estimate that the required accuracy on the visibility of $\simeq$ 1 % can be achieved with the ATs in typically 4 hours on source integration for a significant sample of active T Tauri stars (Figure 4)

The derived characteristic size of the Pa$\beta$ emitting region will be compared to predictions from modelling of the line profile and therefore provide *a critical test of the magnetospheric accretion scenario.* In the systems where the inclination of the rotation axis is known, the derived inclination of the magnetosphere will allow to test for possible mis-alignement of the rotation and magnetic axis. Indeed non-aligned rotation and magnetic axis are suggested to account for the origin of the photometric variability due to hot spots (Johns and Basri, 1995; Bouvier et al., 1999). When possible, comparison of characteristic emitting sizes between HeI (Beristain et al., 2001) and Pa$\beta$ will allow to probe the temperature and density distributions of the funnel flow and further test the magnetospheric accretion scenario. Differential phase measurements between the line and continuum region emissions can be also used to probe asymetries in the magnetospheric flux distribution.

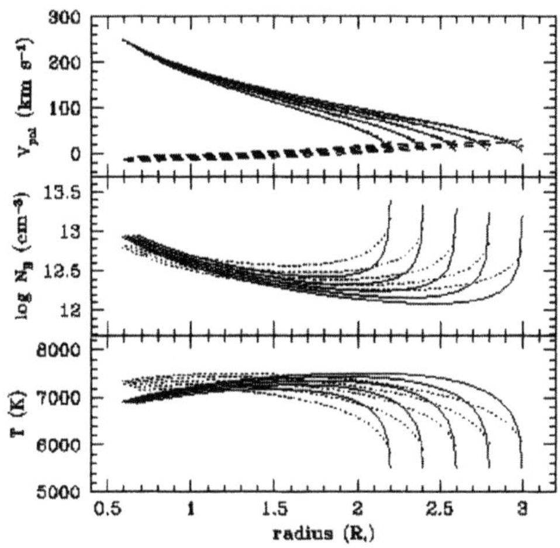


*Figure 5.* Poloidal velocity, density and temperature distributions along field lines for a fiducial magnetospheric flow model (see Muzerolle et al. (2001) for more details). The dashed line in the top panel shows the azimuthal component of the velocity.

## 3. Probing the Kinematics with the UTs at R = 10000

In the magnetospheric accretion framework, the transfer of matter from the disk corotation radius to the star surface occurs along magnetic loops anchored at the magnetic poles. The matter leaves the disk surface at close to zero velocities and reaches the stellar surface at free fall velocities ($\simeq$ 200-300 km/s, Figure 5). Different velocity channels are therefore expected to probe different parts of the magnetosphere, depending on the inclination of the system to the line of sight. In sources where a priori constraints on the inclination are available, velocity resolved visibility profile will therefore provide a crucial test of the magnetospheric flow model. For close to edge-on systems for example, we expect the low-velocity part of the profile to trace the outer boundary of the magnetosphere while the high velocity wings will remain unresolved. On the opposite, if mass-loss significantly contributes to the high-velocity wings of the line profile, as suggested by the H$\alpha$ observations of RU Lup by Takami et al. (2001), these velocity channels will be spatially resolved and their visibility will drop significantly. With AMBER at R = 10000 (ie 30 km/s at 1.28 $\mu$m) the Pa$\beta$ profile will be spectrally resolved (Folha and Emerson, 2001). Although detailed modelling is still required for more accurate estimates, we expect to detect differences in visibilities $\geq$ 15 % and in phases of a few $10^{-2}$ rad across the Pa$\beta$ line (see Figure 2). This program will be preferentially conducted as a detailed follow-up on sources previously resolved at medium resolution with the ATs and will require the UTs. With the UTs and R = 10000, in good seeing conditions (0.5″) we expect to reach S/N > 100 at J < 7.5 in 4 hours on source integration.

## 4. Conclusions

We have shown that AMBER/VLTI provides the first opportunity to directly constrain the extent and kinematics of the magnetosphere in T Tauri stars. Constraints on the magnetospheric size will be derived from medium resolution JH band observations with the ATs and minimum UV coverage (3 baselines minimum). Detailed follow-up studies on selected systems at higher spectral resolution (R = 10000) will allow to probe the kinematics of the line emitting region and further test the accretion flow scenario. No doubt that the interaction region between the star and disk will prove to be far more complex than we envision today and will require true imaging capabilities at sub-milliarcsecond resolution.

## Acknowledgements

PJVG was partially supported by grant POCTI/1999/FIS/34549 approved by FCT and POCTI, with funds from the European programme FEDER.

## References

Beristain, G., Edwards, S. and Kwan, J.: 2001, *ApJ* **551**, 1037.
Bouvier, J. et al.: 1999, *A&A* **349**, 619.
Bouvier, J. et al.: 2002, *A&A*, in preparation.
Folha, D. and Emerson, J.: 2001, *A&A* **365**, 90.
Johns, C. and Basri, G.: 1995, *ApJ* **449**, 341.
Hartmann, L., Hewett, R. and Calvet, N.: 1994, *ApJ* **426**, 669.
Lachaume, R.: 2002, private communication.
Muzerolle, J., Calvet, N. and Hartmann, L.: 1998a, *ApJ* **492**, 743.
Muzerolle, J., Hartmann, L. and Calvet, N.: 1998b, *AJ* **116**, 2965.
Muzerolle, J., Calvet, N. and Hartmann, L.: 2001, *ApJ* **550**, 944.
Takami, et al.: 2001, *MNRAS* **323**, 177.

# UNVEILING THE LAUNCHING REGION OF YSO JETS WITH AMBER

FRANCESCA BACCIOTTI[1], LEONARDO TESTI[1], ALESSANDRO MARCONI[1], PAULO J.V. GARCIA[2], THOMAS P. RAY[3], JOCHEN EISLÖFFEL[4] and CATHERINE DOUGADOS[5]

[1]*I.N.A.F. – Osservatorio Astrofisico di Arcetri, Largo E. Fermi 5, I-50125 Firenze, Italy*
*E-mail:fran@arcetri.astro.it*

[2]*Centro de Astrofísica da Universidade do Porto, Rua das Estrelas, 4150-762 Porto, Portugal*

[3]*School of Cosmic Physics, Dublin Institute for Advanced Studies, 5 Merrion Square, Dublin 2, Ireland*

[4]*Thüringer Landessternwarte Tautenburg, Sternwarte 5, D-07778, Tautenburg, Germany*

[5]*Laboratoire d'Astrophysique de l'Observatoire de Grenoble, BP 53, 38041 Grenoble Cedex, France*

**Abstract.** The interplay between accretion and ejection in the environment of young stellar objects (YSOs) is believed to be a crucial element in the star formation process. Since most of the properties of the models are set up in the first few AUs from the source (below the so-called Alfvèn surface), to validate and constrain the models observationally we need very high angular resolution. With HST (resolution $\sim 0.''1$, i.e. about 14 AU in Taurus) we have been able to access the external border of the acceleration region, for jets in the Taurus-Auriga cloud. Here we see an onion-like kinematic structure in the first 200 AU of the flow, and indications for rotation around the symmetry axis for the resolved low/moderate velocity component. We have now planned observations with AMBER on the VLTI to investigate at 1 mas resolution (in J) the core of the central engine, down to 0.1 AU from the source. Here we describe a joint project by several Institutes in the AMBER consortium dedicated to the study of the morphology and detailed kinematics of a few selected targets. On one hand we will use the large UV coverage of the ATs to explore at medium spectral resolution the structure of the flow. On the other hand the large collection area of the UTs combined with the high resolution mode (R = 10000) of AMBER will allow us to search for interesting kinematic features, among which signatures of rotation around the axis, that would constitute an important validation of the proposed models for the jet launching.

**Keywords:** star formation, Herbig-Haro Jets, interferometry

## 1. YSO Jets and the Need for High Angular Resolution

The interplay between accretion and ejection in the environment of young stellar objects (YSOs) is believed to be a crucial element in the star formation process. In particular, Herbig-Haro jets powered by the central 'engine' should be responsible for the removal of excess angular momentum from the disk, which allows the central star to accrete up to its final mass, and for the final dispersal of the infalling material. Stellar jets are seen primarily in optical forbidden lines (Reipurth and Bally, 2001), but they also have lines in the infrared, as, e.g, Pa$\beta$ and [FeII] lines. Jets show typical radial velocities of 200–400 km s$^{-1}$, and diameters ranging from

*Astrophysics and Space Science* **286:** 157–162, 2003.

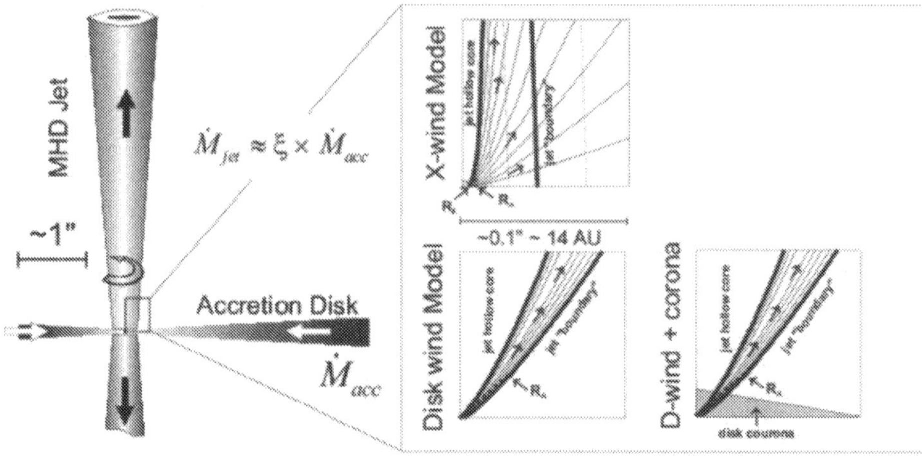


*Figure 1.* Schematic comparison of the Disk-wind model of Ferreira, 1997 and the X-wind model of Shu et al., 2000. Both models predict hollow jets, but the opening angles are very different. The opening angle is observable as an isodensity contour – thick lines. Thin lines are flow surfaces.

unresolved within 0.″1 from the source to 2″ far from it. According to the most popular theoretical scenario magnetic and centrifugal forces act together to launch these jets along magnetic field lines in a 'bead-on-a-wire' fashion, but the models may consider different launching regions: the star/disk interaction at $\leq 0.1$ AU for the *X-wind* (see, e.g. Shu et al., 2000), or the disk up to a few AUs from the source for the *disk wind* (see, e.g. Ferreira, 1997, Garcia et al., 2001). As regards the observational validation of this picture, some progress has been made recently by observing the so-called 'microjets' from unembedded, Class II T Tauri stars (TTSs), for which the flow can be traced back down to the central engine. The jets from DG Tau, RW Aur and CW Tau (at 140 pc) have been observed both from space with the Hubble Space Telescope (HST), with $\sim$ 0.″1 resolution (Bacciotti et al., 2000; Woitas et al., 2002), and from the ground using adaptive optics (AO) techniques (Dougados et al., 2000) at $\sim$ 0.″2 resolution. The flows appear to be well collimated already within 15 AU from the source, with evident asymmetries in the red/blue lobes, and at least in the DG Tau jet the reconstructed velocity channel maps show an onion-like kinematic structure in the first 200 AU from the source.

These studies support the general view of a magneto-centrifugally launched flow, and suggest that the morphological and kinematical structure is indeed extremely rich at the base of the flows, with kinematical features down to a few tens of km s$^{-1}$. The limited spatial resolution, however, does not allow one to see directly the launching region, as with the present limit of 0.″1 we can only access the external border of the acceleration zone, and here the flow is already collimated and has already attained its terminal velocity. Indeed, the core of the central engine lies within the so-called Alfvèn surface, located at a few AUs from the source. This region, yet totally unexplored, can only be investigated with AMBER at the VLTI, that with 1 mas resolution (in J) allows us to probe the circumstellar region down to 0.1 AU in nearby systems. Studies of this type would also constrain the models very efficiently. In fact the various theoretical 'flavours' give different predictions for the morphological and kinematical structure in the first few AUs of the flow (as, e.g., the opening angle of the hollow jet and the density distribution, see Figure 1 and Garcia et al., 2002), while they produce the same observable properties at large distances from the star.

## 2. What Can the VLTI and AMBER Do for Us?

Early studies of the possible application of interferometric techniques to YSO jets considered the measurement of the jet diameter at the base of the flow, in order to constrain the collimation mechanism. In fact available high angular resolution data show that collimation is achieved very early, within 15–20 AU from the source. It was shown by Eislöffel and Dougados (1997) that model visibilities of a jet having two different opening angles (10° and 90°) are clearly distinguishable already with baselines of 50–100 m.

More recently, a joint project has been devised by people in several Institutes of the AMBER consortium to study the morphology, kinematics and excitation properties of the jet base in a number of suitable targets. We have selected the brightest microjet sources where the AO is expected to achieve the best Strehl ratios. All the selected sources are in closeby star formation regions (100–300 pc), and power microjets the properties of which are well known. Our primary targets are the bright southern TTS stars CS Cha and Ru Lupi, at a distance of 150 and 170 pc respectively, which are associated to small jets seen at optical wavelengths and at milliarcsecond scales with spectro-astrometric techniques (Takami et al., 2003). Our next choice, depending on the observation schedule, is the jet from DG Tau (see previous section), that has also been investigated spectro-astrometrically in NIR lines (Takami et al., 2002). The project consists of two parts, which are the bodies of two associated proposals for the AMBER Guaranteed Time.

**Flow morphology.** In the first part of the project we will focus on the determination of the geometrical properties of the flows, by means of medium spectral resolution observations with the ATs in the J band (proposal by R. Foy, P. Garcia et al.). The choice of this band is dictated by higher angular resolution and inclusion of the brightest NIR lines emitted at the base of a flow from TTSs, that is Pa$\beta$, He I 1.08 $\mu$m, and the [Fe II] line at 1.257 $\mu$m. At the wind base the gas can be very dense and hot (up to T$\sim 10^4$ K and n(H)$\lesssim 10^8$cm$^{-3}$), thus forbidden lines may be quenched here. We therefore expect Pa$\beta$ and He I to be the brightest jet lines at the smallest spatial scales sampled by the VLTI. The left panel in Figure 2 shows a typical SNR calculation which assumes that half of the Pa$\beta$ flux comes from the jet. On the other hand [FeII] lines would be the primary NIR jet tracers at larger distances, that provide an important link with AO/HST observations. While

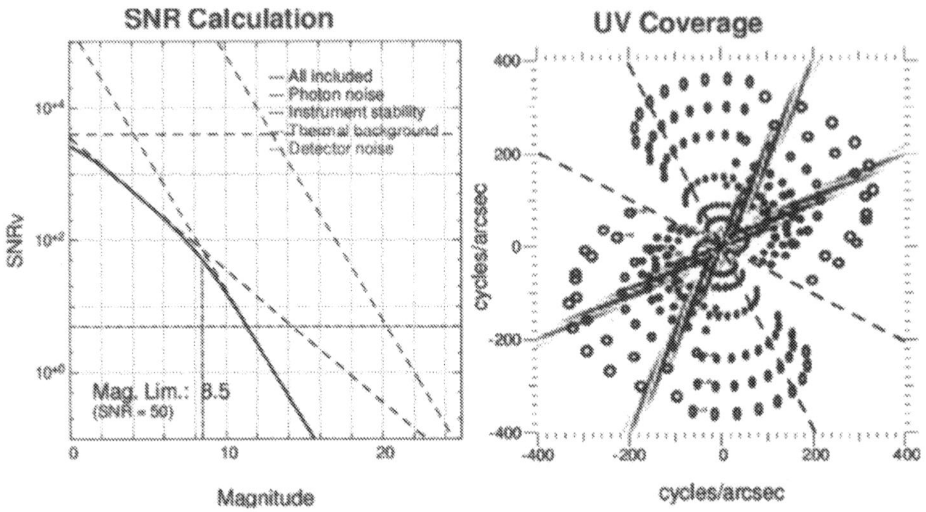


*Figure 2.* **Left:** SNR calculation for AT observations of the jet from RU Lupi, assuming an object visibility of V=0.5 spectral resolution R = 1000, exposure time of 3 × 10 min, fringe tracker on. **Right:** UV points (sampled 1 per hour) for the same observations, superimposed to simulated *visibility* arms of the jet (grey-coloured areas) assuming a 18° opening angle. Dashed lines are the jet in the *image* space.

medium spectral resolution is sufficient to separate the different lines, for the morphological study it is necessary to sample many different baselines, in order to be able to reconstruct good images. This can only be done with the ATs, that provide the best possible coverage of the (U,V) plane (see Figure 2, right panel). An image reconstruction software can then be applied to get images of the jet from the visibility data, to be compared with the model predictions. This tool is described in the contribution by Thiebaut, Garcia and Foy in this volume.

**Flow kinematics.** The second part of the project is dedicated to the determination of the detailed kinematical properties of the flow at AU scales. This will be done by observing the jet in J-band emission lines with the high resolution mode (R = 10000) of AMBER, combined with the large collection area of the UTs (proposal by F. Bacciotti, L. Testi et al.). In particular we want to detect and measure the amount of rotation around the jet symmetry axis at the base of the flow. Rotation is a fundamental element for the launching models, and indeed indications for it could be found with the HST between 20 and 100 AUs from the source in the jet from DG Tau (see Bacciotti et al., 2002, Testi et al., 2002 and Figure 3, left panel). In the acceleration region, i.e. in the first few AUs from the source, the models prescribe a rotational velocity of at least 40 km $s^{-1}$, which is spectrally resolved with the R = 10000 mode. Small centroid displacements in separate wavelengths (as required by rotation) will be identified by taking differential phase measurements, with which we can go down to one tenth of the nominal spatial resolution, i.e. around 0.1 mas.

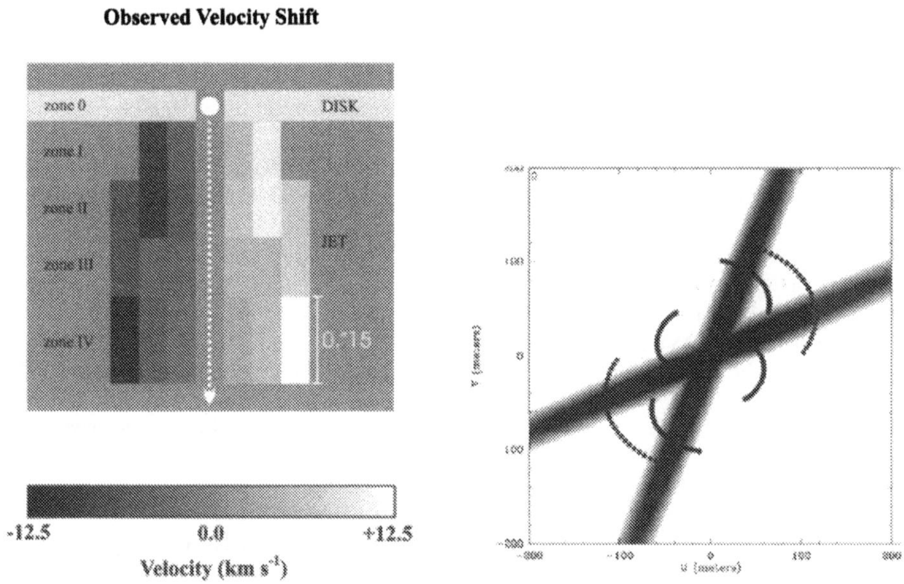


*Figure 3.* **Left** Schematic map of the systematic velocity shifts at the base of the jet from DG Tau derived (at low velocity) from HST/STIS observations (from Bacciotti et al., 2002). **Right:** U,V tracks for UT1-UT3-UT4 for observations of RU Lupi superimposed to simulated visibility maps (grey areas) for the lateral borders of the flow, derived from a toy model of a hollow jet with axis at P.A.=225° and opening angle 45°.

Using UT1,UT3 and UT4, i.e. baselines perpendicular and parallel to the the jet direction, assures reasonable U,V coverage of the object (Figure 3, right). Thus it will be possible to form useful visibility/phase versus observing time plots.

The proposed combined study will allow us to determine for the first time the overall dynamical structure in the first few AUs of a jet flow. We expect to obtain from the observations with the ATs the shape of the flow at its base, the opening angle with respect to the disk surface and the extent of the central hollow core. The derived geometrical parameters will allow us to establish, e.g., if a disk-wind or a X-wind description is more appropriate. Then, the kinematical investigation carried out with the UTs will give us the variation of poloidal and toroidal velocities in the initial jet portion. This in turn will tell us about the efficiency of the acceleration and collimation mechanism, and will allow us to check onto the angular momentum balance between the jet and the star/disk system. In summary, the planned interferometric observations will be a powerful tool to constrain the launching mechanism of both galactic and extra-galactic jets.

## References

Bacciotti, F., Mundt, R., Ray, T.P., Eislöffel, J., Solf, J. and Camezind, M.: 2000, HST/STIS spectroscopy of the optical outflow from DG Tauri: Structure and kinematics on subarcsecond scales, *Astrophys. J. Lett.* **537**, 49, 2000.

Bacciotti, F., Ray, T.P., Mundt, R., Eislöffel, J. and Solf, J.: 2002, HST/STIS spectroscopy of the optical outflow from DG Tauri: Indications for rotation in the initial Jet channel, *Astrophys. J.* **576**, 222.

Dougados, C., Cabrit, S., Lavalley, C. and Ménard, F.: 2000, T Tauri stars microjets resolved by adaptive optics, *Astron. Astrophys.* **357**, 61.

Eislöffel, J. and Dougados, C.: 1997, Outflows from young stars, in: *Science with the VLT Interferometer*, ESO Astrophysics Symposia, Springer-Verlag, Berlin, p. 240.

Ferreira, J.: 1997, Magnetically-driven jets from Keplerian accretion discs, *Astron. Astrophys.* **319**, 340.

Garcia, P.J.V., Ferreira, J., Cabrit, S. and Binette, L.: 2001, Atomic T Tauri disk winds heated by ambipolar diffusion: I-Thermal structure, *Astron. Astrophys.* **377**, 589.

Garcia, P.J.V., Foy, R. and Thiébaut, E.: 2002, Into the twilight zone [C. Bertout, 1989]: Reaching the jet engine with AMBER/VLTI, in: J. Alves and M. McCaughrean (eds.), *The Origins of Stars and Planets: The VLT View*, ESO Astrophysics Symposia, Springer-Verlag, Garching, p. 267.

Reipurth, B. and Bally, J.: 2001, Herbig-Haro flows: Probes of early stellar evolution, *Ann. Rev. Astron. Astrophys.* **39**, 403.

Shu, F.H., Najita, J.R., Shang, H. and Li, Z.-Y.: 2000, X-winds: theory and observations, in: V. Mannings, A.P. Boss and S.S. Russell (eds.), *Protostars and Planets IV*, University of Arizona Press, Tucson, p. 429.

Takami, M., Bailey, J. and Chrysostomou, A.: 2003, A spectro-astrometric study of southern pre-main sequence stars. . ., *Astron. Astroph.* **397**, 675.

Takami, M., Chrysostomou, A., Bailey, J., Gledhill, T., Tamura, M. and Terada, H.: 2002, He I 1.083 $\mu$m emission and absorption in DG Tauri: Line excitations in the Jet, hot wind, and accretion flow, *Astroph. J. Lett.* **568**, L53.

Testi, L., Bacciotti, F., Sargent, A.I., Ray, T.P. and Eislöffel, J.: 2002, The kinematic relationship between disk and jet in the DG Tauri system, *Astron. Astrophys. Lett.* **394**, 31.

Woitas, J., Ray, T.P., Bacciotti, F., Davis, C.J. and Eislöffel, J.: 2002, HST/STIS observations of the bipolar Jet from RW Aurigae: Tracing outflow asymmetries close to the source, *Astrophys. J.* **580**, 336.

# IMAGING STARS AND THEIR ENVIRONMENTS WITH THE VLTI

CHRISTOPHER HANIFF
*Astrophysics Group, Cavendish Laboratory, University of Cambridge, Madingley Road, Cambridge CB3 0HE, UK*

**Abstract.** The opportunity to deliver high-angular resolution model-independent images is one of the most attractive prospects for optical/infrared interferometry. In this paper I use simple imaging simulations to identify some of the practical difficulties that may arise in achieving this goal with the VLTI. For the types of studies investigated here – stellar surface imaging and the mapping of emission line disks – it is likely that the key challenges will be to measure the visibility function on a sufficiently large range of baselines, and to secure accurate and reliable phase information.

**Keywords:** interferometry, synthesis imaging, stellar surfaces, stellar disks

## 1. Introduction

Imaging structurally complex sources in a model-independent manner is one of the 'Holy Grails' of optical/IR interferometry. While existing arrays have only just begun to address this challenge, interferometric observations of stars and their environments have already revealed them as being some of the most attractive targets for study. The processes of convection, rotation, and magnetic activity can all lead to effects that are observable when information on scales comparable to the stellar size becomes available. This has been amply demonstrated by, for example, surface imaging of M supergiants (see, e.g. Buscher et al., 1990 and Tuthill et al., 1997), visibility measurements of rapidly rotating stars (see, e.g. van Belle et al., 2001), and the use of multi-wavelength diameter measurements to probe the atmospheric stratification of cool evolved stars (e.g. Hoffmann et al., 2001).

In a similar vein, interferometric observations of the environments of mass-losing stars have also revealed a wealth of new information. Examples include the discovery of the 'pinwheel nebula' around the Wolf-Rayet system WR 104 (Tuthill et al., 1999), the resolution and monitoring of the highly complex structure of the infrared emission surrounding IRC+10216 (see, e.g. Haniff and Buscher, 1998) and the elaboration of the structures of Be emission line disks through interferometric imaging in the H$\alpha$ line (Quirrenbach et al., 1994).

This paper outlines some of the basic observational challenges for the VLTI in its first phase of operation as it seeks to contribute to these imaging studies. After reviewing the lessons learnt from existing synthesis telescopes, two numerical simulations are used to explore how well the VLTI might be expected to image stellar

*Astrophysics and Space Science* **286:** 163–170, 2003.

disks and their environments. The results are very clear: while imaging will be difficult, it should certainly be possible.

## 2. Requirements for Imaging

In thinking about the potential difficulties of imaging with the VLTI it is reasonable to believe that the following two considerations will play important roles:

- That the most interesting sources for study will be those that display the most complex structure and that are also poorly understood.
- That as the angular resolution increases, the timescale for changes in any source is likely to become smaller.

These statements may appear too obvious to warrant serious comment, but they draw attention to the clear needs for both a *rapid* and *model-independent* imaging capability. Imaging of simple sources, in a mode which requires weeks or months to complete (for example because of a need to reconfigure an interferometer with a small number of array elements), may well be possible, but it is unlikely to capture the interest of the wider astronomical community, and by implication attract further funding.

In the context of stellar imaging, however, there are a number of more positive signs that the task of imaging will be manageable. The first concerns the number of resolution elements required in an interferometric image. The detailed interferometric images obtained using non-redundant aperture masking at 10 m-class telescopes (see, e.g. Tuthill et al., 2000) contain fewer than $10 \times 10$ pixels but have nonetheless been very valuable *scientifically*. This pixel count is much smaller than the $10^6$ that is commonplace in conventional CCD images, and highlights the fact that in many cases even rudimentary imaging will be a powerful tool to investigate topical astrophysical problems.

A second comment is that, at least for stellar sources, the source magnitude and size will be linked. Since stars emit, to first order, as black-bodies, their apparent magnitude will basically depend on their angular size and their temperature. Since the maximum interferometer baseline, $B_{\rm max}$, will determine the smallest source worth studying, this implies that we will know in advance how bright any resolvable targets will be. For the minimum angular size appropriate for the VLTI ($\sim$ 1 milliarcsecond), all stars with surface temperatures greater than 3500 K will have apparent magnitudes in the visible and near-infrared of less than 10, i.e. they will be relatively bright (Baldwin and Haniff, 2002). Stellar sources that are fainter will simply be too small to image!

The above argument does not apply to the imaging of stellar environments which are expected to emit much more weakly than the stellar surface. However, in these cases the circumstellar emission is likely to be present on much larger angular scales than those of the stellar disk, and so it should be possible to use the central star as a point-like reference. This will help both the interferometric measurements and with any adaptive wavefront sensing.

Another approach for assessing how challenging VLTI observing may be is to examine the extent to which it satisfies the general criteria that other interferometric astronomers conventionally use when planning imaging experiments. These 'rules of thumb' are:

1. The total number of visibility data should be greater than, or equal to, the number of filled pixels in the interferometric map. That is, the product of the number of instantaneous baselines and the number of effective array reconfigurations should be greater than or equal to the number of pixels that are expected to contain flux in the final image.
2. The Fourier plane sampling should be as uniform as possible so as to make deconvolution straightforward.
3. The range of baselines sampled, $B_{\rm max}/B_{\rm min}$, should be comparable to the range of angular scales required in the final image.
4. The minimum baseline length should be comparable to the wavelength of observation divided by the maximum angular extent of the source: $B_{\rm min} \sim \lambda/\theta_{\rm max}$.
5. The number and quality of the data should be good enough to achieve the desired dynamic range, i.e. the ratio of the intensities of the brightest and faintest believable features in the final image. The dynamic range will be given by:

$$\frac{\sqrt{n}}{\sqrt{\left[(\delta V/V)^2 + (\delta\phi)^2\right]}}, \tag{1}$$

   where $\delta V/V$ and $\delta\phi$, are the characteristic fractional amplitude and absolute phase errors on the $n$ individual visibility data. Note that this implies that some type of phase measurements are made.

The most obvious impact of these guidelines on the prospects for imaging with the VLTI concerns the number of unit and auxiliary telescopes that will be available at any one time. For routine observations, this may be as low as 4, i.e. the number of currently funded ATs. This, in principle, allows for measurements of up to six visibility amplitudes, four closure phases (using AMBER), and six visibility phases when phase referencing is commissioned with FINITO and PRIMA. However, the current complement of instruments does *not* permit these measurements to be made simultaneously. Hence, the principal shortcoming of the VLTI may well be its inability to secure visibility data that are numerous enough and which cover a wide enough range of spatial frequencies for the task at hand in a usefully short time. Whether we can quantify this concern, at this moment in time, is the subject of the next section of this review.

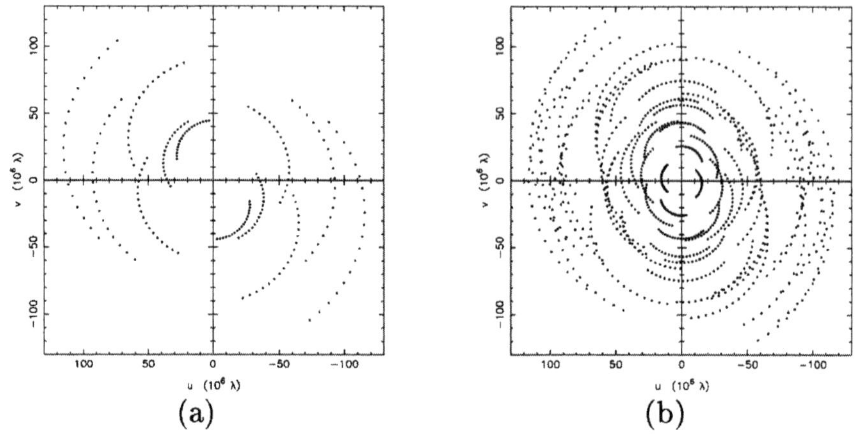


*Figure 1.* The Fourier plane coverage for the two array configurations studied here. The left panel (a) is for a 4-element array with a total of 108 visibility data while the right hand panel shows the coverage for an 8-element array. In this case there are 504 data samples. Note that in both cases, Earth rotation provides essentially no enhancement in the value of the ratio of longest to shortest baseline. Even with 8 telescopes, this ratio is still below 10.

## 3. Examples for the VLTI

One straightforward way to explore the expected imaging capability of the VLTI is to take simulated calibrated visibility amplitudes and closure phases computed for some generic source morphologies and compare the quality of interferometric image reconstructions using differing samplings of these data. This decouples the problems of detecting and calibrating the interferometric signals – which we assume can be performed to some level of precision – and of predicting source brightness distributions – which we reiterate are probably only poorly known at present – from those related to the number and distribution of the visibilities measured.

In the following two subsections, I present results for two 'toy' imaging tests: the first is a resolved stellar surface with two bright hotspots, and the second represents an unresolved star with a edge-on asymmetric disk. Fake complex visibility data were generated with errors equal to $\sim 2\%$ of the total source flux, and sampled so as to mimic observations of a source at declination $-65°$ from Paranal at hour angles between $\pm 3$ hr with measurements made every 20 minutes. For each source, two sets of data were prepared, one corresponding to an observation with 4 array elements at the locations of the VLTI UTs and the other assuming four additional array elements located at four of the available AT stations.

The principal difference between these configurations is evident from their respective Fourier plane coverages (Figure 1): a larger number of array elements immediately allows a wider range of spatial scales to be investigated. These could, of course, be measured sequentially using re-configurations of the 4-element array, but the time to do this would be between five and fourteen times that required for

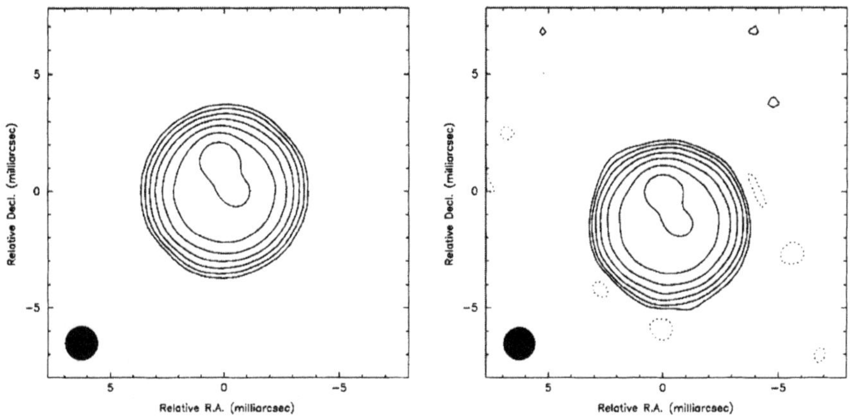


*Figure 2.* Interferometric images of a spotted stellar disk recovered from perfect Fourier data (left) and noisy amplitudes and closure phases (right) from the 8-element array described in the text. It was not possible to map the source reliably with the data from the 4-element array. Contour levels are plotted at −1, 1, 2, 4, 8, 16, 32, and 64% of the peak flux, and in this presentation the images are almost identical. The filled circle at the bottom left hand corner shows the resolution of each map.

the 8-element observation (the longer time corresponds to replicating all possible closure phase measurements). The ability to take advantage of large ($> 5$) numbers of array elements in parallel will clearly lead to large increases in efficiency for the VLTI.

### 3.1. IMAGING STELLAR SURFACES

Here we consider the task of imaging of a stellar surface. With a maximum baseline of 200 m, the most favourable resolved stellar targets for the VLTI will be late-type evolved giants and supergiants. A typical M supergiant with an apparent diameter of 5 milliarcseconds will have an $R$ magnitude of $\sim$ 4 and so will hopefully be bright enough to allow imaging in and out of molecular bands in the optical and near-IR with spectral resolutions of order 100.

To mimic the imaging of such a source, complex visibilities were computed for a model comprising a 5 mas diameter uniform disk and two unresolved 'hotspots' contributing 5% and 10% of the total flux. For this type of target, where a large fraction of the flux from the source is resolved by many of the interferometer baselines, the typical visibility on an arbitrary baseline was low, i.e. well below 30%. On the other hand, the closure phase signals were large, and it is these that carry much of the information on the structure of the target.

The images obtained by inverting the sampled Fourier data are shown in Figure 2. For this resolved source, the Fourier plane sampling provided by the 4-element array did not have a suitable range of baselines to allow *any* type of useful image to be recovered. On the other hand, the map restored from the 8-element data was essentially identical to that which would have been obtained with perfect noise free data. Apart from an uncertainty in the absolute location of the source, the 'truth' and 'test' maps in Figure 2 only differed at the 1% level. The dynamic range in the noisy image was approximately 100:1, i.e. close to the prediction of Eq. (1) for the 2% amplitude and 0.3 radian closure phase errors typical of the visibility and closure phase dataset.

### 3.2. Imaging stellar environments

A quite different type of challenge is presented by a source comprising an unresolved core together with additional distributed emission. As above, model complex visibilities for such a target, comprising an unresolved star and an edge-on asymmetric disk of extent $\pm 4$ mas, were generated with 3% visibility errors. In this case 10% of the total flux was located in the symmetric component of the disk, and 5% in a knot displaced to one side of the star. While these fractions were picked 'at random', this type of brightness distribution is not dissimilar from that expected to be seen in velocity resolved images of the H$\alpha$ disks around Be stars (see, e.g. Stee and Bittar, 2001).

This type of source, with a compact core contributing most of the flux, has a visibility function that remains high ($V \geq 0.8$) on all baselines, and closure phases with quite small amplitudes. Here, for example, $\phi_{\rm cp}$ ranged between $\pm 15°$. For sources such as these, structural information will thus be encoded in the small amplitude modulation of the visibility amplitudes and closure phases as a function of hour angle. This, then, emphasises the need for high precision measurements for reliable imaging, and identifies both signal-to-noise ratio and calibration quality as important goals for an interferometer design.

Images recovered from perfect visibilities and from noisy amplitude and closure phase data are shown in Figure 3. As for the earlier example, the large extent of the circumstellar emission meant that data from the 4-element array did not have a suitable range of baselines to allow useful interferometric image reconstruction. But, again, the 8-element array image was of extremely high quality. In this instance, the smaller phase errors (0.04 radians) meant that a dynamic range of approximately 400:1 was reached, close to the prediction of Eq. (1) for the typical 3% errors in the amplitude data.

## 4. Implications for the VLTI

My conclusions for prospects for imaging with the VLTI can be summarised under three main headings:

- Reliable imaging should certainly be possible if amplitude and phase data can be secured. However, this will require excellent coverage of the Fourier plane and it is likely that multiple reconfigurations of the array will be necessary to guarantee success. The time required for this should not be underestimated.

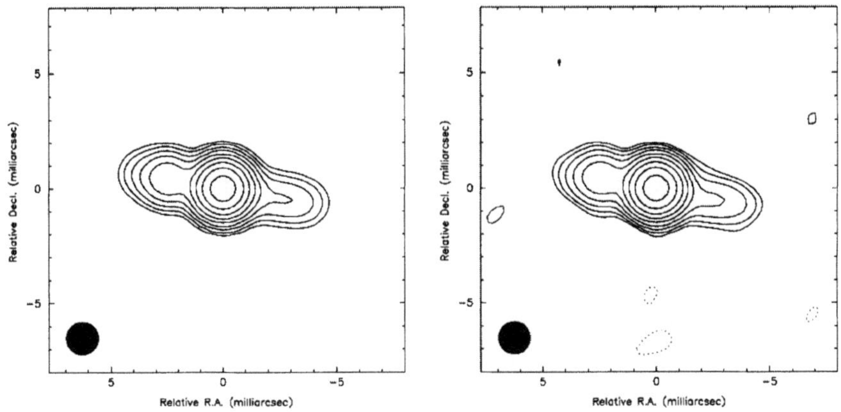


*Figure 3.* Interferometric images of a star plus edge-on disk recovered from perfect Fourier data (left) and noisy amplitudes and closure phases (right) for the 8-element array described in the text. The two images are identical at the 0.25% level. As for the case of a stellar surface, it was not possible to map the source reliably with data sampled by a 4-element array. Contour levels are plotted at −0.25, 0.25, 0.5, 1, 2, 4, 8, 16, 32, and 64% of the peak flux. The filled circle at the bottom left hand corner shows the resolution of each map. This is sufficiently large that the edge on nature of the circumstellar material is not obvious in either image.

- Imaging sources with compact cores will be much more straightforward than imaging fully resolved targets. This should be borne in mind when choosing the targets for the first VLTI imaging studies.
- For resolved sources, there may be additional difficulties associated with the fact that the source visibility, $V$, will be very low on most of the interferometer baselines. Since the instantaneous signal-to-noise ratio scales as $V^2$, it may be difficult to achieve a scientifically useful signal-to-noise unless very long integrations can be realised.

Quite clearly, imaging at high angular resolution will remain a challenge for a sparse array like the VLTI. However, existing prototype arrays have demonstrated that it is possible, and there is every reason to believe that the VLTI will eclipse their performance in due course.

## Acknowledgements

It is a pleasure to thank the organisers for their invitation and generosity hospitality during my attendance at this meeting, and to Dr P.G.One for valuable assistance during the preparation of this manuscript.

## References


Baldwin, J.E. and Haniff, C.A.: 2002, *Phil. Trans. A* **360**, 969–986.

Buscher, D.F., Haniff, C.A., Baldwin, J.E. and Warner, P.J.: 1990, *MNRAS* **245**, 7–11.

Haniff, C.A. and Buscher, D.F.: 1998, *Astron. Astrophys.* **334**, L5–L8.

Hoffmann, K-H., Beckmann, U., Blöcker, T., Coude de Foresto, V., Lacasse, M., Mennesson, B., Millan-Gabet, R., Morel, S., Perrin, G., Pras, B., Ruilier, C., Schertl, D., Schöller, M., Scholz, M., Shenavrin, V., Traub, W., Weigelt, G., Wittkowski, M. and Yudin, B.: 2001 *New Astronomy* **7**, 9–20.

Quirrenbach, A., Buscher, D.F., Mozurkewich, D., Hummel, C.A. and Armstrong, J.T.: 1994, *Astron. Astrophys.* **283**, L13–L16.

Stee, Ph. and Bittar, J.: 2001, *Astron. Astrophys.* **367**, 532–548.

Tuthill, P.G., Haniff, C.A. and Baldwin, J.E.: 1997, *MNRAS* **285**, 529–539.

Tuthill, P.G., Monnier, J.D. and Danchi, W.C.: 1999, *Nature* **398**, 487–489.

Tuthill, P.G., Monnier, J.D., Danchi, W.C. and Lopez, B.: 2000, *Astrophys. J.* **543**, 284–290.

van Belle, G., Ciardi, D.R., Thompson, R.R., Akeson, R.L. and Lada, E.A.: 2001, *Astrophys. J.* **559**, 1155–1164.

# IMAGING WITH AMBER/VLTI: THE CASE OF MICROJETS

E. THIÉBAUT
*AIRI/Observatoire de Lyon (France) and Jean-Marie Mariotti Center, France*
*E-mail: thiebaut@obs.univ-lyon1.fr*

P.J.V. GARCIA
*Dep. Física da Faculdade de Engenharia da Universidade do Porto and*
*Centro de Astrofísica da Universitade do Porto, Portugal*
*E-mail: pgarcia@astro.up.pt*

R. FOY
*AIRI/Observatoire de Lyon, France; E-mail: foy@obs.univ-lyon1.fr*

**Abstract.** The engine that powers pre-main-sequence micro-jets is still unknown and remains a fundamental open question in star formation. The engine can be located solely on the inner disk or in the interaction of the inner disk with the star. In order to ease interpretation problems, imaging the jet engine is the ideal probe to disentangle between the old models and shed evidence for new ones. In this paper, we analyse the feasability of imaging bright southern targets, and show that even at low SNR, accurate image reconstruction is still possible with high contrast. However, the small number of ATs requires a fast reconfigurable array.

**Keywords:** Amber, VLTI, PMS, microjets, image restoration

## 1. Introduction

The VLTI with its four movable ATs and four UTs has an enormous potential for interferometric imaging. The focal instrument AMBER, adds to the imaging capability moderate to very high spectral resolution. In particular at intermediate resolutions $R = 1\,000$ a broad spectral coverage is available allowing not only the combined study of of several spectral lines but also wavelength bootstrapping.

This paper addresses the issue of imaging with the VLTI in the context of pre-main-sequence (PMS) star micro-jets. Fundamental open questions remain in this region of astrophysics. In the literature models for the engine that powers the jet can be found. Those who present observational predictions lie in two classes: pure disk engines, and star-disk engines. These models share one property, because most of the mass is ejected in the inner disk or star-disk interface, jets are hollow. This property was noticed by (Garcia, 2002) to produce well defined visibility 'arms' in the spatial frequencies space, and therefore showed a context within which visibility measurements could be interpreted. A companion paper in these proceedings (Bacciotti et al., 2002) describes in detail this project as well as it's astrophysical context.

*Astrophysics and Space Science* **286:** 171–176, 2003.

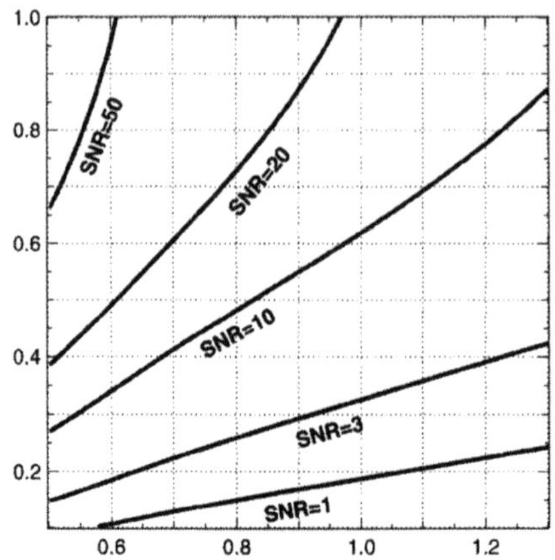


*Figure 1.* SNR isocurves, for RU Lupus, in $V^2$ – seeing space.

Here the next step is analyzed, namely if images can be produced by AMBER for such objects. With no a priori model for the object brightness distribution, images free us of the degeneracy often connected to model dependent data interpretation. AMBER/VLTI allows the recombination of 3 telescopes which provide, in addition to visibilities ($V$) or power spectra ($V^2$), phase closures. If enough spatial frequencies are obtained it will be possible to reconstruct an image. This allows one to identify the structure of the jet's engine without assuming a particular physical model. On the other hand for the relatively small number of AT's available the required spatial frequency coverage is obtained at the expense of moving some AT's around. In the next section we address the problem of signal to noise ratio obtained in one visibility measurement, then we introduce our image reconstruction algorithm and apply it to a PMS micro-jet, finally we end with the conclusions.

## 2. Signal to Noise Ratio Estimation

The brightest southern PMS harbouring a micro-jet is RU Lupus with $m_J = 8.6$. The Pa$\beta$ emission line is probably the most interesting one, very near the star densities are expected to be high enough so that H recombination dominates emission. The line profile peaks at 2.5 times the continuum. We used a calculator written by Malbet (Malbet, 2000) to estimate the signal to noise ratio (SNR). The flux entering AMBER fibers critically depends on the Strehl ratio (SR). We computed the expected SR for the ATs assuming tip-tilt correction. Our star being very bright at visible is not limiting the AO compensation, the final SR is limited only by the seeing. We assumed that a fringe tracker was able to freeze the fringes for 1 minute, and that 30 such frames were obtained. We selected the medium resolution ($R = 1000$) configuration. Figure 1 shows the expected SNR per wavelength bin.

## 3. The Multi-Aperture Image Reconstruction Algorithm

MIRA (Multi-aperture Image Reconstruction Algorithm) is the imaging software devoted to optical interferometry data and currently under development at the JMMC (Jean-Marie Mariotti Center).

Owing to the small number of telescopes, optical interferometers are only able to measure very few Fourier frequencies at the same time and even less Fourier phases because of the lack of phase referencing capability at this time. Unless a large number of nights are spent to collect the data, optical/IR aperture synthesis imaging is much more challenging than what can be routinely done in radio-astronomy. The solution adopted in MIRA algorithm is to directly fit the phase closure and power spectrum data without explicitly rebuilding the missing Fourier phases (i.e. no explicit self calibration). The image reconstruction in MIRA is regularized in order to cope with missing data and to avoid artifacts due to the sparse/non-even sampling. Various kinds of regularization (entropy, generalized Tikhonov, . . .) can be used by the algorithm. MIRA states the image reconstruction as an inverse problem solved by minimizing a so-called penalty function. Owing to the specific relationship between the object brightness distribution and the data (phase closures and power spectra), the penalty to minimize cannot be guaranteed to be convex. The problem is then very difficult to solve because global optimization is required to find the best solution among the multiple local minima. At this time, the VMLM-B non-linear constrained optimization algorithm (Thiébaut, 2002) is used in MIRA. Since VMLM-B is mainly a descent algorithm, it can only provide a local optimum. We therefore expect the results obtained with this preliminary version of MIRA to be somewhat worse than what is achievable when global optimization will be implemented.

## 4. Application to PMS's Micro-Jet

In order to assess the imaging capability of Amber/VLTI, we used an image of a micro-jet in Pa$\beta$ computed with the model by (Garcia, 2001) (see Figure 2). In the simulation, at the wavelength of Pa$\beta$, the integrated brightness of the jet roughly equates to that of the PMS star. Nevertheless, since the jet covers a large area, the ratio of the mean jet brightness over a $0.643 \times 0.643$ marcsec$^2$ pixel divided by the star brightness is about 1/100 which is quite unfavorable. As a consequence, imaging of the micro-jet should require high quality data.

We simulated Amber/VLTI data of RU Lupus (phase closures and power spectra) for 6 different configurations involving 3 auxiliary telescopes (AT's, see Figure 2). In our simulations, we considered various SNR for the power spectra*: 100, 50, 20 and 10; we took the standard deviation of the noise of the phase closures

* One limitation of these simulations is that the SNR was assumed independent of the value of $V^2$. As can be seen from Figure 1 this is not the case.

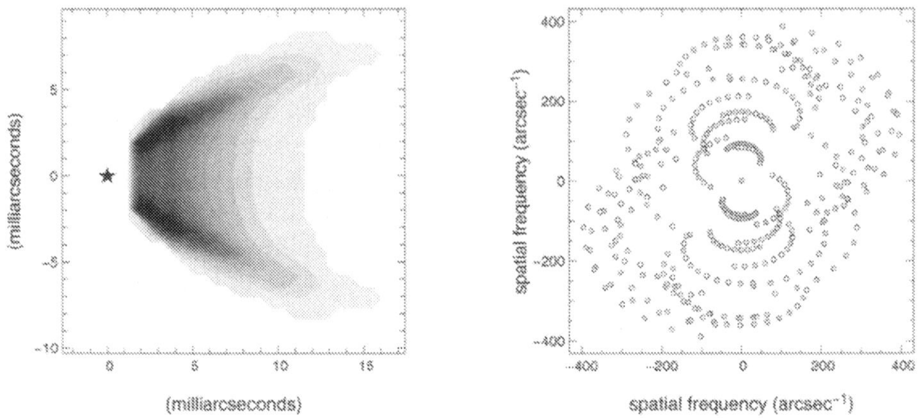


*Figure 2.* **Left:** Micro-jet emitted by a PMS. The *pixel* size is 0.643 marcsec. The point-like star contribution has been removed (to display the jet brightness distribution) and replaced by a star symbol. The isocontours are evenly spaced. **Right:** $(u, v)$ coverage in our simulations. The configurations were J3-I1-K0, D1-J2-K0, C2-G2-K0, J2-I1-H0, D0-G1-K0, and G2-J1-B5.

to be $1/\mathrm{SNR}$ that is: 0.6, 1.1, 2.9 and 5.7 degrees respectively. Assuming that the star (RU Lupus) is observable in the $\pm 4$ hours range from transit and that one measurement is obtained every 12 minutes**, we got 11 measurements per night. For the 6 simulated observing nights, the total number of measurements is 199 power spectra and 66 phase closures.

Using MIRA and starting the image restoration iterations with a point source, we reconstructed the images shown in Figure 3 for the various SNR considered. These results show that even for the lowest SNR: (i) the micro-jet can be detected and (ii) the open angle can be estimated. The main artifacts are due to the fact that we choose a very small pixel scale (0.643 marcsec) which is 4 times smaller than the resolution achieved by the largest of the bases in our configurations.

## 5. Conclusions

We have shown that it is possible to obtain an image of the close surroundings of PMS's with no a priori (real imaging) and sufficient resolution ($\sim$ 0.1 AU at 140 pc) with VLTI/Amber instrument. Our simulations demonstrate that 6 configurations (6 observing nights) with SNR $\sim$ 10 (i.e. using 3 AT's) is sufficient to recover an image of the jet good enough to measure its opening angle.

It should be noted that these results were obtained from a preliminary version of MIRA the image reconstruction software under development by one of us (ET)

** At this point we are not allowing for reference calibration. In the considerable wavelength coverage of J–H, we however expect that the continuum is unresolved and can serve as a proper calibrator.

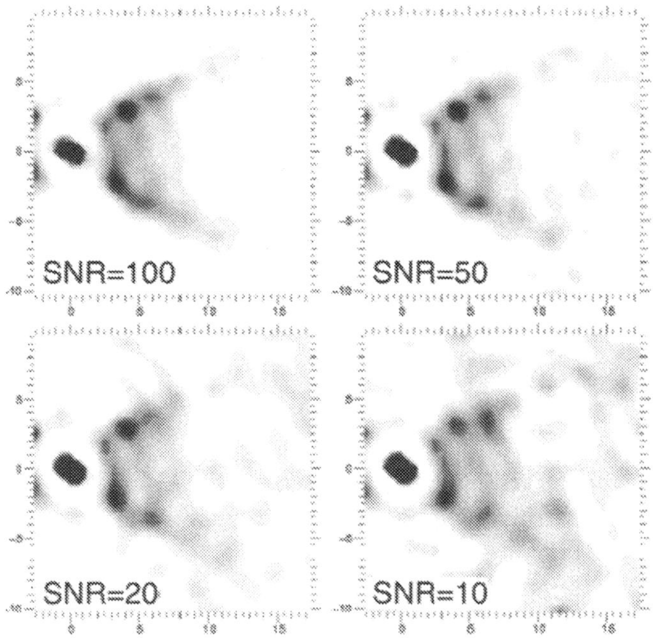


*Figure 3.* Reconstructions achieved from data with different signal to noise ratios (SNR). Intensities have been clipped to display the jet brightness distribution, the PMS star therefore appears as a saturated area at coordinates (0, 0).

within the framework of JMMC. In particular the current version of the algorithm lacks of global optimization capability (i.e. only a local optimum image is recovered not the best one over all possible images). Our derived requirements in terms of sufficient SNR and $(u, v)$ coverage may therefore be rather pessimistic. With global optimization, the algorithm could cope with much worse data (lower SNR and/or more sparse $(u, v)$ coverage resulting in substantially less observing time). Optimization of telescope configurations could also reduce the number of required observing nights.

Nevertheless, in order to recover worthwhile images with VLTI, some basic rules must be followed: (1) since the object is unknown (which is likely at the spatial resolution achieved by VLTI), no a priori brightness distribution can be used in the image reconstruction process and $(u, v)$ coverage must be as uniform as possible; (2) even if some Fourier interpolation/extrapolation is allowed thanks to the constraints and the regularization, the maximum baseline extension $B_{\max}$ is related to the effective spatial resolution: $\Delta\Omega \simeq \lambda/B_{\max}$ whereas the mean $(u, v)$ sampling $\Delta B/\lambda$ is related to the reconstructed field of view: $\Omega \simeq \lambda/\Delta B$; the number of interferometric measurements needed to reconstruct an image (with no a priori) will therefore be about $(\Omega/\Delta\Omega)^2 \simeq (B_{\max}/\Delta B)^2$. Image reconstruction will always require a not so small number of different interferometric configurations (with a small number, let us say 3 or 4, of telescopes) requiring several observing nights.

Another point to not forget is that at the small spatial scales achieved by interferometry, the observed object is more likely to be variable. This add the requirement that all the observations (of the same object) be made within the shortest time as possible (no longer than a few days), i.e. the telescope configuration should be changed every night. At this time, optical interferometers only have a small number of telescopes (says 3 or 4), imaging objects at the spatial resolutions achievable by these instruments therefore requires a versatile array of telescopes that are easy/fast to reconfigure.

## Acknowledgements

PJVG work was supported by grants POCTI/1999/FIS/34549 and PRAXIS XXI/BPD/20179/99 approved by FCT and POCTI.

## References

Bacciotti, F. et al.: 2002, Unveiling the launching region of YSO jets with AMBER, in: *JENAM proceedings*.

Garcia, P.J.V., Ferreira, J., Cabrit, S. and Binette, L.: 2001, *A&A* **377**, 589.

Garcia, P.J.V., Foy, R. and Thiébaut, E.: 2002, in: Alves and McCaughrean (eds.), the proceedings of the ESO workshop *The Origins of Stars and Planets: The VLT View*, Springer-Verlag, Germany, p. 267.

Malbet, F. et al.: 2000, AMBER – Instrument Analysis Report.

Thiébaut, E.: 2002, Optimization issues in blind deconvolution algorithms, in: *Telescopes and Instrumentation* SPIE proceedings **4847-22**, Hawaii.

# PULSATING STARS

BENJAMIN F. LANE
*Department of Geological and Planetary Sciences, 150-21 California Institute of Technology, Pasadena, CA 91125, USA*

**Abstract.** Modern long-baseline interferometers are capable of resolving stellar diameters in the range of one to several milli-arcseconds with measurement precision approaching a few percent. This level of precision allows astronomers to directly resolve diameter changes associated with pulsation for various classes of stars, including Cepheids and Miras. For several nearby Cepheids it is relatively straightforward to apply a Baade-Wesselink analysis and hence directly measure the distance to the Cepheid with a high degree of precision. This field is still quite new; I expect that in the near term several new interferometers will provide direct Cepheid distances to perhaps a few tens of Galactic Cepheids. However, equally important will be direct measurements of the atmospheric properties (in particular limb darkening) of these stars. Such measurements should help reduce the considerable systematic uncertainties that remain.

**Keywords:** interferometry, fundamental stellar parameters, Cepheids

## 1. Introduction

The past decade has seen a number of long-baseline stellar interferometers enter service (GI2T, COAST, IOTA, NPOI, PTI). These instruments typically operate in the optical or near infrared (400 nm to 2.2 $\mu$m), and have baselines on the order of 20–400 meters, with corresponding angular resolutions as high as 0.2 mas. However, until recently all but two of these instruments were only capable of single-baseline operation, and hence were limited to observations of simple source morphologies such as single and binary stars. Fortunately, there are several interesting scientific questions that can be addressed with this capability; here I discuss one such application – diameter measurements of pulsating stars.

There are two classes of stars that have proved particularly amenable to study with stellar interferometers: Cepheids and Miras. Here I will concentrate on Cepheids.

## 2. Cepheids

Possessing a well-behaved period-luminosity relation, the class of pulsating stars known as Cepheids is a cornerstone in determining the distances to nearby galaxies. These stars are massive and thus intrinsically very luminous, making it possible to observe Cepheids located in very distant galaxies (Tanvir, 1999). Because of the

*Astrophysics and Space Science* **286:** 177–183, 2003.

usefulness and fundamental importance of Cepheids, it is important to calibrate their period-luminosity relation. However, there are no Cepheids close enough for a high-precision parallax measurement with current instrumentation (including Hipparcos), and hence a number of indirect methods have been used, primarily Baade-Wesselink (Wesselink, 1946; Bersier, Burki and Kurucz, 1997) and surface brightness (Laney and Stobie, 1995; Fouque and Gieren, 1997; Ripepi et al., 1997) methods. However, it is expected that in the next decade both the SIM (Boden, Unwin and Shao, 1997) and GAIA (Perryman , 2002) space missions will be able to obtain high-precision Cepheid parallaxes. Nevertheless, the period-luminosity relations used currently have uncertainties on the order of 0.09 mag (Feast, 1999), which in turn make up a significant portion of the systematic uncertainty in estimates to the Large Magellanic Cloud.

Cepheids have been observed by a number of interferometers, including GI2T, IOTA, NPOI and PTI. The first mean diameter measurement was obtained by GI2T in 1997 (Mourard et al., 1997), followed by NPOI (Nordgren et al., 2000) and IOTA (Kervella et al., 2001). Resolving the diameter change with pulsation required considerable precision in the measured fringe visibilities (on the order of 1%) and has only recently been achieved (Lane et al., 2000).

### 2.1. Diameter and Distance Measurements

Determining the distance and radius of a Cepheid via the Baade-Wesselink method requires comparing the measured changes in angular diameter to the expansion of the Cepheid photosphere measured using radial velocity techniques. In effect, given measurements of radial velocity as a function of phase, one infers a model apparent angular diameter as

$$\theta(\phi) = \frac{R_0 + p\int_0^{\phi} v(\phi')d\phi'}{d} \tag{1}$$

where $R_0$ is the mean stellar diameter, $d$ the distance to the Cepheid, $v(\phi)$ is the radial velocity as a function of pulsational phase, and $p$ is a projection factor; it accounts for the difference between physical expansion velocity and apparent radial velocity. PTI has measured the changes in angular diameter of two Cepheids, $\eta$ Aql and $\zeta$ Gem. When combined with previously published radial velocity data one can derive the distance and mean diameter to the Cepheids. Lane et al. (2002) find $\eta$ Aql to be at a distance of $320 \pm 32$ pc with a mean radius of $61.8 \pm 7.6 R_{\odot}$ while $\zeta$ Gem is at a distance of $362 \pm 38$ pc, and has a mean radius of $66.7 \pm 7.2 R_{\odot}$. These results are in good agreement with previous work.

### 2.2. Surface Brightness Calibrations

Interferometrically measured Cepheid diameters can also be used to calibrate surface brightness relations, e.g. of the form

$$F_V = a + b(V - K) \tag{2}$$

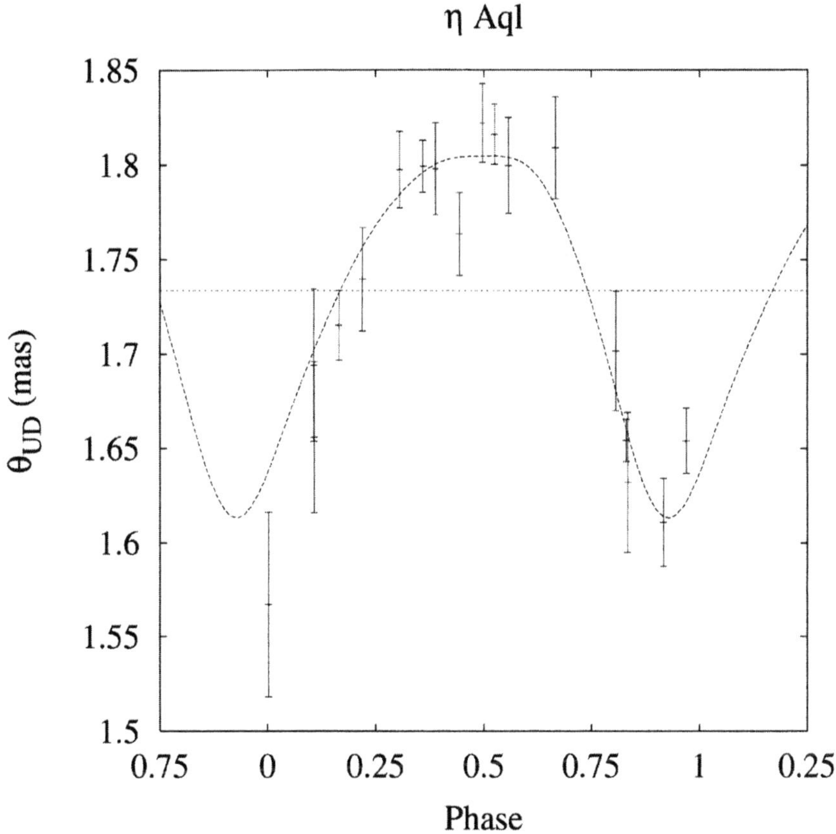


*Figure 1.* The angular diameter of $\eta$ Aql as a function of pulsational phase, together with a model based on radial velocity data, but fitting for distance, mean radius and phase shift. From Lane et al., 2002.

which can then be used with

$$F_i = 4.2207 - 0.1m_i - 0.5\log(\theta_{LD}) \tag{3}$$

where $F_i$ is the surface brightness in magnitudes in passband $i$, $m_i$ is the apparent magnitude in that band, and $\theta_{LD}$ is the apparent angular diameter of the star. With the above relations one can determine the angular diameter based on photometry alone. Nordgren et al. (2002) have used available interferometrically determined Cepheid diameters to derive surface brightness relations with good precision; these should be of considerable value to future work.

### 2.3. Limitations

The precision achieved in published Cepehids distances and radii is $\sim 10\%$ in the parameters; further improvement is at present limited by both statistical and systematic error sources. In particular the details of limb darkening, projection factors

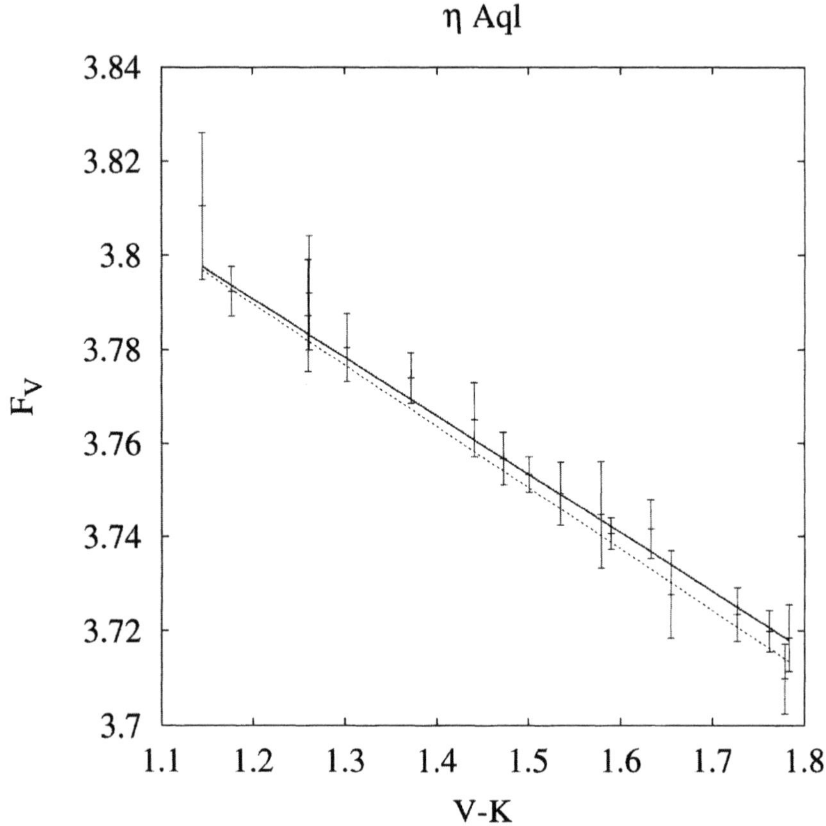


*Figure 2.* Dereddened surface brightness relation $F_V$ vs. $V - K$ for $\eta$ Aql. The solid line is the weighted linear least-squares fit to the data. The dashed line represents the relation from (Fouque and Gieren, 1997), and the dotted line represents the (Nordgren et al., 2002) result.

(p-factor), and optical depth effects need to be understood. The latter are a source of potential systematic error stemming from the fact that the photosphere seen by the infrared interferometer is different from that detected by the visible-light spectrometers used for measuring radial velocities. Clearly it is important to obtain more data; both radial velocity measurements in the near-IR and interferometric diameter measurements at visible wavelengths will be useful.

However, the largest sources of systematic uncertainty are from the radial velocity p-factor and limb darkening, both of which depend on the detailed atmospheric structure of the Cepheids. There is a considerable range in the estimates of the p-factor in the published literature, values from 1.31 to 1.41 can be found (Hindsley and Bell, 1986; Albrow and Cottrell, 1994). In addition, the p-factor should not really be treated as a constant, but allowed to vary as a function of pulsational phase.

Clearly, there is a need for reliable Cepheid atmospheric models that can provide good values for the p-factor and limb darkening. Although considerable work has been done in this area, it has been difficult to observationally verify such models. However, a small number of observations of Cepheid limb darkening would serve to constrain existing atmospheric models and drastically reduce this systematic error. This is where interferometry can provide uniquely valuable information.

### 2.4. FUTURE DEVELOPMENTS

There are two areas where long-baseline stellar interferometers can improve our current understanding of Cepheids. The first is simply by obtaining many more measurements of Cepheid diameters, particularly if they have sufficient precision to resolve the pulsational changes. This will require significant investments in observing time at those facilities that have sufficiently high angular resolution (on the order of 1-2 milli-arcseconds).

However, there is a second area where a few targeted observations can provide a very tight constraint on currently used models, and hence significantly reduce the level of systematic uncertainty in Cepheid distance determination. Measuring limb darkening with an interferometer is a challenging observation that has to date only been done on a few stars ((Hanbury Brown et al., 1974; Quirrenbach et al., 1996; Wittkowski et al., 2001)). It typically requires an approach called baseline bootstrapping, outlined below.

Given the ability to combine three interferometric baselines at a time, one can choose two of the baselines to be relatively short and hence have a high Signal-to-Noise Ratio (SNR is proportional to fringe amplitude; see Figure 3). However, the third baseline can be up to twice as long as the others, thus providing twice the resolution. Although the SNR is necessarily lower, the fact that the baselines form a closed loop implies that by tracking the fringe location on the two high-SNR baselines, one automatically knows the location of the fringes on the low-SNR baseline. This allows one to integrate coherently, rapidly recovering the SNR. This technique is perfect for measuring stellar limb darkening, where all the useful information is encoded in the fringes measured at very long baselines, such that they would be very difficult to observe without coherent integration.

## 3. Summary

In summary, optical/IR interferometry has recently begun to provide valuable observations of Cepheids, including both direct distance measurements and calibrations of surface brightness relations. In time these observations should help reduce uncertainties in Cepheids distance scales. However, there remains a lot of work to be done. In particular, interferometry can provide a useful observational constraint on models of Cepheid atmospheres by directly measuring limb darkening.

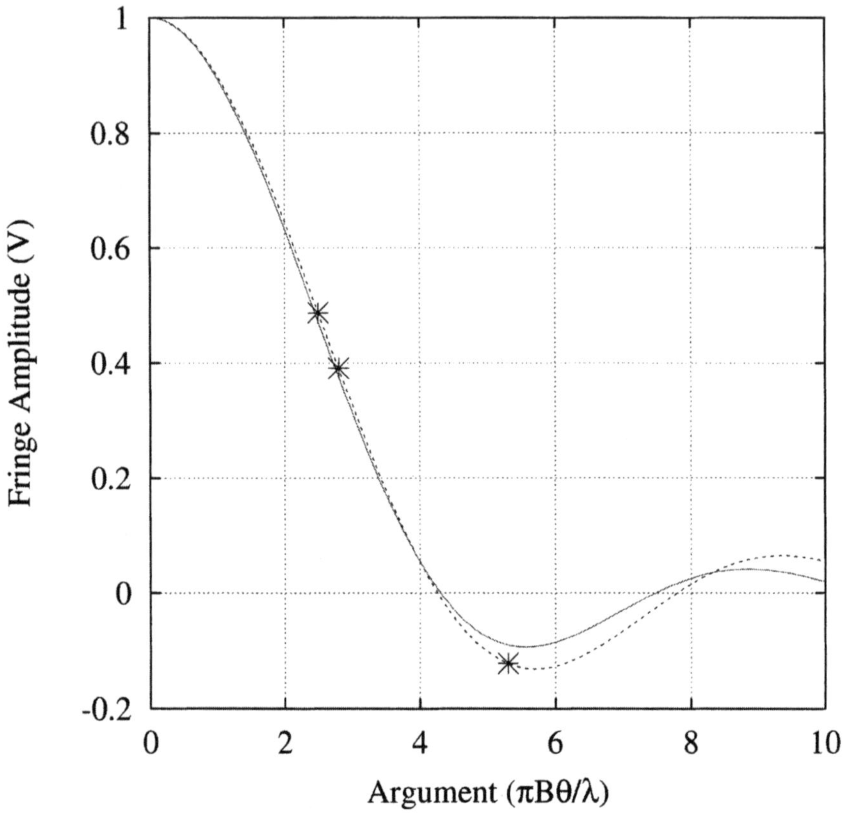


*Figure 3.* An illustration of baseline bootstrapping. Plotted is the fringe amplitude for a uniform disk (dashed) and a 10% larger limb-darkened (solid) disk model of a single star. Also shown are the points sampled by three available baselines. Parameters are chosen to represent use of 200-meter class baselines and a 1.5 milliarsecond source. Only at long baselines is the degeneracy between size and limb darkening broken.

## Acknowledgements

I would like to thank D. Sasselov and M.M. Colavita for valuable comments. Observations with PTI are only made possible through the efforts of the PTI collaboration. Funding for the development of PTI was provided by NASA under its TOPS (Toward Other Planetary Systems) and ASEPS (Astronomical Studies of Extrasolar Planetary Systems) programs, and from the JPL Director's Discretionary Fund. I also gratefully acknowledge the support of NASA through the Michelson fellowship program.

## References

Albrow, M.D. and Cottrell, P.L.: 1994, Line profiles of Cepheid variables – Projection factors and metallic cline asymmetries, *MNRAS* **267**, 548–556.

Bersier, D., Burki, G. and Kurucz, R.L.: 1997, Fundamental parameters of Cepheids. IV. Radii and luminosities, *A&A* **320**, 228–236.

Boden, A., Unwin, S. and Shao, M.: 1997, Global astrometry with the space interferometry mission, in: B. Battrick (ed.), *Hipparcos Venice '97*, (ESA SP-402; Noordwijk: ESA), pp. 789–794.

Colavita, M.M. et al.: 1999, The Palomar testbed interferometer, *ApJ* **510**, 505–521.

Feast, M.: 1999, Cepheids as distance indicators, *PASP* **111**, 775–793.

Feast, M.W. and Catchpole, R.M.: 1997, The Cepheid period-luminosity zero-point from HIPPARCOS trigonometrical parallaxes, *MNRAS* **286**, L1–L5.

Fouque, P. and Gieren, W.P.: 1997, An improved calibration of Cepheid visual and infrared surface brightness relations from accurate angular diameter measurements of cool giants and supergiants, *A&A* **320**, 799–810.

Hanbury Brown, R., Davis, J., Lake, R.J.W. and Thompson, R.J.: 1974, The effects of limb darkening on measurements of angular size with an intensity interferometer, *MNRAS* **167**, 475–484.

Hindsley, R. and Bell, R.A.: 1986, An investigation of photoelectric radial-velocity spectrometers as used in the analysis of Cepheid variables, *PASP* **98**, 881–888.

Kervella, P. Coudé du Foresto, V., Perrin, G., Schöller, M., Traub, W.A. and Lacasse, M.G.: 2001, The angular diameter and distance of the Cepheid $\zeta$ Geminorum, *A&A* **367**, 876–883.

Lane, B.F., Creech-Eakman, M. and Nordgren, T.E.: 2002, Long-baseline interferometric observations of Cepheids, *ApJ* **573**, 330–337.

Lane, B.F., Kuchner, M.J., Boden, A.F., Creech-Eakman, M. and Kulkarni, S.R.: 2000, Direct detection of pulsations of the Cepheid star $\zeta$ Gem and an independent calibration of the period-luminosity relation, *Nat.* **407**, 485–487.

Laney, C.D. and Stobie, R.S.: 1995, The radii of galactic Cepheids, *MNRAS* **274**, 337–360.

Mourard, D. Bonneau, D., Koechlin, L., Labeyrie, A., Morand, F., Stee, P., Tallon-Bosc, I. and Vakili, F.: 1997, The mean angular diameter of $\delta$ Cephei measured by optical long-baseline interferometry, *A&A* **317**, 789–792.

Nordgren, T., Armstrong, J.T., Germain, M.E., Hindsley, R.B., Hajian A.R., Sudol, J.J. and Hummel, C.A.: 2000, Astrophysical quantities of Cepheid variables measured with the navy prototype optical interferometer, *ApJ* **543**, 972–978.

Nordgren, T., Lane, B.F., Hindsley, R.B. and Kervella, P.: 2002, Calibration of the Barnes-Evans relation using interferometric observations of Cepheids, *AJ* **123**, 3380–3386.

Perryman, M.A.C.: 2002, GAIA: An astrometric and photometric survey of our galaxy, *Ap&SS* **280**, 1–10.

Quirrenbach, A., Mozurkewich, D., Buscher, D.F., Hummel, C.A. and Armstrong, J.T.: 1996, Angular diameter and limb darkening of Arcturus, *A&A* **312**, 160–166.

Ripepi, V., Barone, F., Milano, L. and Russo, G.: 1997, Cepheid radii and the CORS method revisited, *A&A* **318**, 797–804.

Tanvir, N.R.: 1999, Post-Hipparcos cosmic candles, *ASSL* **237**, 17+.

Wesselink, A.J.: 1946, The observations of brightness, colour and radial velocity of $\delta$ Cephei and the pulsation hypothesis, *BAN* **368**, 91+.

Wittkowski, M., Hummel, C.A., Johnston, K.J., Mozurkewich, D., Hajian A.R. and White, N.M.: 2001, Direct multi-wavelength limb-darkening measurements of three late-type giants with the Navy Prototype Optical Interferometer, *A&A* **377**, 981–993.

# CEPHEID LIMB DARKENING MODELS FOR THE VLTI

MASSIMO MARENGO, MARGARITA KAROVSKA, DIMITAR D. SASSELOV and COSTAS PAPALIOLIOS
*Harvard-Smithsonian Center for Astrophysics, Cambridge, MA, USA*

J.T. ARMSTRONG
*Remote Sensing Division, Naval Research Laboratory, Washington, DC, USA*

TYLER E. NORDGREN
*Department of Physics, University of Redlands, Redlands, CA, USA*

**Abstract.** We present a new method to compute wavelength- and phase-dependent limb darkening corrections for classical Cepheids. These corrections are derived from hydrodynamic simulations and radiative transfer modeling with a full set of atomic and molecular opacities. Comparison with hydrostatic models having the same stellar parameters show a larger limb darkening for most phases in our models, and temporal variations related to the hydrodynamics of the stellar pulsation. We assess the implications of our results with respect to the geometric Baade-Wesselink method, which uses interferometric measurements of Cepheid angular diameters to determine their distances and radii. The relevance of the hydrodynamic effects predicted by our model on the limb darkening of pulsating Cepheids is finally discussed in terms of the predicted capabilities of the VLTI.

## 1. Introduction

The period-luminosity (PL) relation of Classical Cepheids, since its discovery by Leavitt (1906), has played an important role in setting the extragalactic distance scale. The reliability of this relation, however, depends on the accurate calibration of its zero-point. This is difficult to accomplish, due to the low density distribution of Cepheids in our Galaxy. The determination of the PL relation zero point from geometric parallaxes has been attempted using the database of the Hipparcos mission (Feast and Catchpole, 1997), but large uncertainties do not allow an unambiguous interpretation of the results (Madore and Freedman, 1998). The same is also true for the classical Baade-Wesselink (BW) method (Baade, 1926; Wesselink, 1946), which suffers from large uncertainties related to the source colors and metallicity.

The recent advances in interferometry, however, have finally allowed the direct detection of pulsations in nearby Cepheids (Lane et al., 2000; Kervella et al., 2001; Lane et al., 2002). This result have opened the way for a *geometric* variant of the BW method (Sasselov and Karovska, 1994), allowing an independent distance measurement of pulsating stars, based on the interferometric determination of their radial pulsations.

*Astrophysics and Space Science* **286:** 185–190, 2003.

As pointed-out by Sasselov and Karovska (1994), the feasibility of the geometric BW method requires accurate predictions for the limb darkening (LD) of the pulsating star, necessary to convert the measured interferometric visibilities into angular diameters. In this paper we present our time- and wavelength-dependent LD models, based on hydrodynamic models of the classical Cepheids, and we discuss their importance for future VLTI observations of this type of pulsating stars.

## 2. Dynamic Models for Pulsating Cepheids

In Marengo et al. (2002) we presented a new method for computing accurate time- and wavelength-dependent center-to-limb brightness distributions for classical Cepheids. This method is based on hydrodynamic simulations of the Cepheid atmosphere, performed in non-LTE conditions with a simplified radiative transfer (Sasselov and Lester, 1994). The pulsations are induced by perturbing the model atmosphere with a piston, parametrized by matching the radial velocity curve, phase lags and other stellar observables. On top of the hydrodynamic atmosphere, we solve the full LTE radiative transfer problem, using the complete set of atomic and molecular opacities provided by the ATLAS/SYNTHE program (Kurucz, 1970, 1979, 1993). From the SYNTHE emergent spectral intensity, we derive the LD of the pulsating star at a number of wavelengths and pulsational phases.

The center-to-limb intensities from which the LD is derived are computed with SYNTHE for a model atmosphere having (1) the effective temperature $T_{eff}(\phi)$ obtained as a function of the pulsational phase from observations (Krockenberger et al., 1997), (2) $\log g$, $L/H$, metallicity, $v_{turb}$ and the atmospheric Rosseland opacity from a hydrostatic model approximating the hydrodynamic atmosphere at each pulsational phase, and (3) $T(r)$, $n_e(r)$ and $P(r)$ from the hydrodynamic model, excluding the chromospheric region. The main limitation of this procedure resides in the opacities being computed with a hydrostatic atmosphere in LTE conditions. However, the resulting limb darkened intensities are still much improved with respect to competing models, usually derived for static yellow supergiants, since the thermodynamic state of the atmosphere is completely set by the hydrodynamics.

The radial profile of the hydrodynamic atmosphere can be very different from the structure of a hydrostatic atmosphere with equivalent $T_{eff}$ and $\log g$. This is especially true for pulsational phases characterized by the propagation of hydrodynamic instabilities or shocks, opposed to other phases where the structure of the pulsating atmosphere is quasi-static. The fact that the atmosphere structure changes as the Cepheid pulsates, leads to different amount of limb darkening for different pulsational phases. Figure 1 shows how the photosphere of the classical Cepheid $\zeta$ Gem appears to be more limb darkened near the minimum radius, and less limb darkened as a consequence of a shockwave perturbing its atmosphere.

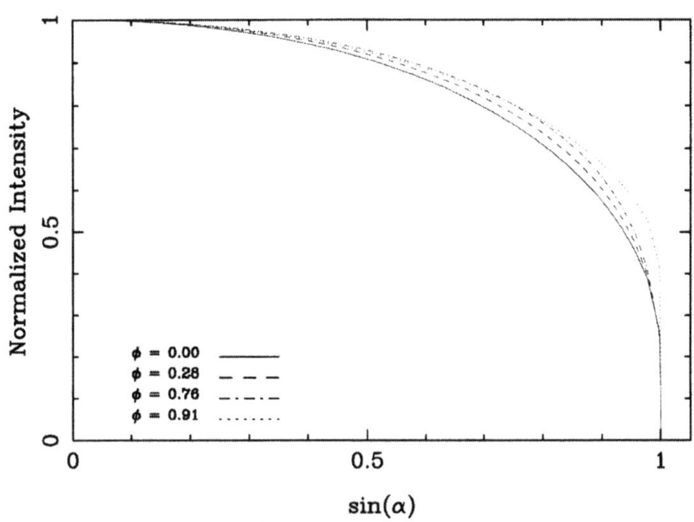


*Figure 1.* Limb-to-center profiles for $\zeta$ Gem at dynamical phases 0 (minimum radius), 0.28 (maximum luminosity), 0.76 (minimum luminosity) and 0.91 (passage of a shockwave through the photosphere). The profiles are computed for $\lambda = 500$ nm. The x-axis coordinate sin($\alpha$) is the projected radial coordinate of the star.

The center-to-limb intensities shown in Figure 1 can be used to derive the LD correction $k(\lambda, \phi)$ for the modeled Cepheid. This correction is defined as the ratio between the angular diameter measured by fitting the interferometer visibilities with a uniform brightness disk (UD) model, with the *true* diameter of the limb darkened star (the 'limb darkened diameter'): $k(\lambda, \phi) = \theta_{UD}(\lambda, \phi)/\theta_{LD}(\phi)$. The LD correction is wavelength and phase-dependent, as a consequence of the wavelength dependence of the atmospheric opacities, and because of the changes in the atmospheric structure as the star pulsates.

As an example, Figure 2 shows the spectral dependence of the LD correction for our model of $\zeta$ Gem at minimum radius. The limb darkening is larger at UV and optical wavelengths than in the infrared. Neglecting to correct for LD in the near-IR, however, will still produce an error of the order of $\sim 2\%$ with respect to the UD diameter of the star. Note that the LD correction $k$ is very sensitive to the photospheric spectral lines, which can thus be used as an effective diagnostic tool to better constrain the LD with interferometric observations.

Figure 3 shows the variations of the LD corrections with the pulsational phase, in the visible and at near-IR wavelengths (H-band). The dotted lines are the LD corrections for of equivalent hydrostatic atmosphere, having the same $T_{eff}$ and $\log g$ than the hydrodynamic models. The hydrostatic LD corrections closely follows the changes of $T_{eff}(\phi)$ with the pulsational phase. In the hydrodynamic atmosphere, however, $T_{eff}$ is not the dominant parameter. The hydrodynamic LD correction is instead characterized by sudden variations due to shocks traveling through the

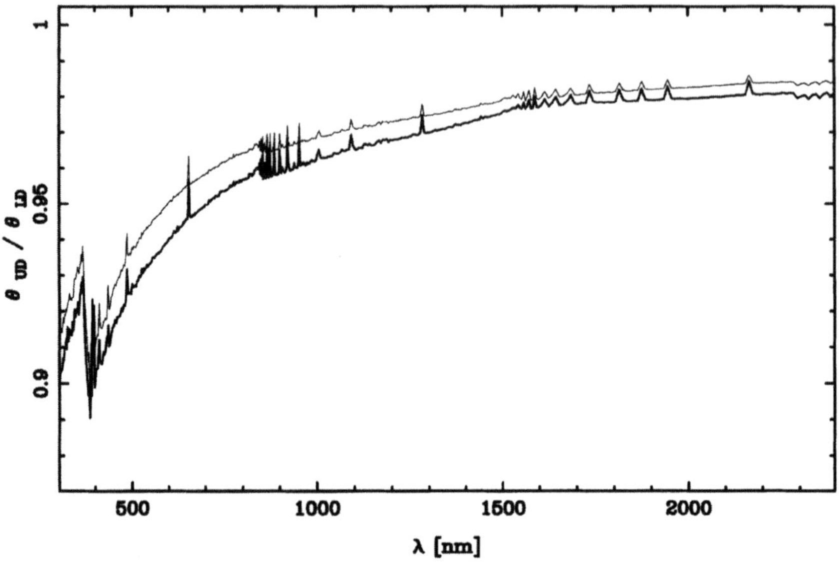


*Figure 2.* Wavelength dependence of the LD correction $k(\lambda)$ for $\zeta$ Gem at minimum radius. The solid thick line is $k(\lambda)$ from our hydrodynamic model; the thin line is derived from a hydrostatic model having $T_{eff}$ and $\log g$ from observations.

photosphere. Even when shocks are absent, the LD correction is still strikingly different from the one predicted by an hydrostatic model, as it is fully determined by the time-dependent structure of the expanding/contracting atmosphere. This dependence from the hydrodynamics results in a larger LD of the hydrodynamic models with respect to hydrostatic ones.

## 3. Limb Darkening Corrections for the VLTI

The VLTI interferometer, equipped with the near-IR camera AMBER will be especially suited to measure accurate distances of southern Cepheids with the geometric Baade-Wesselink method. To obtain the required accuracy, however, the interferometric measurements should be accompanied by a detailed modeling of the observed stars, taking into account the hydrodynamic effects induced by the pulsations.

The unprecedented sensitivity of AMBER will result in high accuracy visibilities ($\sim 10^{-3}$). This will allow the direct detection of the phase dependence of LD for nearby Cepheids. Figure 3 shows that the variation of the LD in the near-IR continuum is of the order of 0.2% for $\zeta$ Gem. AMBER will be able to detect these variations for similar Cepheids at a distance of 300–500 pc. The spectral capabilities of AMBER, coupled with the sensitivity offered by the Unit Telescopes (UT,

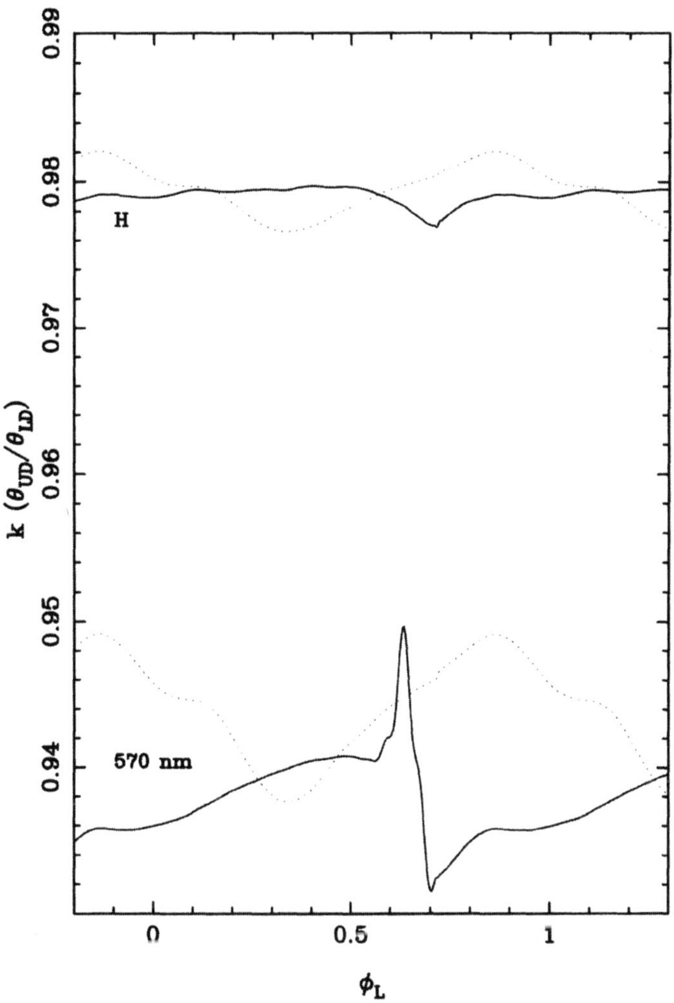


*Figure 3.* Phase dependent $k(\phi_L)$ for $\zeta$ Gem in the near-IR H-band and in the visible. The solid line is $k(\phi_L)$ from our hydrodynamic model, while the dotted line is derived from a hydrostatic model having $T_{eff}$ and $\log g$ from observations.

providing an expected limiting magnitude K$\sim$14 for $R \sim 1,000$) will also allow to observe the LD in spectral lines, where the hydrodynamic effects are much larger.

The VLTI with AMBER will be the ideal instrument to directly measure the LD of pulsating Cepheids, and set strong constrains on the models. This will result in more accurate predictions of the LD corrections to use for the geometric BW method. The large aperture of the VLTI telescopes and the large maximum baseline will allow to detect the pulsation of a large sample of classical Cepheids. The limiting factor of the interferometers currently operating in the northern hemisphere is their limiting magnitudes (e.g. V$\sim$5 for NPOI in the present configuration, and

K∼4.5–5 for the PTI). The 1.8m aperture of the Auxiliary Telescopes (AT) and the 8m diameter of the UT (necessary to resolve the spectral lines of Cepheids at $R \sim 1,000$) will extend the sample of Cepheids that can be observed in the southern hemisphere. Although the LD will not be directly measured for the most distant sources with the maximum 200 m baseline, the accurate predictions from the models calibrated with nearby Cepheids will still allow reliable distance determinations even when the stellar disk is not fully resolved. This in turn will result in a better calibration of the PL relation, and thus in a much improved distance scale from classical Cepheids.

## Acknowledgements

This work was partially supported by NSF grant AST 98-76734. M.K. is a member of the Chandra Science Center, which is operated under contract NAS8-39073, and is partially supported by NASA.

## References

Baade, W.: 1926, *Astron. Nachr.* **228**, 359.
Feast, M.W. and Catchpole, R.M.: 1997, *MNRAS* **286**, L1.
Kervella, P., Coudé du Foresto, V., Perrin, G., Schöller, M., Traub, W.A. and Lacasse, M.G.: 2001, *A&A* **367**, 876.
Krockenberger, M., Sasselov, D.D. and Noyes, R.W.: 1997, *ApJ* **479**, 875.
Kurucz, R.L.: 1970, *SAO Special Rep.* **309**.
Kurucz, R.L.: 1979, *ApJS* **40**, 1.
Kurucz, R.L.: 1993, in: E.F. Milone (ed.), *Light-Curve Modeling of Eclipsing Binary Stars*, Springer, New York, p. 93.
Lane, B., Kuchner, M.J., Boden, A.F., Creech-Eakman, M. and Kulkarni, S.R.: 2000, *Nature* **407**, 485.
Lane, B.F., Creech-Eakman, M. and Nordgren, T.: 2002, *ApJ* **573**, 330.
Leavitt, H.S.: 1906, *Ann. Astron. Harvard Coll. Obs.* **60**, 87.
Madore, B.F. and Freedman, W.L.: 1998, *ApJ* **492**, 110.
Marengo, M., Sasselov, D.D., Karovska, M.K., Papaliolios, C. and Armstrong, J.T.: 2002, *ApJ* **567**, 1131.
Sasselov, D.D. and Karovska, M.: 1994, *ApJ* **567**, 1131.
Sasselov, D.D. and Lester, J.B.: 1994, *ApJ* **423**, 795.
Wesselink, A.: 1946, *Bull. Astron. Inst. Netherlands* **10**, 91.

# PROSPECTS OF HIGH ANGULAR RESOLUTION MEASUREMENTS OF BINARY STARS PROPERTIES WITH VLTI

S. CORREIA
*European Southern Observatory, Karl-Schwarzschild-Str. 2, 85748 Garching, Germany*
*UMR 6525 Astrophysique, Université de Nice-Sophia Antipolis, Parc Valrose, 06108 Nice Cedex 2, France; E-mail: scorreia@eso.org*

A. RICHICHI
*European Southern Observatory, Karl-Schwarzschild-Str. 2, 85748 Garching, Germany*
*E-mail: arichich@eso.org*

M. SCHÖLLER
*European Southern Observatory, Casilla 19001, Santiago 19, Chile; E-mail: mschoell@eso.org*

**Abstract.** We present a program for observations of spectroscopic binaries with VLTI. The aim of such observations is to obtain high accuracy measurements of stellar parameters, in particular masses, combining the visual orbit with spectroscopic orbital elements. We selected a sample of spectroscopic binaries, including pre-main sequence stars, using the estimated angular nodal separation as a criterion for feasibility. A cross-check with the 4th catalog of measurements of binary stars (Hartkopf 2001) was carried out and we discuss the statistical properties of the sample.

**Keywords:** binaries, spectroscopic binaries, pre-main sequence, interferometry

## 1. Binary Stars as Targets for Interferometry

Binary stars are traditional targets for interferometric studies since the morphology of these kind of objects is simple. When the single components are not well resolved, the visibility curve is well determined with only a few measurements. Another advantage in this case is that points of high visibility are available for the longest baselines. However, more visibility measurements are needed if the single components are resolved (Figure 1), in order to determine without ambiguity the five parameters of the binary (separation, position angle, brightness ratio and radii of the single components). In such a case, the availability of multiple baselines or the use of earth rotation is necessary.

## 2. Determining Masses from Spectroscopic Binaries

The mass of a star is, with its chemical composition, the key parameter of its evolution. It is therefore of fundamental importance to measure the mass of stars of different types for which the metal abundance is known with enough accuracy in

*Astrophysics and Space Science* **286:** 191–196, 2003.

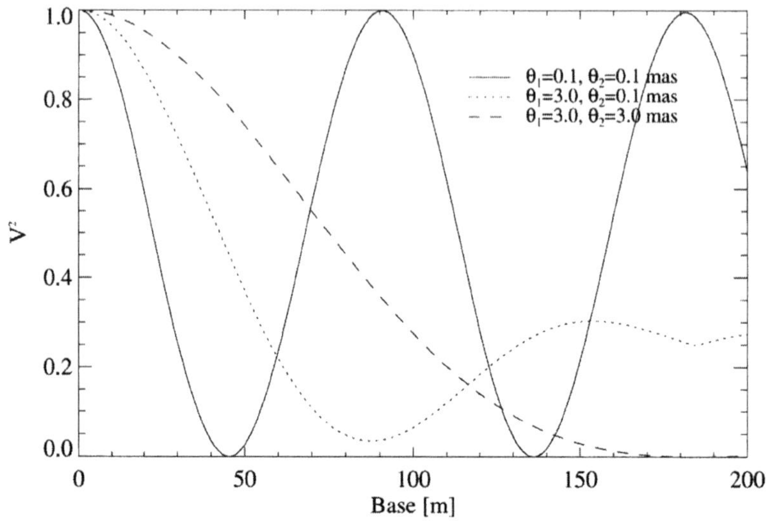


*Figure 1.* Exemple of squared visibility curves as functions of the projected baseline length and along the direction of the separation for a binary star observed in K band ($\lambda$ = 2.2 $\mu$m) and for different uniform disk diameters $\theta_1$ and $\theta_2$ of the single components (separation 5 mas, brightness ratio unity).

order to provide critical tests for the models of stellar evolution. Andersen (1991, 1998) defined a criterion of 1–2% on the accuracy of the mass measured to be useful for this aim. So far a handful of accurate stellar masses has been provided using double-lined eclipsing spectroscopic detached binaries (Andersen 1991). However it can be noted that this sample contains mainly early type main sequence stars.

The combination of the orbital solution obtained using radial velocity measurements with the visual orbit provided by interferometry leads also to the single component masses. This method provides also the *orbital parallax* which is a geometric measurement of the distance and consists of a substantial improvement in accuracy with respect to Hipparcos distances. Until now 16 visual orbits of double lined spectroscopic binaries (SB2s) have been measured using the first generation of interferometers (see Quirrenbach 2001, Hummel et al., 2001; Torres et al., 2002a). However, the accuracy provided by these interferometers was limited and reached Andersen's criterion in only a few cases. In addition, the systems were essentially the brightest accessible from the northern hemisphere.

VLTI with its larger collecting area and better accuracy on the visibility measurements (goal down to 0.1%) will provide the possibility to obtain accurate masses of fainter objects. Its location will also allow access to the southernmost objects.

## 3. Preliminary Sample

We selected a preliminary sample of spectroscopic binaries (SBs) from the eighth catalogue of orbital elements of spectroscopic binaries of Batten et al. (1989). To this sample of 1469 systems we added 44 systems mainly coming from Duquennoy and Mayor (1991). Only the southern systems ($\delta < 30$ deg.) have been retained. In order to obtain an estimate of the angular separation between the two components of the SBs we computed the so-called *nodal separation*, which is the separation at the node of the orbit i.e. when the two stars are in the plane of the sky. This idea was originally proposed by Alden (1946) to provide a better estimate of the angular separation of a SB than the division of $a_1 sin(i)$ (or $a_1 sin(i) + a_2 sin(i)$ for a SB2) by the distance, and was subsequently applied to the SBs of the sixth and seventh catalogue of Batten by McAlister (1976) and Halbwachs (1981) respectively. Following Halbwachs (1981) the nodal separations are given by replacing the true anomalies with the values $-\omega$ or $\pi - \omega$ in the polar equation of the ellipse of the orbit. The maximum separation is then given by:

$$sep = a(1 - e^2)/(1 - e\,|cos\omega|), \tag{1}$$

where the semi major axis is deduced from the third Kepler law:

$$a = (M_1 + M_2)^{1/3} P^{2/3} r^{-1}, \tag{2}$$

where r is the distance of the system. If the masses are expressed in solar masses, P in years and r in parsecs, then a is in arcseconds. The masses $M_1$ are estimated from the mass-spectral type relations of Schmidt-Kaler (1982), and concerning the SB2s the masses $M_2$ are derived from $M_1$ and the mass ratio $M_1/M_2$. For the SB1s, $M_2$ is set to zero. The Figure 2 presents the distribution of nodal separation of the sample considering both SB1s and SB2s and for a range between 1 and 60 mas. This results in 289 SB1s and 93 SB2s. A comparison with the 4th catalog of measurements of binary stars (Hartkopf 2001) shows that 17 SB1s among the 289 and 12 SB2s among the 93 have been already resolved.

Figure 3 shows that the complete visual orbit can be measured in a reasonable period of time considering the orbital periods of these systems.

The sample of SBs is well within the limiting magnitude of the VLTI+AMBER as shown in Figure 4. We point out that such a program will need only the Auxiliairy Telescopes (ATs) with which the multiple baselines provided are a clear advantage.

## 4. Pre-Main Sequence Spectroscopic Binaries

It is of fundamental importance to measure with accuracy masses of pre-main sequence (PMS) stars with the aim of validating and constraining the models of stellar structure evolution during the PMS phase. We compiled a preliminary sample

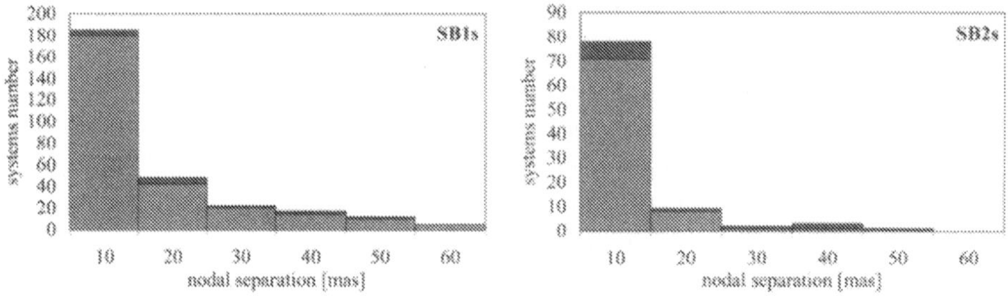


*Figure 2.* Distribution of estimated angular separation for the sample of 289 SB1s and 93 SB2s between 1 and 60 mas. The systems found in the Fourth Catalog of Interferometric Measurements of Binary Stars are represented in dark grey.

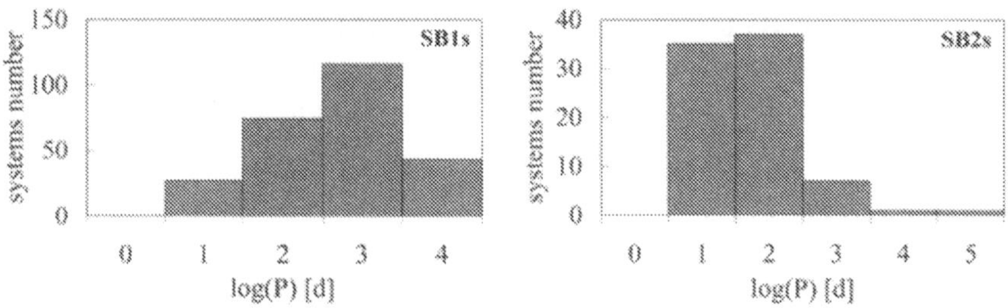


*Figure 3.* Distribution of the orbital period for the sample of SB1s and SB2s.

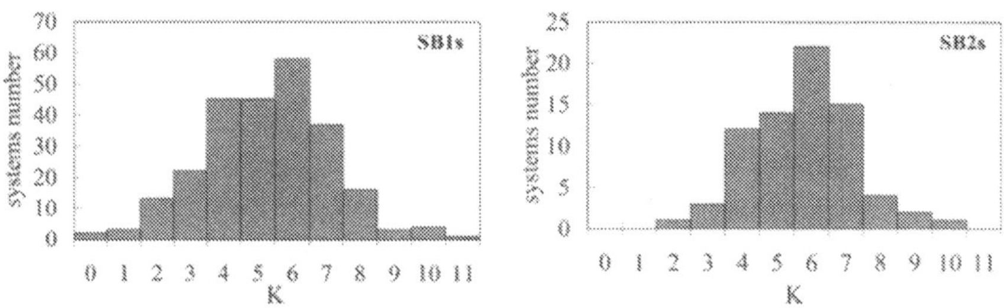


*Figure 4.* Distribution of the main component K magnitude for the sample of SB1s and SB2s.

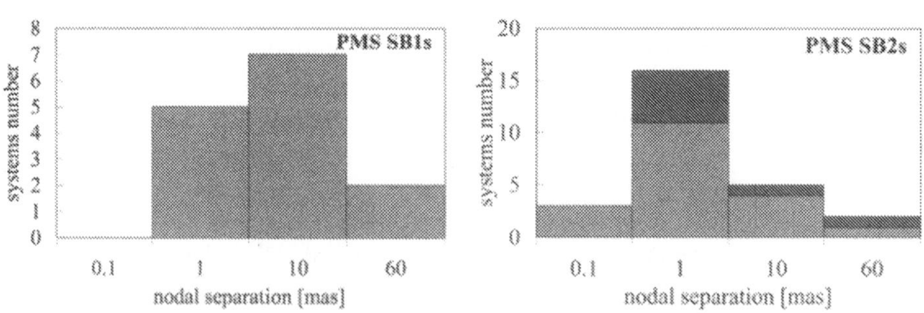


*Figure 5.* Distribution of estimated angular separation for the sample of PMS SB1s and PMS SB2s. The eclipsing systems are represented in dark grey.

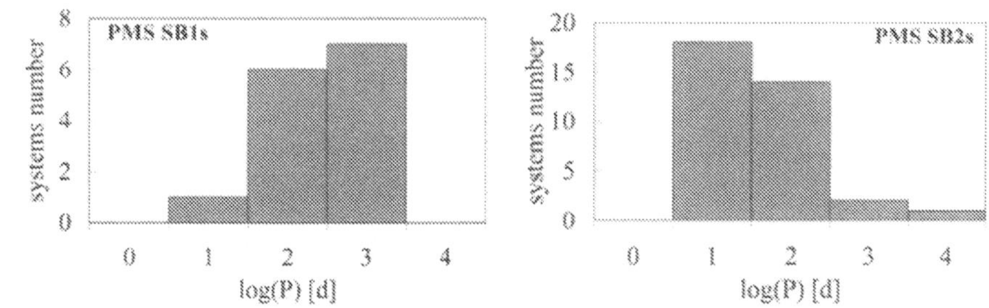


*Figure 6.* Distribution of the orbital period for the sample of PMS SB1s and PMS SB2s.

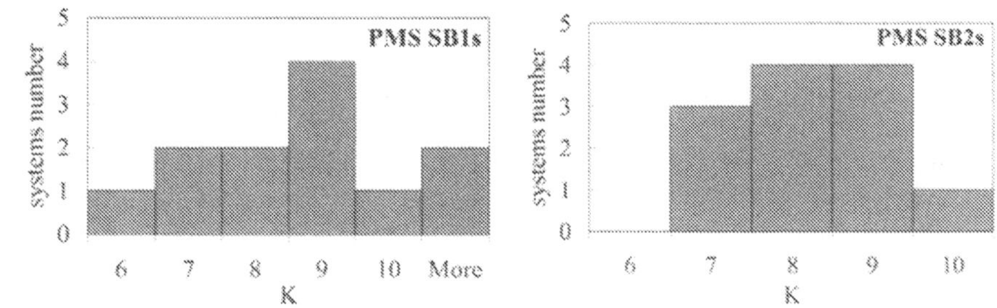


*Figure 7.* Distribution of main component K magnitude for the sample of PMS SB1s and PMS SB2s (combined magnitude).

of the PMS SBs discovered recently from the literature (Melo et al., 2001; Torres et al., 2002b; Prato et al., 2002) and we provide the same feasibility studies as described before for the sample of non-PMS SBs. The PMS SB sample will probably grow with the results of radial-velocity surveys in progress (see e.g. Guenther et al., 2001) which are sensitive to longer orbital period systems. The sample consists of 14 PMS SB1s and 33 PMS SB2s. Among the 14 SB1s, 4 have been converted into SB2s in the near-infrared but without orbits (Prato et al., 2002). Among the 33 SB2s, one has been resolved astrometrically by the HST (Steffen et al., 2001), leading to a mass accuracy of 10%, 6 others are eclipsing systems (with masses measured with 1 to 5% accuracy) and 9 have no orbits or the PMS nature is not determined yet (and are discarded from the sample). Figure 5 shows the distribution of estimated angular separation and in Figures 6 and 7 the distributions in orbital period and in K magnitude respectively. Although the angular resolution needed to resolve some of these systems, as well as the sensitivity, is at the edge of the VLTI+AMBER+ATs capability, the real difficulty with these interferometric measurements will probably be to take into account properly the presence of circumstellar (possibly circumbinary) disks in the derivation of the binary parameters.

## References


Alden, H.L.: 1946, *AJ* **52**, 37.

Andersen, J.: 1991, *A&ARv* **3**, 91.

Andersen, J.: 1998, in: T.R. Redding, A.J. Booth and J. Davis (eds.), *Fundamental Stellar Properties: The Interaction between Observation and Theory*, Proc. of IAU Symp. **189**, Kluwer, Dordrecht, 99.

Batten, A.H. et al.: 1989, Pub. of the Victoria Dominion Astrophysical Observatory.

Duquennoy, A. and Mayor, M.: 1991, *A&A* **248**, 485.

Guenther, E.W. et al.: 2001, in: H. Zinnecker and R.D. Mathieu (eds.), *The Formation of Binary Stars*, Proc. of IAU Symp. **200**, p. 165.

Halbwachs, J.L.: 1981, *A&AS* **44**, 47.

Hartkopf, W.I. et al.: 2001, *AJ* **122**, 3480.

Hummel, C.A. et al.: 2001, *AJ* **121**, 1623.

McAlister, H.A.: 1976, *PASP,* **88**, 317.

Melo, C.H.F. et al.: 2001, *A&A* **378**, 898.

Prato, L. et al.: 2002, *ApJ* **569**, 863.

Quirrenbach, A.: 2001, *ARA&A* **39**, 353.

Schmidt-Kaler, T.: 1982 , in: Landolt-Börnstein Num. Data and Funct. Rel. in Science and Tech., New Series, Group VI, **2b**, Springer-Verlag, p. 1.

Steffen, A.T. et al.: 2001, *AJ* **122**, 997.

Torres, G. et al.: 2002a, *AJ* **124**, 1716.

Torres, G. et al.: 2002b, *AJ* **123**, 1701.

# INTERFEROMETRY: THE TOOL TO STUDY GIANT, SUPERGIANT AND MIRA STARS

GUY PERRIN
*LESIA, UMR 8109, Observatoire de Paris, 5 place Jules Janssen, 92190 Meudon, France*

**Abstract.** Late-type giant stars have been traditional targets for infrared interferometers. They are bright and big and are therefore easy targets to resolve and to detect. Considerable progress has been and is being made on the spatial structure of these objects thanks to existing interferometers. Beyond the classical measurement of their diameters, pulsations have been directly detected, spatial intensity distributions are more and more understood and more important, consistent scenarios for both spectroscopic and interferometric measurements are on the verge to be validated. All this has been possible with prototype instruments having a small number of baselines and very limited spectral capabilities. AMBER and MIDI will surely open a new era with high spectral resolution, high efficiency and imaging capabilities. This is not an exhaustive review of all the work done in the field but rather a presentation of the context. A recent review of Mira star observations with interferometers was recently written (Scholz, 2003). A brief science case is first introduced in this paper. Achievements with high angular resolution single-telescope techniques are presented. The contributions of optical and infrared interferometers are then explained. Eventually, some hints about the possible progress with VLTI are discussed.

**Keywords:** AGB, giants, supergiants, VLTI, interferometry

## 1. A Brief Science Case

Depending on their initial mass on the Main Sequence, stars will have different fates during the later stages of their evolution. The different evolutionary tracks split after the Red Giant phase. Stars with the larger masses over 5 to 8 $M_{\odot}$, depending on evolution models, will turn into supergiants and will or may end-up in a dramatic supernova explosion leaving a degenerate neutron star or a black-hole in the center of an expanding element-rich envelope. Stars with lower masses will reach the Asymptotic Giant Branch (AGB) to continue their evolution at a much slower pace. The next stage of their evolution is a protoplanetary nebula, a central degenerate white dwarf surrounded by a relatively slowly expanding envelope of ionized material.

In both types of scenarii, a major fact of stellar evolution is the mass loss phenomenon. Cool evolved stars are molecule and dust factories. Mass loss has become far more efficient (up to a few $10^{-6} M_{\odot}/yr$) compared to what it was on the Main Sequence for example for solar type stars ($10^{-15} M_{\odot}/yr$) and shells surrounding the stars are created. Although the mechanisms for mass loss are not fully understood, it can be believed that supergiants lose their mass because of their

*Astrophysics and Space Science* **286:** 197–212, 2003.

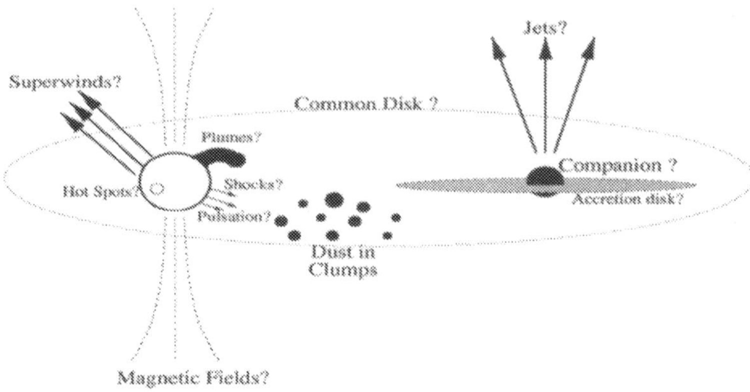


*Figure 1.* Possible exhaustive scenario of an evolved mass-losing star and of its environment. Courtesy of B. Lopez and P. de Laverny.

very high intrinsic luminosity that exceeds the Eddington limit hence gravity is not strong enough to retain the material. The mechanism for AGB stars is somewhat different. A possible scenario is that they undergo a regime of pulsations that can reach a good fraction of the star size in amplitude. During a pulsation cycle, material is lifted in the upper atmosphere, molecules cool down and dust can form. Radiation pressure is then efficient enough on dust grains that they flow away from the star. Dust drags gas molecules which also contribute to the mass loss. AGB stars contribute to more than 75% of the mass loss in the galaxy with an estimated cumulated rate of $0.55 M_{\odot}/yr$ (Sedlmayr, 1994). Apart from potential planet suites, the environment of stars on the Main Sequence looks quite simple compared to how very complex it becomes as soon as enough material has been generated around the stars. Understanding this environment to understand the secrets of mass loss and evolution is the goal of high angular resolution observations of evolved stars. A sketch of a possible environment for both low and high mass stars is displayed in Figure **??**. The general case of a binary system is considered as this may be the explanation for some particular sources.

The field of cool evolved stars is very rich and is addressed with many different techniques which share many goals. Some of the major goals can be summarized as follows. As dust contributes to the enrichment of the galaxy in evolved material and in the mass-loss process of AGB stars for example, it is an important matter to discover where, when and how dust forms and how it couples to the gas to induce the gas-loss, a hundred times larger than the dust-loss. In the same way, at least for AGB stars, pulsations are crucial to explain the mass-loss process and a similar series of questions can be raised about these: why are stars pulsating, how do they pulsate, when do pulsations start, when do they stop, what kind of regularity can be expected? As the star itself cannot be considered separately from its environment, one may wonder what is the connection between the observations at large angular

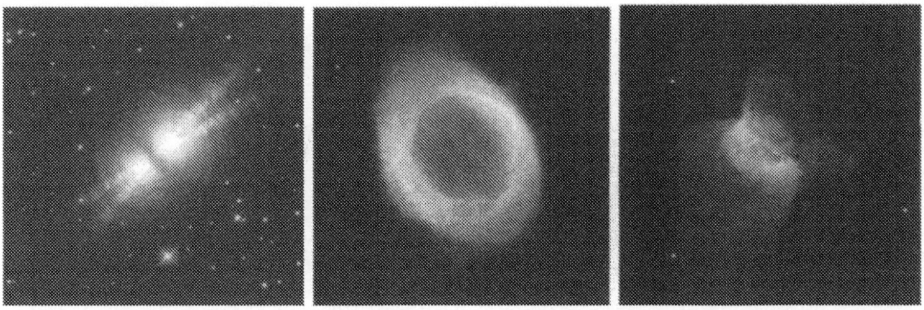

*Figure 2.* From left to right: the Egg Nebula, a post-AGB source, M57 and NGC2346, two planetary nebulae. Credit: NASA/STScI.

scales and the stellar photosphere. What is its smoothness, does it have spots, if so, how many, what are the time and spatial scales for these and are they generated by convection, is the dust formation process connected to the existence of spots on the stellar surface? As the dust and molecule factory probably takes place in the lower atmosphere, what is the link with the atmospheric height scale, can this be guessed from a study of limb-darkening for these objects? Is the scenario for mass-loss in AGB stars based on pulsation correct? In the end, can the outflows and planetary nebulae shapes be connected to some earlier stages of the star history. These points will be addressed in this review as they may have benefitted from high angular resolution observations or because VLTI has the potential to bring answers.

## 2. Contribution of Single-Telescope High Angular Resolution Techniques

The improvement of the quality of optical telescopes, the increase of their diameter and the advent of techniques to fight the limitations due to atmospheric turbulence have allowed to access to new informations on the environment of evolved stars. The images of post-AGB stars and planetary nebulae taken with the Hubble Space Telescope at short wavelengths in the visible or the ultraviolet show structures with similarities between the ionized material of planetary nebulae and the light scattered by dust in the flow of post-AGB stars. The examples of Figure 1 may be illustrative of this. The Egg Nebula is a post-AGB source whose active mass loss has stopped. Several symmetries are visible at short wavelengths. A spherical mass loss superimposed to an outflow possibly collimated by a disk clearly visible as it obscures the equatorial plane of the nebula. Those symmetries can be found in the two planetary nebulae in the same figure: M57 and NGC2346. The material ejected around M57 is very close to spherical symmetry whereas that around NGC2346 is clearly collimated. In the case of the Frosty Leo Nebula, images taken with several large telescopes equipped with adaptive optics systems show an asymmetric collimated flow (see e.g. Roddier et al. (1995)). The pinched-waist shape is also due

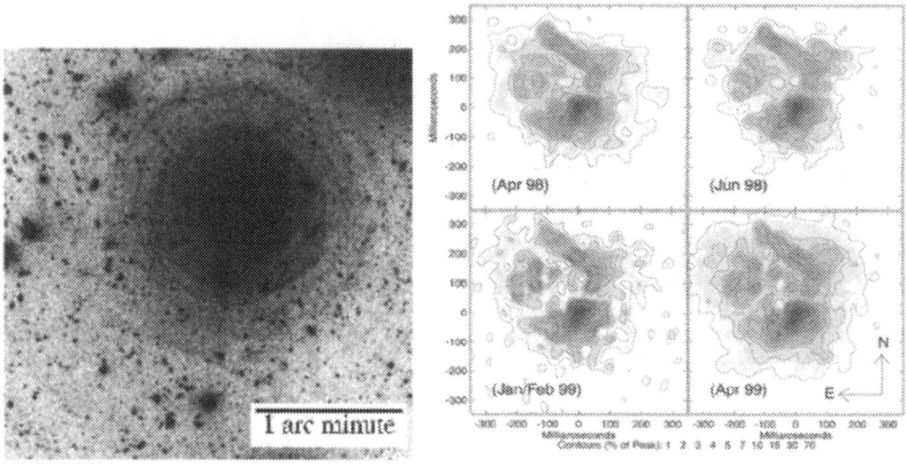


*Figure 3.* Left: images of IRC+10216 obtained at the VLT in the visible on a two arcminute scale (de Laverny, 2002). Right: images of IRC+10216 obtained with the aperture masking technique at Keck in the K band on a half arcsecond scale (Tuthill et al., 2000).

to a central obscuring disk. The asymmetric flow could be due to a binary system embedded in the central disk. No clear direct detection of this binary system has been achieved yet. Yet, binarity is a serious hint to explain the large scale shapes of some post-AGB stars (Mastrodemos and Morris, 1999).

The situation on large scales seems manageable as simple symmetries may have a relatively simple cause and as large scales are also synonymous of large periods of time to average small time scale phenomena. When looking at smaller scales one notices that the situation is somewhat different. A good example for that is the very famous carbon star IRC+10216 as both large and small scales observations of this source are available. De Laverny et al. (2002) have made observations in the visible with the VLT with a resolution better than 1 arcsec on a two arcminute scale (cf. Figure **??**). The long exposure unveils the almost spherically and steadily emitted rings of matter, a trend very similar to large scale observations. Yet, these good resolution data show that dust emission is not strictly symmetric and that departure from circular symmetry is very common on short spatial scales. It is then interesting to connect these observations of light scattered by dust to observations of regions very close to the star in the infrared. Observations carried out at Keck with the aperture masking technique (Tuthill et al., 2000) and at the SAO 6 m telescope with the speckle technique (Osterbart et al., 2000) clearly show a very complex source at the 100 mas scale (cf. Figure **??**). The possible position of the star is obscured by the thick material. Interpretations of the small field images is still under study (position of the star for example). The bright spots may be interpreted as clumps of dust linked to the mass loss process either as a source or a consequence. The position of the clumps is seen to evolve with time and move

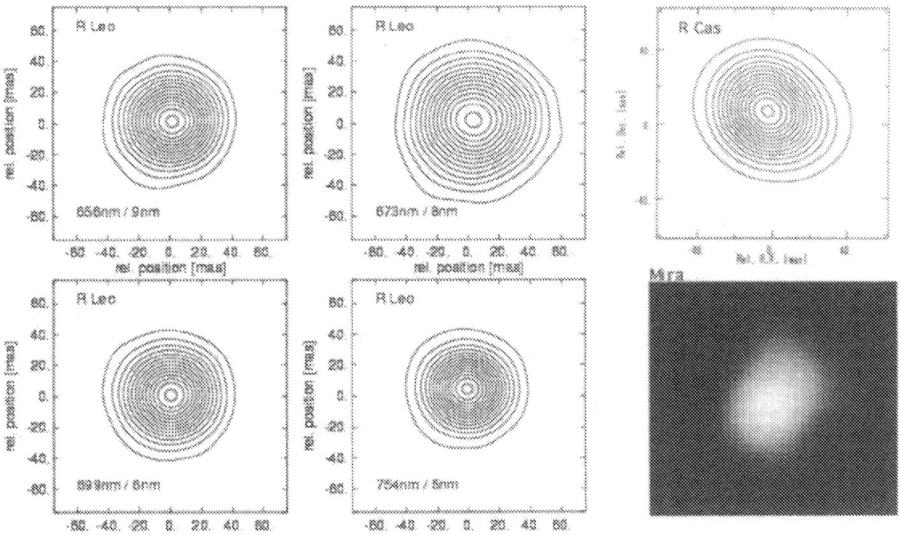


*Figure 4.* Images of the Mira-type stars R Leo and R Cas obtained at the 6 m SAO telescope with the speckle technique (R Leo: Hofmann et al. (2001); R Cas: Weigelt et al. (1996)) and of Mira obtained with HST (Karovska et al., 1997).

away from the center. The differences between the large and small scale images are striking and it becomes clear, at least on the example of IRC+10216, that there is a lot to understand with high angular resolution.

IRC+10216 is a quite peculiar source as the amount of dust in its environment prevents from directly seeing the central engine. Oxygen rich Mira stars are also good targets for single-telescope high angular resolution techniques but have far less material around them allowing to see deeper in the atmosphere. Observations with the speckle technique and with HST allow to resolve the disk of the biggest stars at short wavelengths. Examples are displayed in Figure **??**. Hofmann et al. (2001) have observed R Leonis in the TiO bands and in the continuum and have found a round-shaped star whereas in the same TiO band Weigelt et al. (1996) found a clear departure from circular symmetry for R Cassiopeiae. The same result was achieved by Karovska et al. (1997) with HST in the ultraviolet on Mira itself. The asymmetry of Mira may be due to the presence of its companion. But no companion is known to R Cassiopeiae and another mechanism must be invoked. Although this is still very speculative, non-radial pulsations could explain the asymmetric shape (see Scholz (2003) for other possible mechanisms). Another potential explanation is the presence of spots at the surface of the star which convolved with the instrument response could produce such shape. R Leonis has quite similar characteristics but does not so clearly show asymmetry. Tuthill et al. (1999b) found it slightly asymmetric at 833 nm in 1992 (axis ratio of $89 \pm 3\%$) but did not detect clear evidence for an asymmetric shape a year later in the TiO band

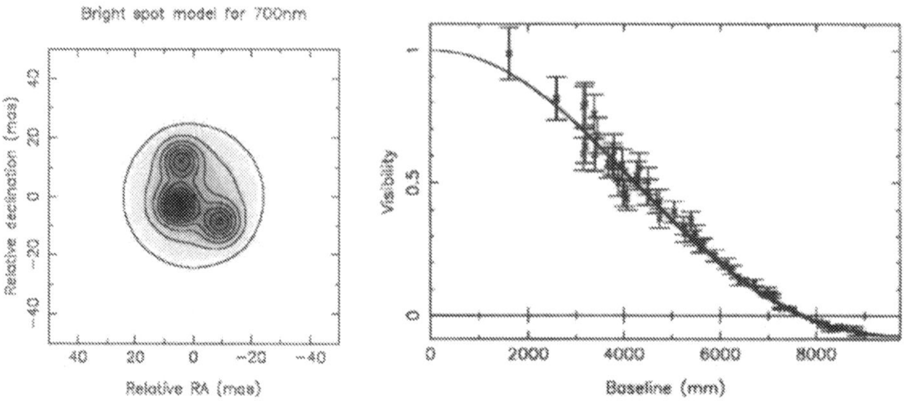


*Figure 5.* Left: image of Betelgeuse obtained at the W. Herschel Telescope with the aperture masking technique at 700 nm (Wilson et al., 1997). Right: Visibilities of Betelgeuse measured with the COAST interferometer at 1290 nm (Young et al., 2000b).

at 710 nm and in the adjactent continuum at 700 nm. Asymmetries are therefore a difficult issue not yet solved.

The study of high mass stars has also benefitted from high angular resolution techniques. Recently, first images of a Wolf-Rayet star with the Keck aperture masking system have revealed the interaction of the wind of a Wolf-Rayet star and of its close companion creating a spiral of dust heated by a shock (Tuthill et al., 1999a). The biggest and brightest supergiant stars Betelgeuse and $\alpha$ Herculis have been observed several times with both the pupil masking technique and HST (Gilliland and Dupree, 1995; Burns et al., 1997; Tuthill et al., 1997; Wilson et al., 1997; Young et al., 2000a). In all cases, the stars were found asymmetric and modeled by spots on their surfaces (see Figure **??** for a sample image measured at 700 nm). The distribution of the spots and their number changed from one observation to the other. Based on theoretical arguments, a classical explanation for the spots is convection cells. Larger wavelengths data obtained with long baseline interferometers are necessary to understand the physical characteristics of these spots as will be shown in the next section.

Although important results have been obtained, single-telescope observations left most of the questions of Section 1 unanswered. Dust has been observed directly distant from and close to stars, and spots have been detected on the brightest supergiant stars. But other questions are left open.

## 3. The Discoveries of Interferometry

Interferometry was first essentially used to measure stellar diameters and binary stars. With a technique becoming more and more mature, the quality of data has

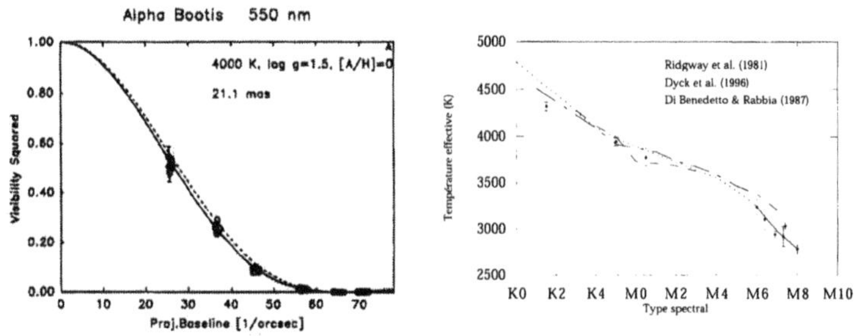


*Figure 6.* Left: visibilities of Arcturus measured with the Mark III interferometer at 550 nm (Quirrenbach et al., 1996). The dashed-line is the best uniform disk fit. The continuous line is the best limb-darkened disk fit. Right: effective temperature scale of late type giants (Perrin et al., 1998).

improved and some higher order information can now be searched for. Diameter measurements have allowed to directly compare the predictions of theory to the size and effective temperature of non-pulsating giants. We will also see in the next paragraphs how optical interferometry has caused a quantum leap in the study of Mira stars and Supergiants.

### 3.1. THE STUDY OF GIANT STARS

The problem of directly measuring the diameter of giant stars has been addressed by optical interferometry but also by the lunar occultation technique. When combined to bolometric flux measurements, diameters yield the effective temperature of stars, a radiometric quantity useful to represent the global properties of a star population as is needed for example to synthesize the spectrum of galaxies. The first exhaustive effective temperature scale of giants was obtained with lunar occultation measurements (Ridgway et al., 1980). The giant star effective temperature scale was completed for the later types by interferometric measurements (Figure 3.1) which confirmed the increasing difference between color and effective temperatures for giant stars later than K4 (Perrin et al., 1998; Van Belle et al., 1999).

The advent of interferometry has allowed to measure diameters more accurately and to study the photospheric structure of stars. First direct measurements of stellar limb-darkening, a consequence of the stratification of temperature in the atmosphere, were obtained with the intensity interferometer in the visible (Hanbury Brown et al., 1974). Modern interferometers measure limb-darkening more routinely on giant stars despite the small departure from uniform disk (Quirrenbach et al., 1996; Hajian et al., 1998) and limb-darkening models can now be better tested against the multi-color measurements obtained. An example of limb-darkening measurement on Arcturus in the visible is displayed in Figure 3.1. These measurements need to be accurate to discriminate between different models of

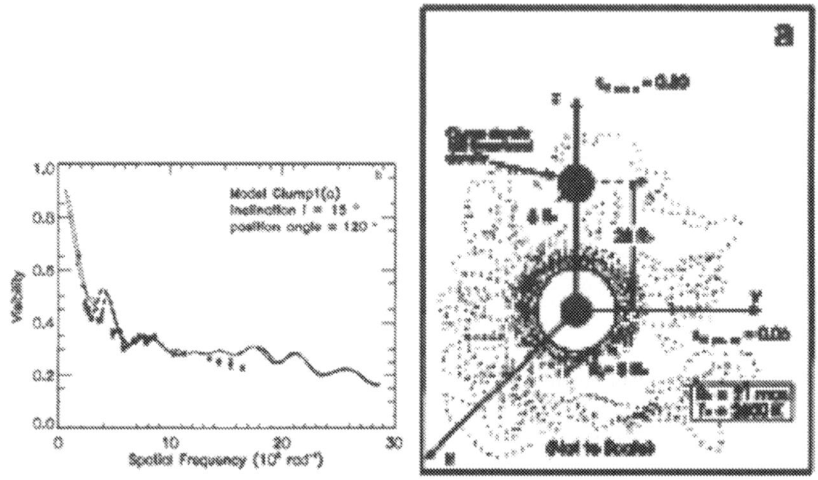


*Figure 7.* Left: visibilities of Mira measured with ISI at 11.15 $\mu$m (Lopez et al., 1997). The dust shell, peak close to zero frequency, is fully resolved whereas the stellar disk is barely resolved. Right: proposed model of a dust shell and of a clump to model the data.

limb-darkening. The comparison shows good agreement between data and theory for the hydrostatic model atmospheres for non pulsating giants.

### 3.2. Mira stars

Dust has been detected and studied around Mira stars with diverse techniques in the infrared. The dust environment is better constrained with the spatial information provided by interferometry and is an excellent complementary information for the interpretation of Spectral Energy Distributions. The measurements by Danchi et al. (1994) with the ISI interferometer at 11.15 $\mu$m allowed to characterize several parameters such as the distance, the inner radius and grain distribution of the dust. This gave a view of the condensation radius where dust forms, a few stellar radii above the photosphere at a temperature about 1200 K consistently below the sublimation temperature of the grains. For stars observed at different phases, the characteristics of the shells were changing as dust around Mira is created at minimum phase. Yet, a later study by Monnier et al. (2000) showed that the creation of dust was not necessarily correlated to pulsation for the carbon star IRC+10216. Although oxygen-rich and carbon-rich Miras have different chemistries, this illustrates the complexity of the mass loss phenomenon. The dust creation process may stop and resume despite the continuing pulsation cycle. The characteristics of the creation of dust varies with the type of star and its age. Systematic studies will therefore help understand the history of mass-loss hence the shape of nebulae as that of IRC+10216 imaged with the VLT.

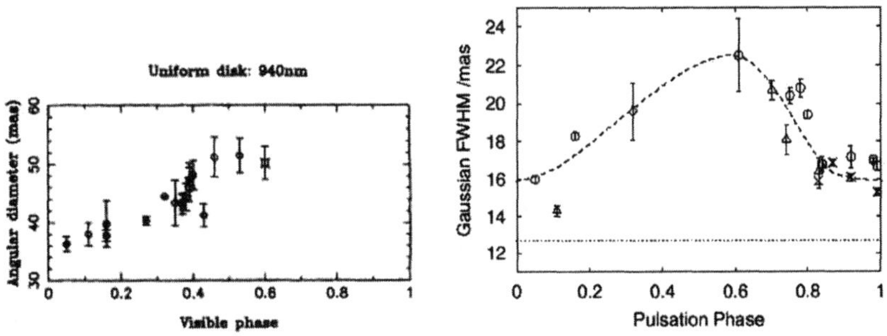


*Figure 8.* Left: Apparent diameter variations of R Leo measured with COAST and WHT at 940 nm (Burns et al., 1998). Right: Apparent diameter variations of $\chi$ Cyg measured with COAST at 905 nm (Young et al., 2000a).

Lopez et al. (1997) have studied the dust around Mira with ISI on a time scale covering several periods and found that the dust shell characteristics are changing. The spatial distribution of dust may be compatible with several models comprising more than a single shell and asymmetries were found to play a major role (Figure 3.2). The hypothesis of the presence of clumps in the shell was favoured by the data and raised the question of dust formation. Does dust form in clumps and could these clumps be connected with the presence of spots in the star atmosphere?

The pulsation mechanism is important in Mira stars as it provides the mechanical energy necessary to generate the enhanced mass-loss. Although theoreticians favor the fundamental mode of pulsation in Mira stars, there is no *a priori* clear evidence that this pulsation mode should be favored from direct observations. Mira stars are known to pulsate from their varying light curves and also from varying cyclic velocity variations detected in the atmosphere. As the environment of Mira stars is quite complex with layers in the atmosphere containing molecular species as shown by the first measurements of Labeyrie and collaborators (Labeyrie et al., 1977) it is difficult to disentangle the real pulsation of the photosphere from the motion of some atmospheric layers observed in molecular bands as TiO in the visible or CO and $H_2O$ at near-infrared wavelengths. The first measurements of the pulsation of a Mira star, R Leo, were achieved by Burns et al. (1998) at 940 and 830 nm partly with the COAST interferometer and with the WHT telescope. The stellar size variations were about 70% in amplitude over 2 periods (Figure 3.2). Young et al. (2000a) followed the pulsation of the S star $\chi$ Cyg over several pulsation cycles in TiO bands and were able to reproduce consistent cycle-to-cycle 30% apparent atmospheric diameter variations. Because of the molecular bands, it is not clear whether the measured effect is due to a physical motion of the layers above the photosphere or to a change in opacity of the absorbing layers as shock waves sweep through them. Observing in less contaminated bands at longer wavelengths could potentially provide an answer. A first direct measurement of the star size of R Leo very close to the photopshere was performed in the K band by Perrin et al. (1999) at two different phases and its apparent size varied by about 8%. The K band is *a priori* cleaner than the bands at lower wavelengths and the flux emitted by the star is not as much scattered by the molecular layers as it is at shorter wavelengths hence the Gaussian intensity distribution and the much bigger size compared to the K band (about 50% larger in average). Yet, the visibility curve measured on R Leo in the K band showed a clear departure from classical limb-darkening profiles and the hypothesis of a strong contamination of the visibilities by the molecular environment was proposed. A direct consequence is a smaller radius for the photosphere (although the photosphere is difficult to define in the case of Mira stars as it is very likely wavelength-dependent) compatible with both the spectral type and the fundamental mode pulsation. Measurements of R Tri with PTI by Thompson et al. (2002a) at four different phases also showed a variation of the star size. Departure from the uniform disk model was also detected but was interpreted as an asymmetry. Measurements carried out in the L band and in molecular (CO and $H_2O$) and continuum bands in K (Mennesson et al., 2002; Perrin et al., 2003a) clearly favor the hypothesis of contamination by molecules. Similar measurements were also carried out at PTI in the K band by Thompson et al. (2002b). The K and L band data are well modeled by a molecular layer located 1 or 2 $R_\star$ above the photosphere with a typical temperature of 1500 K. It is remarkable that the parameters found for the molecular layer by the interferometric technique are in perfect agreement with the $H_2O$ layers found by the ISO satellite (e.g. Yamamura et al., 1999). In the case of Mira, the K&L band model is also consistent with the visibility curve measured with ISI by Weiner et al. (2000). The K band data and the molecular layer model are presented in Figure 3.2. The apparent size of Mira stars measured by fitting visibilities by a uniform disk model is in fact intermediate between that of the molecular layer and the real size of the photosphere. Using the molecular layer model, the photosphere diameter of Mira for example is found to be on the order of 23 mas depending on phase and the effective temperature is about 2500 K, in agreement with the very late type of the star (Perrin et al., 2003a).

### 3.3. THE ENVIRONMENT OF RED SUPERGIANT STARS

Fewer red supergiant stars have been studied interferometrically compared to Mira stars. Yet, the brightest ones have quite large angular diameters and were ideal targets for the pre-VLTI generation of interferometers. Among them, $\alpha$ Orionis, the star with the largest apparent diameter seen from the Northern hemisphere, has been one of the first targets of interferometers as early as the pioneering ages of interferometry at the end of the nineteenth century. Michelson measured its diameter quite accurately in the visible. $\alpha$ Orionis and $\alpha$ Herculis are by far the most observed supergiant stars by interferometers because of their large size and their diameters have been measured from the visible range up to 11.15 $\mu$m (see e.g.

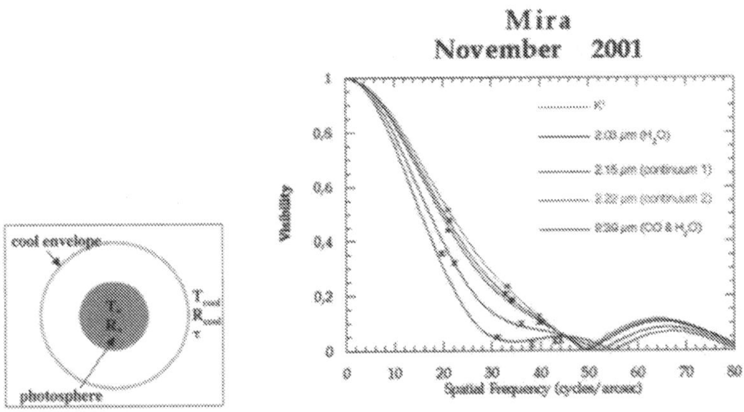


*Figure 9.* Visibilities acquired with FLUOR at IOTA in molecular and continuum bands in K (Perrin et al., 2003a). The chromatic variations are well explained by a simple model of a photosphere surrounded by a thin shell. The same model allows to explain L band and ISI observations.

Benson et al., 1991; Mozurkewich et al., 1991; Dyck et al., 1992; Mennesson et al., 1999; Young et al., 2000b; Weiner et al., 2000; Chagnon et al., 2002). As for Mira stars, the diameters are found to decrease from the visible to the near-infrared and then to increase again towards the mid-infrared. Short wavelengths are sensitive to scattering by the atmosphere whereas the dusty environment shows up in the thermal infrared.

The properties of the inner parts of dust shells around supergiants have been studied with ISI by Danchi et al. (1994). Apart from VY CMa, dust is found relatively far from the star photosphere, up to a few tens of stellar radii. This indicates sporadic emission of dust on time scales of a few years. The larger time scale compared to Mira stars correlates quite well with the longer period of luminosity variations of these semi-regular variable stars, on the order of 2000 days. Later, Bester et al. (1996) detected a recently formed closer dust shell 5 stellar radii above the photosphere of Betelgeuse. As the theory of Mira stars links dust formation to pulsation, the irregular dust formation in supergiant stars is quite puzzling. The natural cause for mass-loss being the excessive luminosity of the star, one would expect a regular flow of dust.

Clues for alternative scenarii for the formation and emission of dust are probably to be found in interferometric data. Among other possible leads, spots may be an interesting clue. They are signatures of inhomogeneities of temperature in the atmosphere of evolved stars. Lower temperature patches could be a favorable nest where dust could condense. We have seen that spots on supergiants have been studied in the visible with single-telescope instruments. Young et al. (2000b) have extended this study to the near-infrared with the COAST interferometer (Figure ??).

No spots were detected at 1290 nm contrary to shorter wavelength observations. This change of regime between the visible and the near-infrared is explained by the fact that the temperature contrast between the spots and the photosphere leads to a decreasing brightness ratio from the visible to the infrared due to the properties of the Planck distribution of the blackbody radiation. This behaviour was somewhat confirmed by Perrin et al. (2003b) with observations in the K band which, although sparse, do not show evidence for departure from circular symmetry. Unfortunately, no correlation has been made so far between spots and dust events. Coordination would be necessary to achieve this and some exciting science can be expected from interferometers at least in this particular field. In parallel to the study of dust, the study of the size and structure of the atmosphere of red supergiants in molecular bands is starting. Using K and L band observations, Perrin et al. (2003b) have attempted to explain the ISI 11.15 $\mu$m observation of the disk of Betelgeuse very similar to a limb-darkened disk but with a larger diameter than found in K and L where they are the same. The possible explanations are either that limb-darkening is very strong in the near-infrared and the star appears smaller at K and L, or that the bigger diameter detected by ISI is due to a warm molecular layer. The second hypothesis is favored by the K band observations as there is no evidence of very strong limb-darkening in this band. The real nature of this molecular layer cannot be directly inferred from the interferometric observations as no molecular band observations are available for Betelgeuse. Yet, such observations have been carried out in CO and $H_2O$ bands in K on another supergiant star, $\mu$ Cep, and show a clear increase of diameter compared to the continuum (Verhoelst et al., 2003) which is compatible with the MOLsphere detected spectroscopically by Tsuji (2000). The MOLsphere of Tsuji has a temperature of 1500$\pm$500 K in excellent agreement with the 2000 K temperature needed to consistently explain the K, L and 11.15 $\mu$m observations. This suggests atmospheric structures very similar for Mira-type and supergiant stars. The question is: are there also similarities for the process of dust formation?

### 3.4. Summary of Progress Made with Interferometers

Concerning the two main topics, the nature of late-type stars and the mass loss process, a lot of progress has been made with the contribution of interferometers which would have not been possible otherwise. The structure of dust shells very close to the atmosphere is now well studied and some features of the process of dust creation such as the motion of successive dust shells are addressed. But the formation of dust is still an open issue. Models based on clumps of dust have been suggested and need to be tested on a larger sample. A better time coverage of the sources is needed to see the dust being created and follow its migration away from the star.

The link between pulsation and dust formation in Mira stars is not very clear yet. At least in IRC+10216, dust formation and pulsation are not necessarily connected. The pulsation of long-period variables has been detected and the diameter variations of a few stars have been monitored. It is clear that a good understanding of the atmosphere structure is necessary to well characterize the type of pulsation. Models with warm molecular layers tend to decrease the measured photospheric diameter and favor the fundamental mode pulsation scenario as the high layer increases the apparent diameter and correcting for it decreases the deduced diameter. A global model of Mira stars consistent with both dust and warm gas is now required to proceed further.

In order to validate the models of cool evolved stars, one of the first necessary step is to validate the models on regular non-pulsating giants. Direct measurements of limb-darkening are certainly helping a lot in this respect and measurements of stratified atmospheres in multiple bands conducted at NPOI are certainly of an important value for this. Combined measurements of limb-darkened diameters and flux also allow to refine the effective temperature scales of giant stars.

The problem of spots and asymmetries is now well addressed with imaging interferometers. Results achieved on supergiants for which the physical characteristics of spots can be studied are contributing a lot to the understanding of these objects. The question of the connection between the spots and the environment of the stars is still open and more extensive and promising studies are now on their way.

## 4. Breakthroughs with VLTI

In this section some basic characteristics of VLTI relevant to our subject are reviewed and three examples of sources of breakthroughs in the field of evolved stars are presented.

One of the basic needs is to be able to cover a wide wavelength range. With AMBER and MIDI, VLTI can be operated from 1 to 2.5 $\mu$m and from 8 to 12 $\mu$m with a possible extension to 20 $\mu$m (Q band). No instrument is anticipated as of now for the L and M bands, two regions intermediate between the continuum and dust emmission wavelengths. Spectral resolution is needed to study molecules, shocks and dust grains. AMBER will have up to 10000 resolution and MIDI up to 230. Images (more exactely visibility phase) will be required to study the departures from circular symmetry. This will be possible with both AMBER which is a three telescope beamcombiner and with MIDI using the phase referencing capability of PRIMA. In order to be able to observe stars with the largest spectral resolution or at very low visibilities, the best sensitivity will be required and will be provided by the PRIMA fringe tracker. Last but not least, an excellent uv plane coverage will be needed to study the most complex objects. It will be granted by the four Unit Telescopes and the four relocatable Auxilliary Telescopes.

**Source of breakthrough ♯1: spectral resolution**
Physics comes from spectra and performing a spectral analysis of visibilities is therefore important. The classical approach to model spectra and explain the spatial characteristics of a source is to fit the observed Spectral Energy Distribution (SED). Unfortunately this is not enough and some spatial information is required. A better way will be to fit the Spatial and Spectral Energy Distribution (SSED) to get a fully consistent model of a source. With this, it becomes possible to perform a direct 3D (wavelength dimesion and two spatial dimensions) study of stratified envelopes and to connect the spectroscopic studies in the infrared to the interferometric observables. As an example, the temperature of a particular layer will be both constrained by the properties of spectral features pertaining to this layer and by its distance to the photosphere. Up to now, either one of these informations is missing and a unique solution may be difficult to find. Fitting of SSEDs will help solve ambiguities. Some applications of spectral resolution are for example the validation of model atmospheres on quiet giants before modeling complex objects with extended atmosphere. The study of more evolved objects and of shock waves in the wind of Mira stars are obviously excellent applications.

**Source of breakthrough ♯2: uv coverage and phase measurements**
The ability of VLTI to realize good quality images will have a strong impact on the study of objects with complex structures. A good example is the study of spots on supergiants. As we have seen before, spots are a good proxy for convection in giants and supergiants and also may be connected to the process of dust formation. A simple calculation shows that imaging of the largest spots with a size of a stellar radius requires about 16 pixels on the stellar disk at Nyquist sampling. This can be achieved with at least two baselines and is relatively easy on bright sources which can be observed with the relocatable ATs as will be the case for the brightest which are also the most resolved stars. As a consequence, the surfaces of a good sample of supergiants (with diameters larger than 5 to 10 mas) should be imaged in J, H and K at least. The surface of a few big giants (smaller spots) will also be imaged.

**Source of breakthrough ♯3: VLTI is not a prototype interferometer**
Contrary to previous generation interferometers, VLTI benefits from high quality infrastructure and of a lot of engineering power. It will therefore almost always work and will provide tens to hundreds of visibility points per source. This has to be compared to usually sparse data acquired by prototype interferometers. This will make a dramatic change in the interpretation of data and will reduce the share of speculative explanations. Besides, thanks to the generalized use of spatial filtering, nice accuracies (typically 1%) will be achieved on visibilities. As a consequence many more stars will be observed as the interferometer will be more efficient and many more high quality data will be acquired per star. This will help build larger samples of observed stars per class of object and perform more systematic studies, a real new feature in interferometry. In the same vein, being mostly operated by

ESO, VLTI will provide an excellent temporal coverage of cyclic or time varying phenomena, a feat rarely achieved up to now. This characteristic of VLTI is very important for all the matters we have reviewed in this paper.

## 5. Conclusion

As a conclusion, it is obvious to say that VLTI will do much more and better than what has been achieved so far in the field of evolved stars. Pre-VLTI interferometers have provided very encouraging results and have opened the way to the solution of some of the big questions. Considering the general science case, VLTI has the potential to solve most if not all issues known so far and will very likely raise some new ones. But at least most of the questions we have now will find an answer during the VLTI era and this is the announcement of very exciting science.

## Acknowledgements

I wish to thank the Scientific Organizing Committee of the conference for inviting me to give this review. I am very grateful to S.T. Ridgway and P. de Laverny for reading and commenting on this article.

## References

Benson, J.A. et al.: 1991, *AJ* **102**, 2091, 1991.
Bester, M. et al.: 1996, *ApJ* **463**, 336.
Burns, D. et al.: 1998, *MNRAS* **297**, 462–466.
Burns, D. et al.: 1997, *MNRAS* **290**(1), L11–L16.
Chagnon, G. et al.: 2002, *AJ*.
Danchi, W.C. et al.: 1994, *AJ* **107**(4), 1469.
De Laverny, P.: 2002, in: *Mass-losing Stars and their Circumstellar Matter held in Sendai, Japan, May 13–16, 2002*, in press.
Dyck, H.M. et al.: 1992, *AJ* **104**, 1982.
Gilliland, R.L. and Dupree, A.K.: 1995, **463**, L29.
Hajian, A.R. et al.: 1998, *ApJ* **496**, 484.
Hanbury Brown, R. et al.: 1974, *MNRAS* **167**, 475.
Hofmann, K.H. et al.: 2001, *A&A* **376**, 518–531.
Karovska, M. et al.: 1997, *ApJ* **482**, L175.
Labeyrie, A. et al.: 1977, *ApJ* **218**, L75–L78.
Lopez, B. et al.: 1997, *ApJ* **488**, 807–826.
Mennesson, B. et al.: 1999, *A&A* **346**, 181–189.
Mennesson, B. et al.: 2002, *ApJ* **579**, 446–454.
Monnier, J.D. et al.: 2000, *ApJ* **543**, 861.
Mastrodemos, N. and Morris, M.: 1999, *ApJ* **523**, 357.
Mozurkewich, D. et al.: 1991, *AJ* **101**, 2207–2219.
Osterbart, R. et al.: 2000, *A&A* **357**, 169–176.

Perrin, G. et al.: 1998, *A&A* **331**, 619–626.
Perrin, G. et al.: 1999, *A&A* **345**, 221–232.
Perrin, G. et al.: 2003a, in preparation.
Perrin, G. et al.: 2003b, submitted to *A&A*.
Quirrenbach, A. et al.: 1996, *A&A* **312**, 160–166.
Ridgway, S.T. et al.: 1980, *ApJ* **235**, 126–137.
Roddier, F. et al.: 1995, *ApJ* **443**, 249.
Scholz, M.: 2003, *SPIE* **4838**, 163.
Sedlmayr, E.: 1994, *Held at Copenhagen, Denmark, May 24–29, 1993* **428**, p. 163.
Thompson, R.R. et al.: 2002a, *ApJ* **570**, 373–378.
Thompson, R.R. et al.: 2002, *ApJ* **577**, 447–456.
Tsuji, T.: 2000, *ApJ* **538**, 801.
Tuthill, P.G. et al.: 2000, *ApJ* **543**, 284–290.
Tuthill, P.G., Monnier, J.D. and Danchi, W.C.: 1999a, *Nature* **398**, 487.
Tuthill, P.G., Haniff, C.A. and Baldwin, J.E.: 1999b, *MNRAS* **306**, 353.
Tuthill, P.G., Haniff, C.A. and Baldwin, J.E.: 1997, *MNRAS* **285**(3), 529–539.
Van Belle, G. et al.: 1999, *AJ* **117**(1), 521–533.
Verhoelst, T. et al.: 2003, in preparation.
Weigelt, G.: 1996, *A&A* **316**, L21–L24.
Weiner, J. et al.: 2000, *ApJ* **544**, 1097–1100.
Wilson, R.W., Dhillon, V.S. and Haniff, C.A.: 1997, *MNRAS* **291**, 819.
Yamamura,I., de Jong, T. and Cami, J.: 1999, *A&A* **348**, L55–L58.
Young, J.S. et al.: 2000a, *MNRAS* **318**, 381–386.
Young, J.S. et al.: 2000b, *MNRAS* **315**, 635.

# SYNTHETIC RADII AND VISIBILITIES FOR PULSATING RED GIANTS

JOSEF HRON and WALTER NOWOTNY
*Institut für Astronomie der Universität Wien, Vienna, Austria*

RITA GAUTSCHY-LOIDL
*Rüti, Switzerland*

SUSANNE HÖFNER
*Department of Astronomy and Space Physics, Uppsala University, Sweden*

**Abstract.** We present synthetic radii and visibilities in the IR, based on hydrostatic and dynamic model atmospheres of carbon stars. The differences between these two sets of input models and to uniform disc models are discussed and compared to the expected capabilities of the VLTI.

**Keywords:** AGB stars, red giants, Interferometry, visibilities, synthetic

## 1. Introduction

Red giants have always been prime targets for optical interferometry since they have large diameters and are very bright in the red and infrared (e.g. Labeyrie, 1978; Perrin, this volume). Besides that, the atmospheres of these stars are astrophysically highly interesting due to the following reasons: (i) the cool temperatures allow the formation of molecules and dust; (ii) pulsations can lead to strong temporal and radial variations in the structure of the atmosphere and (iii) a stellar wind, driven by the pulsations and radiation pressure on dust, causes very extended atmospheres.

Due to the complexity of the involved processes, modelling the atmospheres of pulsating red giants remains one of the major challenges for stellar astrophysics. Very good progress has been made in this field over the last decade (see Section 2) and the best constraints for the models come from high angular and high spectral resolution data. Thus the existing and new interferometric data will be crucial for improving the theoretical models.

In this paper we will concentrate on carbon stars for three reasons: (i) these stars are of special relevance for stellar and galactic evolution due to the dredge-up of carbon into the atmospheres and the consequent enrichment of the ISM with this element; (ii) the dust formation process is much better understood in the case of a carbon rich atmosphere than for oxygen rich (i.e. M-type) stars and as a consequence the available model atmospheres are more advanced and (iii) until now only very few carbon stars have been observed but with the VLTI several 'typical' objects are accessible.

*Astrophysics and Space Science* **286:** 213–218, 2003.

## 2. Model Atmospheres for Red Giant Stars

Red giant model atmospheres may be grouped into three categories: (i) 'classical', i.e. hydrostatic and dust-free models; (ii) 'dynamic', i.e. taking into account the effects of pulsation and (iii) 'dusty', i.e. one of the previous including circumstellar dust. An overview of existing models and problems can be found in Plez (1999) and Höfner (1999).

For this work, we used various hydrostatic model atmospheres for C-rich giants, taken from Gautschy (2001). They were computed with an improved version of the MARCS code (Jørgensen et al., 1992).

As the stars evolve towards the upper part of the AGB, they become unstable against large amplitude radial pulsations. These trigger shockfronts, which propagate outwards. In cooler regions molecules and even dust can form. The capability of hydrostatic models in reproducing observable properties of these complex and temporally variable atmospheric structures is very limited. Hydrostatic models give good results (spectra from the visual to mid IR, narrow band colours) for red giants with relatively high $T_{eff}$, low $\Delta m_{bol}$ and small $\dot{M}$ (Hron et al., 2002). Dynamic models, including a combined and self-consistent solution of hydrodynamics, frequency-dependent radiative transfer and a detailed time-dependent treatment of dust formation (Höfner et al., 2003), are often better suited for stars with lower $T_{eff}$, higher $\dot{M}$ and a more pronounced variability (Miras).

A major difficulty is the treatment of dust (formation, absorption and emission). For M-type stars (O-rich chemistry), the formation process is still poorly known, which is one of the limitations of the models (Hron et al., 2002). Our understanding of the behaviour of dust is much better for C-type stars (C-rich chemistry), a detailed description using the 'moment method' (Gauger et al., 1990) is included in the dynamic models.

For this work we used a dynamic model for a 'typical' C-rich Mira, from Höfner et al. (2003). The stellar pulsation is simulated by applying a variable inner boundary (piston) moving sinusoidally with a velocity amplitude $\Delta u_p$ and a period $P$. The luminosity variation follows from the variable inner radius and the assumption of constant flux in the innermost mass shell. The parameters of the used C-rich model are: $L=10^4\,L_\odot$, $T_{eff}$=2800 K, C/O=1.4, $M=1\,M_\odot$, period=$525^d$, $u_{piston}$=4 km/s. The resulting mass loss rate and outflow velocity are $\dot{M}=2.5 \cdot 10^{-6}\,M_\odot$/yr and $u_{terminal}$=14 km/s. The bolometric lightcurve and spatial structures at two phases are shown in Figure 1, one model at maximum and another one at minimum luminosity were used for the calculations. Pronounced differences can be seen between different phases (moving shells with enhanced density and dust, multiple shock fronts).

For both, the selected phases of the dynamic model and the hydrostatic models, molecular abundances are computed on the basis of equilibrium chemistry. Opacities are treated in the opacity sampling approximation (Jørgensen et al., 2000,

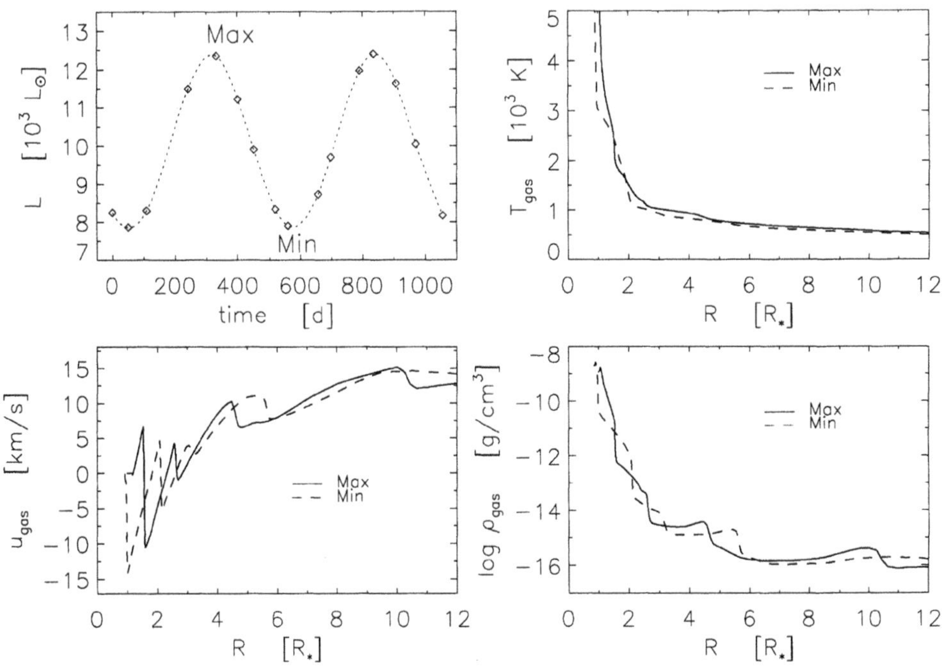


*Figure 1.* Dynamic model for a 'typical' C-rich Mira. The upper, left panel shows the change in luminosity, together with the two phases selected for synthesizing the visibilities ('Max' and 'Min'). The other plots show the changing atmospheric structure (T, $\rho$, u) for these phases. Note the extension of the atmosphere, 1 $R_* \approx 1.65$ AU.

2001), they include basically all the molecular species relevant for the range of pressures and temperatures, i.e. CO, CH, $C_2$, CN, $C_3$, HCN, $C_2H_2$ and CS.

The radiative transfer is done in spherical geometry, the frequency-dependent transfer equation is integrated along characteristic rays. This results in intensities for a variety of impact parameters across the projected stellar disk and for every frequency point of our calculations. Convolving them with different filter curves gives intensity profiles at different wavelengths. Visibilities were then synthesized by applying a Fourier transform (which reduces for spherical symmetric distributions to the Hankel transform) to these intensity profiles.

## 3. Synthetic Intensity Profiles and Visibilities

While several publications dealing with synthetic visibilities of pulsating M-type giants exist (e.g. Hofmann et al., 1998), the few corresponding works on carbon stars only covered stars with very high mass loss rates (e.g. Winters et al., 1995). The models presented here all concern carbon stars with zero or intermediate mass loss.

The intensity profiles and visibilities were computed in several broad and narrow infrared wavelength regions between 1 and 13$\mu$m, including the standard near IR filters J, H, and K. For the narrow bands, the central wavelengths were chosen according to the presence or absence of molecular features, the width was a few 0.01$\mu$m and the passbands were assumed to be rectangular. We also computed the radii where an optical depth of unity was reached ($R_{\tau=1}$), taking into account the finite width of our passbands by a proper weighting procedure.

### 3.1. Hydrostatic Models

Carbon stars with bolometric amplitudes below a few $0.1^{mag}$, temperatures above 3000 K and small mass loss can be modelled well with hydrostatic atmospheres, except for the region beyond 10$\mu$m (Jørgensen et al., 2000; Hron et al., 2002). To get an idea about the behaviour of the intensity profiles and the deviations from uniform disks (UD), we used several representative models with C/O=1.4, M=1 $M_{\odot}$, $T_{eff}$=2650 and 3000 K and luminosities between 3000 and 7000 $L_{\odot}$.

Using $R_{\tau=1}$ and the first zero of the visibility to define the radius, the radius difference between 1 and 13$\mu$m was around 10% for a given model, and up to 50% between the extreme values of temperature and luminosity. The deviations from UD-like visibilities were less than 2% in the regions of strong $C_2H_2$ absorption, i.e. around 3.2$\mu$m and 14$\mu$m. For the J, H and K-bands the intensity profiles and visibilities were quite similar due to the similar absorption by CN, $C_2$ and CO in all three bands. Visibility differences to a UD of about 5% were found and the intensity profiles showed also a slight local depression in the outer part. This feature could be associated with a local temperature minimum around 2000 K, related to the effect of molecular cooling.

### 3.2. Dynamic Models

The main results for the selected dynamic model atmosphere are summarized in Figure 2. The total extension of the 'photosphere' is around 8 AU or 8 mas at a distance of 1 kpc. The shock fronts cause sub-structures on scales of 1 to 3AU.

As for the hydrostatic models, the profiles and visibilities for J, H and K are quite similar to each other but differ significantly from those in the narrow wavelength regions of strong $C_2H_2$ absorption. Figure 2 demonstrates the importance of observations around 3.2$\mu$m (i.e. in the L-band region) as a diagnostic tool to study the extension and sub-structures of dynamic atmospheres. The visibilities of the dynamic model differ by more than 10% from those of hydrostatic or UD models and this difference is well pronounced at spatial frequencies beyond the first zero where the dynamic models show significantly lower visibility. It should be noted at this point that there is no clear correspondence between a dynamic model and a hydrostatic or UD model. One possibility would be choosing the same $R_{\tau=1}$ for each model. In terms of visibilities one could also choose models having the same location of the first zero.

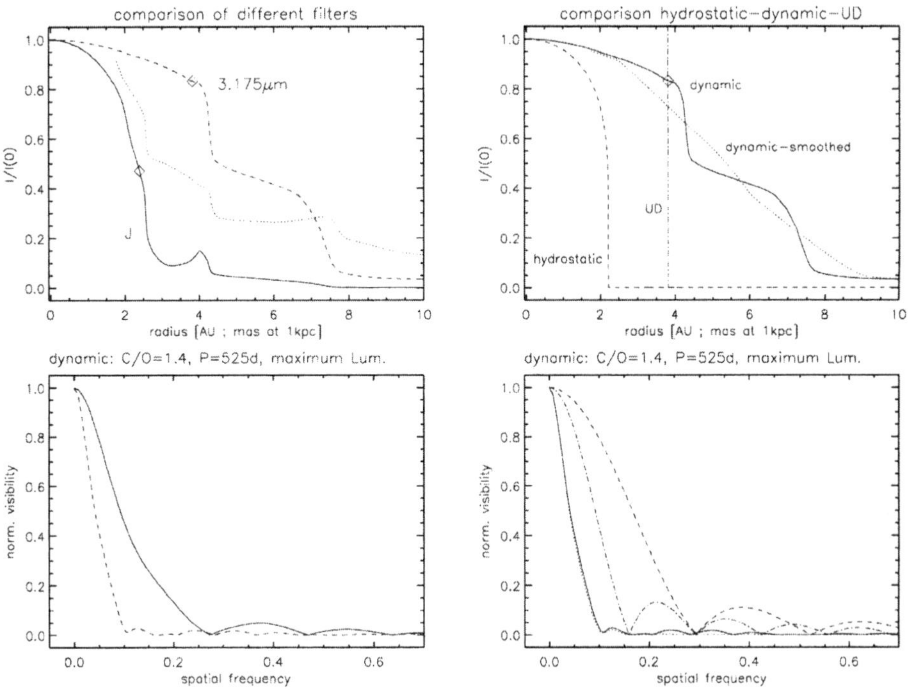


*Figure 2.* Radial intensity profiles (upper panels) and corresponding normalised visibilities (lower panels) for different models of carbon stars. The diamonds in the upper panels mark the radii where $\tau$=1. *Left:* intensity and visibility for the Max-phase of the dynamic model in the J-band (full) and a narrow band at 3.175$\mu$m (dashed). The dotted line in the upper left panel shows the radial gas density profile for comparison (the logarithm of the density was scaled to fit into the plot). *Right:* intensity and visibility for the Max-phase of the dynamic model at 3.175$\mu$m (full) are compared with the corresponding quantities for a uniform disk (having the same integrated intensity and $\tau$=1 radius, dash-dotted) and our most extended hydrostatic model (dashed). The dotted lines correspond to a smoothed version of the dynamic intensity profile.

In general, the differences from smooth but extended profiles only cause visibility differences of less than 5%. An effect of the same order is also caused by the temporal variability.

## 4. Conclusions

It has been shown that dynamic models for carbon stars show clear visibility differences from hydrostatic and UD models, all being on the order of up to 10%. Thus these effects should be detectable with the VLTI, especially at shorter wavelengths. There are pronounced differences between the J/H/K-bands and narrow regions of strong $C_2H_2$ absorption, i.e. around 3.2$\mu$m and 14$\mu$m. Given the higher angular resolution at shorter wavelengths, the L-band is better suited than the mid-IR to study the radial structure of dynamic atmospheres. An extension of the current VLTI instrument sensitivities to this region thus appears to be very useful.

## Acknowledgements

This work has been supported by the Austrian FWF under project P14365-PHY.

## References

Gauger, A., Sedlmayr, E. and Gail, H.-P.: 1990, *A&A* **235**, 345.

Gautschy, R.: 2001, *Ph.D. Thesis*, University of Vienna.

Hofmann, K.-H., Scholz, M. and Wood, P.R.: 1998, *A&A* **339**, 846.

Höfner, S.: 1999, in: T. Le Betre, A. Lebre and C. Waelkens (eds.), *Proc. of 'Asymptotic Giant Branch Stars'*, IAU Symp. **191**, p. 159.

Höfner, S., Gautschy-Loidl, R., Aringer, B. and Jørgensen, U.G.: 2003, *A&A* **399**, 589.

Hron, J., Aringer, B., Gautschy-Loidl, R., Höfner, S. and Jørgensen, U.G.: 2002, *A.S.P Conf. Ser.* **274**, 210.

Jørgensen, U.G., Johnson, H.R. and Nordlund, Å.: 1992, *A&A* **261**, 263.

Jørgensen, U.G., Hron, J. and Loidl, R.: 2000, *A&A* **356**, 253.

Jørgensen, U.G., Jensen, P., Sørensen, G.O. and Aringer, B.: 2001, *A&A* **372**, 249.

Labeyrie, A.: 1978, *ARA&A* **16**, 77.

Plez, B.: 1999, in T. Le Bertre, A. Lebre and C. Waelkens (eds.), *Proc. of 'Asymptotic Giant Branch Stars'*, IAU Symp. **191**, p. 75.

Winters, J.M., Fleischer, A.J., Gauger, A. and Sedlmayr, E.: 1995, *A&A* **302**, 483.

# FIRST VLTI OBSERVATIONS OF MIRA STARS

ANDREA RICHICHI and MARKUS WITTKOWSKI
*European Southern Observatory*

**Abstract.** We present first observations of Mira stars obtained at the Very Large Telescope Interferometer (VLTI) at 2 microns, equipped with the VINCI instrument, using different baselines with both the test siderostats and the 8.2 m unit telescopes. These observations, collected in the course of the VLTI commissining program, have targeted so far about sixty cool giant stars. In this talk, we list and describe the measurements obtained for 14 Mira stars, many of them representing first-time determinations of the visibility. In particular, we devote special attention to a study of Mira itself, for which very accurate determinations of the visibility at several baseline lenghts and orientations could be obtained. We find that a two-component model is more consistent with the data than a single stellar disc. Further observations are needed for a better understanding of this source. The VLTI will constitute an ideal instrument for such studies in the future. In particular it will permit detailed investigations of southern AGB stars, such as accurate measurements of surface structure parameters (diameters, diameter variations, asymmetries, center-to-limb variations, special features like hot spots) and of circumstellar envelopes.

**Keywords:** Techniques: interferometri, Stars: fundamental parameters, Stars: individual: Mira

## 1. Introduction

The ESO Very Large Telescope Interferometer (VLTI) is located on Cerro Paranal, in Chile. This facility consists of the four fixed 8.2 m Unit Telescopes (UT), and of four 1.8 m Auxiliary Telescopes (AT) which can be moved over an array of 30 stations. At present, small siderostats are being used routinely, mainly for commissioning purposes. Also, all UTs have been successfully combined and used during several runs. The ATs will become available starting from 2003. Two facility instruments will be available initially at the VLTI: MIDI and AMBER, operating in the thermal and near-IR respectively. MIDI has achieved first light in December 2002, while AMBER will follow shortly after in 2003. For the observations detailed in this paper, only the 40 cm siderostats were used.

The Mira star phenomenon is a fundamental stage of stellar evolution for stars of solar mass or slightly higher. Not only do Mira stars represent a critical check for evolutionary models, but they also shed light on important processes such as stellar pulsation, stellar chemical composition, and the dust enrichment of the Galaxy. With their extended, cool atmospheres, they have comparatively large angular diameters and substantial infrared fluxes. As such, they have been traditionally a key target for observations by high angular resolution techniques.

*Astrophysics and Space Science* **286:** 219–224, 2003.

In this report, we illustrate first observations of Mira stars carried out at the VLTI, and present preliminary results on Omicron Ceti, i.e the star after which the whole Mira class is named.

## 2. Observations

A first sample of Mira stars observed with VLTI/VINCI is reported in Table I. In the table, the last column reports the number of entries present in the CHARM catalogue (Richichi and Percheron, 2002). It can be seen that apart from a couple of rather famous stars, most of the objects observed by the VLTI had been only scarcely studied before. In fact, for five stars the VLTI data represent the first observations ever. This highlights the fact that the VLTI is ideally positioned to expand significantly our knowledge in this field of research. The log of the observations is reported in Table II, in which the first and last date of observation are reported, along with the number of measurements and the baselines on which they were obtained. The actual projected baseline length and orientation of each observation was of course determined by the circumstances of the observation.

It can be seen that a total of 290 measurements were collected on 14 sources. Three different baselines were used, although in practice the 16 m baseline was used in most of the cases. The Mira phenomenon is characterized by significant pulsations, with periods which are typically of one year or longer. At this preliminary stage, our observations do not span a sufficient time span to characterize the pulsation characteristics of the sources.

## 3. Results

We concentrate on Omicron Ceti, i.e. Mira itself. A large number of observations were collected on this star using the 16 m baseline between October 2001 and January 2002. In this period, the phase of the source varied between 0.1 and 0.3. A first step was to convert the visibility measurements into equivalent angular diameters, using a simple model of a uniform circular disk. This is shown in Figure 1.

Two conclusions can be drawn already, even if this approach is not formally correct since an asymmetric disc could not be uniformly illuminated. Firstly, the VLTI data are of remarkable accuracy and show consistency on a night-by-night basis. Secondly, they show a marked increase of the angular diameter with time. This is expected, since as the phase increases, the temperature drops and the diameter increases.

However, it can also be appreciated that the equivalent angular diameter seems to be a function of the position angle of the baseline. This is indicative of the fact that the assumption of a uniform disk is not adequate. Better models might be either an asymmetric disk, or a two (or more) component model for the stellar disk. This conclusion is highlighted very clearly in Figure 2.

TABLE I
List of observed Mira stars

| Name | V | K | Spec. | P(d) | CHARM |
|---|---|---|---|---|---|
| T Eri | 9.3 | 2.6 | M5/M6IIIe | 252 | 0 |
| R Leo | 6.0 | −3.0 | M8IIIe | 310 | 20 |
| S Scl | 6.2 | 0.4 | M7IIIe | 363 | 0 |
| R Lep | 8.1 | 0.1 | CIIe | 427 | 1 |
| R Hya | 6.4 | −2.5 | M7IIIe | 389 | 1 |
| U Her | 8.5 | −0.3 | M7III | 406 | 4 |
| T Lep | 9.9 | 0.3 | M6evar | 368 | 0 |
| *o* Cet | 3.0 | −2.4 | M7IIIe | 332 | 27 |
| UU For | 9.8 | 1.1 | M9 | - | 0 |
| R Peg | 8.1 | 0.3 | M7e | 378 | 2 |
| U Ori | 6.7 | −0.5 | M8III | 368 | 7 |
| R Cnc | 7.1 | −0.7 | M7IIIe | 362 | 1 |
| X Hya | 8.5 | 1.0 | M7e | 301 | 0 |
| R Aqr | 7.7 | −1.2 | M7IIIpevar | 387 | 11 |

TABLE II
Summary of observations

| Name | first | last | n | bas. |
|---|---|---|---|---|
| T Eri | 03-Dec-01 | 03-Dec-01 | 5 | 102.5 |
| R Leo | 01-Apr-01 | 10-Feb-02 | 46 | 16+66 |
| S Scl | 21-Oct-01 | 14-Nov-01 | 23 | 16 |
| R Lep | 22-Oct-01 | 12-Jan-02 | 38 | 16 |
| R Hya | 01-Apr-01 | 01-Apr-01 | 3 | 16 |
| U Her | 01-Apr-01 | 01-Apr-01 | 1 | 16 |
| T Lep | 21-Oct-01 | 20-Jan-02 | 43 | 16 |
| *o* Cet | 22-Oct-01 | 20-Jan-02 | 55 | 16 |
| UU For | 24-Oct-01 | 24-Oct-01 | 1 | 16 |
| R Peg | 05-Nov-01 | 15-Nov-01 | 5 | 16 |
| U Ori | 05-Nov-01 | 22-Jan-02 | 18 | 16 |
| R Cnc | 05-Dec-01 | 08-Jan-02 | 32 | 16 |
| X Hya | 26-Jan-02 | 26-Jan-02 | 2 | 16 |
| R Aqr | 21-Oct-01 | 22-Jan-02 | 18 | 16 |

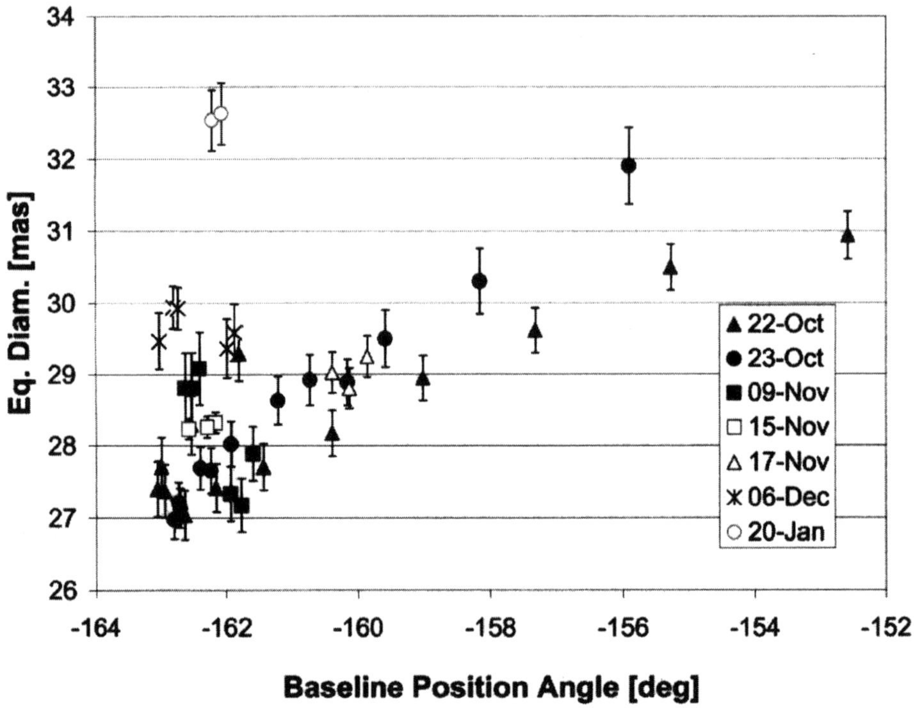


*Figure 1.* Measurements of Mira obtained using the VLTI on a 16 m baseline, coded by epoch. The visibility measurements have been converted, in a first approximation, to equivalent angular diameters.

If one investigates further the hypothesis of a possible asymmetry of the stellar disc, a rather strong convex shape is found for the stellar limb. However, this explanation cannot be supported convincingly by the data, since the span in position angle is very limited. Furthermore, it is well known from observations at other wavelengths as well as in the near-IR, that Mira is surrounded by circumstellar emission (Quirrenbach et al., 1992; Tuthill et al., 1995). In order to test this possibility, we present in Figure 3 a fit obtained with a simple-minded two-components model wihich assumes a circular disk of 30 milliarcseconds and a gaussian component about 7 times larger having 20% of the total flux. Of course, these assumptions are totally ad hoc and will need to be quantified in a more detailed study, possibly taking into account also other baselines. However, one can see that the visibility curve of such a model is much more consistent with the observed data than that of a homogenous circular disc.

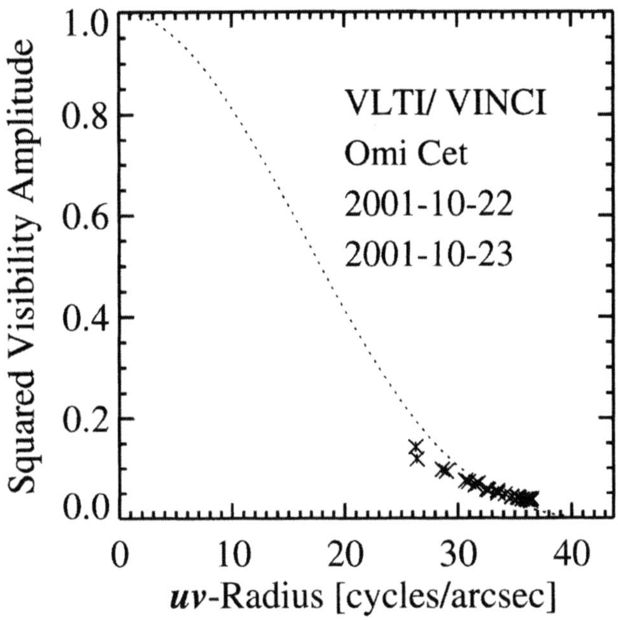


*Figure 2.* Global fit of the visibilities with a single uniform disk model, showing inconsistency of the model with the data.

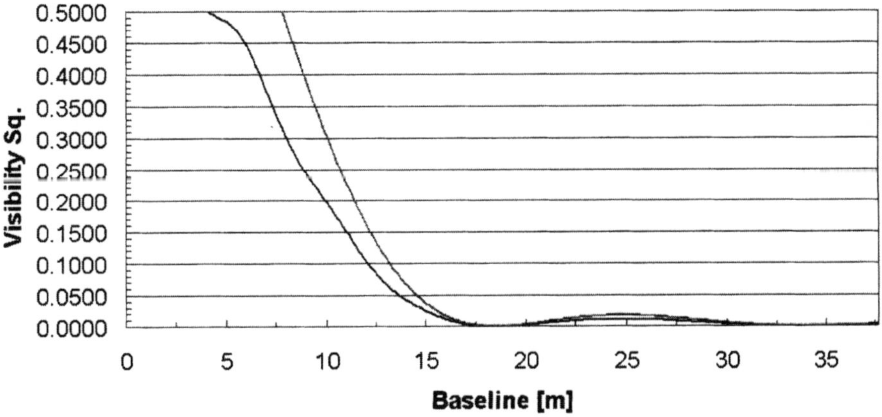


*Figure 3.* Comparison between the visibility plot of a model with a single uniform disk (top), and a model with two components (bottom). This latter is better suited to fit data such as those of Figure 2.

## Acknowledgements

We thank the whole VLTI commissioning team for their work which has made possible the observations detailed in this paper.

## References

Quirrenbach et al.: 1992, *A&A* **259**, L19.
Richichi, A. and Percheron, I.: 2002, *A&A* **386**, 492.
Tuthill et al.: 1995, *MNRAS* **277**, 1541.

# ACTIVE HOT STAR SHELLS

PHILIPPE STEE
*Observatoire de la Côte d'Azur, Site de Grasse-Roquevignon, Avenue Copernic, 06130 Grasse – France; E-mail: Philippe.Stee@obs-azur.fr*

**Abstract.** Interferometry in the visible provides milliarcsecond spatial resolution and thus has been used for studying the circumstellar environment of active hot stars. In this paper I will illustrate how the visibility modulus and phase can be used to better constrain the physics of Be disks through results from the VLA, the MkII and the GI2T interferometers. I will insist on the importance and the potential of coupling high angular resolution with high spectral resolution to the study of Be shells. Finally I will present a possible study of the circumstellar disk of Be stars using the VLTI.

**Keywords:** $\gamma$ Cas, $\phi$ Per, $\psi$ Per, $\eta$ Tau, 48 Per, $\zeta$ Tau, $\beta$ CMi, $\beta$ Lyr interferometry, visibility, angular diameter, GI2T/REGAIN, MkIII

## 1. Introduction

Be stars are hot and fast rotating stars surrounding by an extended circumstellar envelope of gas. They are an important source of UV photons that are ionizing the circumstellar and interstellar environment up to a few hundred of parsecs. Thus, they are playing an important role in the heating of the gas in the arms of galaxies as well as the formation of radiative shocks in the interstellar medium. Moreover, in a given stellar field approximately 20% of the B stars are in fact Be stars. This percentage can be much higher in some young clusters where up to 60–70 % of the B stars are showing the Be phenomenon, i.e. Balmer lines in emission and infrared excess due to free-free and free-bound emission. These stars are very bright and over luminous compared to B 'normal' stars due to the presence of their circumstellar envelope. Thus, in young clusters, with many Be stars, the luminosity function may seem to contain too massive stars, leading to an artificially top-heavy Initial Mass Function (IMF).

What causes B stars to become Be stars is not yet well understood. The proposed mechanisms include the wind compressed disc model of Bjorkman and Cassinelli (1993), the SIMECA code by Stee et al. (1994; 1995) based on the bi-stability effect proposed by Lamers et al. (1995), non-radial pulsations, magnetic activity and binarity. The optical/infrared and the ultraviolet observations of Be stars have been widely interpreted as being the evidence for the existence of two quite distinct regions in the circumstellar envelopes of these objects. The optical emission lines and the infrared excess are most probably produced in a dense rotating equatorial region, often called 'equatorial disc'. The asymmetric UV lines seem to be formed

*Astrophysics and Space Science* **286:** 225–236, 2003.

in a much more rarefied region which expands above and below the equatorial disc with velocities that may reach 1500 km/s also called the 'polar wind' (e.g. Marlborough, 1987). An important aspect of this traditional picture is that the equatorial region would be characterized by a very low radial expansion, as considered by Poeckert and Marlborough (1978) and by Waters and co-workers (1986, 1987, 1992).

Interferometry in the visible has been used successfully to study the circumstellar environment of Be stars (Thom et al., 1986; Quirrenbach et al., 1993, 1994; Stee et al., 1995; Quirrenbach et al., 1997). See also the review given by Stee (2000). It is now clear that the matter responsible for the visible emission lines and the continuum is not spherically distributed.

In the visible the star appears very bright and thus it is more difficult to 'see' the circumstellar matter. In the near-IR, the central star is very faint and the surrounding matter can be seen at very large distances. It is clear that first interferometric measurements of Be stars at near-IR wavelengths, even with a limited number of spatial frequencies will definitly estimate the envelope extension even if details of the disc structure itself will require refine measurements and models.

## 2. First Result Using a Large (u,v) Plane Coverage: the Be Star Ψ Per Observed with the VLA Radio Interferometer

These observations were done with the VLA radio interferometer by Dougherty and Taylor (1992). The synthesized beam full width at half maximum (FWHM) of the VLA interferometer was $176 \times 144$ mas. The high projected rotational velocity of $\psi$ Per (300 kms$^{-}$1) indicates that the rotation axis of the star lies closely to the plane of the sky. The detection of centimeter wavelength radio emission due to free-free emission demonstrates that the gas extends to very large radii. The major axis found at 15 Ghz is more than 34 time its diameter at visible wavelengths. Assuming that their data represents the diameter of the optically thick emitting region, they found a linear size to the radio emission of 3700 $R_{\odot}$ at a distance of 155 pc. The angular size of the source allowed them to determine a lower limit to the brightness temperature of 870 K which is also a lower limit to the kinetic temperature of the electrons. The reconstructed map is shown on Figure 1. Note that the new distance for Ψ Per determined from Hipparcos ($215 \pm 34$ pc) leads to a linear size to the radio emission larger than 5000 $R_{\odot}$.

## 3. Science with 2.5 cm Siderostats: the Mk III Interferometer

The Mk III interferometer was a Michelson interferometer with two 2.5 cm siderostats on a single north-south baseline between 3.0 and 31.5 m located at Mount Wilson (U.S.A) (see Shao et al., 1988, for a detailed description of the instrument).

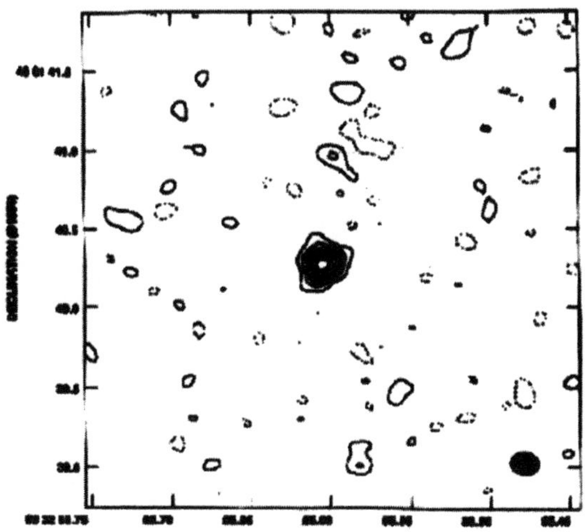

*Figure 1.* Ψ Per observed with the VLA radio interferometer. The FWHM of the beam (black dot) at the lower right corner is 176 × 144 mas (Dougherty, with permission).

They exploited the capability of this instrument to cover a large (u,v) plane by using the 'super-synthesis' effect and definitively proved that Be stars envelopes are not spherical. For instance in Quirrenbach et al. (1993), they found that $\gamma$ Cas envelope is elongated to the East from $20^o$ with respect to the North-South direction. They also found that, if the envelope is symmetric with respect to the rotation axis, the inclination angle of this axis must be at least $42^o$. These results confirm the usual picture for the origin of the linear polarization by electron scattering in a flattened envelope (see Figure 2).

They have also reconstructed an intensity map, using a maximum of entropy algorithm, of the Be star $\zeta$ Tau (Quirrenbach et al., 1994) showing for the first time how flattened the envelope can be. The disk of $\zeta$ Tau is certainly seen edge-on and they were able to derive an inclination angle of at least $73^o$ assuming an axially symmetric rotation axis (see Figure 2).

In Quirrenbach et al. (1997), they obtained constraints on the geometry of the circumstellar envelopes of seven Be stars using optical interferometric observations in the H$\alpha$ emission line from the Mk III interferometer and polarimetric data from the PBO spectropolarimeter at the University of Wisconsin. Their results are summarized in Table I (Note that the photometric diameters are not interferometric measurements but are from Barnes-Evans relation). The envelope diameters are smaller than 5 mas for the 7 Be stars. Among these 7 stars, 4 were not circularly symmetric and showed evidence of elongation. The result (3.47 mas) for $\gamma$ Cas agrees well with the diameter in the N-S direction of $3.25 \pm 0.4$ mas measured by Thom et al. (1986) with the I2T interferometer.

The position angles of the major axis of the elongation were in agreement with the disk orientation determined from polarization data and also with the result from

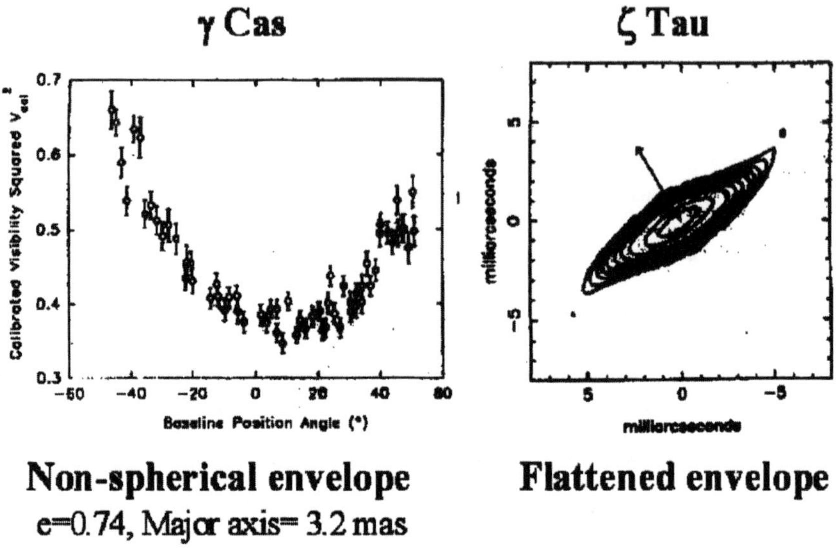


*Figure 2.* $\gamma$ Cas and $\zeta$ Tau observed with the Mk III interferometer (Quirrenbach, with permission).

TABLE I

Photospheric and disk diameters from the MkIII interferometer (From Quirrenbach et al., 1997)

| Star | V | photosphere (mas) | envelope (mas) | env/photo | W(H$\alpha$) Å |
|---|---|---|---|---|---|
| $\gamma$ Cas | 2.47 | 0.56 | 3.47 | 6.2 | 20.1 |
| $\phi$ Per | 4.07 | 0.39 | 2.67 | 6.8 | 53.8 |
| $\psi$ Per | 4.23 | 0.35 | 3.26 | 9.4 | 38.0 |
| $\eta$ Tau | 2.87 | 0.71 | 2.65 | 3.8 | 6.0 |
| 48 Per | 4.04 | 0.34 | 2.77 | 8.0 | 19.7 |
| $\zeta$ Tau | 3.00 | 0.39 | 4.53 | 11.6 | 22.0 |
| $\beta$ CMi | 2.85 | 0.74 | 2.65 | 3.6 | 5.4 |

radio observations for $\psi$ Per. It means that there is no misalignement between the larger scale envelope from the radio observation ($\sim$ 100R$_*$), the intermediate scale from H$\alpha$ emission ($\sim$ 10 R$_*$) and the inner regions where the polarization is produced.

Nevertheless, we must be cautious with the diameters of the circumstellar envelopes given in Table I since the Mk III interferometer observations were done using a narrow filter (10 Å) centred on the H$\alpha$ line. In fact, most of the 7 stars were showing H$\alpha$ line profile widths larger than 10 Å (see last column in Table II).

This is an important point since line flux is lost not only in the wings of the profile but through spatial filtering. Thus, there is potential confusion due to 'cross-talk' of flux produced from iso-velocity regions and from other regions which emit line flux at wavelengths outside the bandpass limits of the H$\alpha$ filter.

## 4. Interferometry with 'Large' Telescopes: The GI2T Interferometer

The GI2T/REGAIN interferometer is a Michelson interferometer with two 1.5 m telescopes on a single north-south baseline between 12.0 and 65.0 m located at Caussols (France) (see Mourard et al., 1994a, b and 1998) for a detailed description of the instrument). This instrument uses the dispersed fringes technique (X-$\lambda$ mode) which allows to study the circumstellar environment of Be stars with an angular resolution of $\sim$ 1 mas in the visible and a spectral resolution of $\sim$ 1 Å. Thus it is possible to study the size of an emitting region at different wavelengths but also within a given spectral line.

In a paper by Stee et al. (1998) they report the first spectrally resolved observations of $\gamma$ Cas in the He I $\lambda$6678 and H$\beta$ emission lines. They concluded that the H$\beta$ emitting region must be smaller than 8.5 stellar radius and close to 2.3 stellar radii in He I $\lambda$6678. This is, for He I $\lambda$6678, smaller than the nearby continuum extent. These results confirmed the $\gamma$ Cas basic parameters for this star obtained by Stee et al. (1995).

In the previous sections only the modulus of the visibility has been exploited. With the GI2T interferometer it is also possible to follow the photocenter $\vec{\epsilon}(\lambda)$ displacement as a function of wavelength ($\lambda$) and/or time since it is related to the visibility phase. This technique is called 'differential interferometry'. For instance if you take the continuum emission as a reference source (which is not really the best reference since the envelope can contribute up to 20% of the total continuum emission, see Stee et al. (1994)) you can nevertheless follow the photocenter displacement as a function of the doppler shift across the H$\alpha$ line profile at different epochs.

### 4.1. 'One-armed' oscillations in the Be Shell star $\zeta$ Tau and in the Be star $\gamma$ Cas

Observing the Be shell star $\zeta$ Tau on October 93, Vakili et al. (1998) found that a high density pattern occurs around a radial velocity of +130 kms$^{-1}$ and has a north-south projection position of 0.7 mas to the south of the continuum source (i.e at 3.6 R$_*$). For October 94, this density pattern has rotate prograde and occurs at a radial velocity of –70 kms$^{-1}$. Thus, for 94 It is located at 0.5 mas to the north of the contiuum source which corresponds to 2.6 R$_*$. These observations are the first detection of an axi-asymmetric envelope around a Be star that is interpreted as a direct evidence for a prograde one-armed oscillation of its equatorial disk.

After $\zeta$ Tau, $\gamma$ Cas is the second Be star for which interferometric observations provide direct evidence of a prograde one-armed oscillations of its equatorial disk (see Figure 3). This prograde precesssion agrees with the Okazaki's model (Okazaki, 1997) of the one-armed (m = 1) oscillation confined by the radiative effect. In their paper on $\gamma$ Cas Berio et al. (1999) have investigated the possibility of such oscillation to occur within the equatorial regions of a latitude dependent radiative wind model developed by Stee and Araújo (1994). In this model, the star is distorted due to its high rotation. The centrifugal force makes the effective gravity and the brightness temperature decrease from pole to equator and thus the corresponding radiative force will depend on the stellar latitude. Finally the star has a highly ionized fast wind in the polar regions and a slow wind of low ionization, with a higher density near the equator. In the inner ($\leq 10R_*$), cool, dense equatorial regions the kinematics is dominated by the Keplerian rotation, the radiative wind is due to optically thin lines where the one-armed oscillations develop. The global period of the oscillation remains about 7 years, which is consistent with the mean value found by Hirata and Hubert-Delplace (1981) from spectroscopic observations.

### 4.2. Interferometric observations of $\beta$ Lyr international multitechnique campaign in 1994

Be interacting binaries have also benefit from high angular resolution observations. In the paper by Harmanec et al. (1996), they confirm the spatial orientation of the projection of the vector of the orbital angular momentum on the sky ($\theta = 160^o$ from the North). Moreover, interferometric, spectroscopic, photometric and polarimetric data confirm that the bulk of the H$\alpha$ and HeI 6678 emission seems to originate from jets perpendicular to the orbital plane. These jets are associated with the more massive (primary) component of the binary. They may originate from a 'hot-spot' in the disk where the gas stream is flowing through the Roche lobe of the second component (B6-8 II) toward the primary. It seems that some contribution to the emission may also comes from an intermediate region between the two stars. This result agrees well with the analysis of UV line polarization.

## 5. From Be to B[e] Stars: HD 62623

The peculiar bright star HD 62623 (3 Puppis, HR 2996; $\alpha(2000) = 07^h43^m48^s.47$, $\delta(2000) = -28^\circ57'17''.4$) is classified as an A2Iabe supergiant in the *Yale Bright Catalogue* (Hoffleit and Jaschek, 1982). Allen (1973) found an IR excess due to circumstellar dust in HD 62623. This star was presented as photometrically not variable by Trams et al. (1991). Trams et al. (1990) proposed that it is a massive star with a low mass unseen companion. Plets et al. (1995) interpreted the significant IR excess observed for HD 62623 as due to the presence of circumstellar dust in a shell that extends from 35 to 5000 stellar radii where the density falls as $R^{-1.3}$.

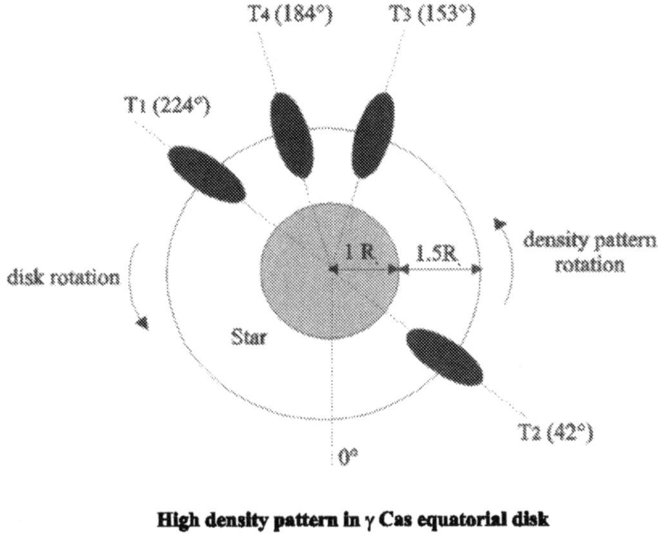


*Figure 3.* Schematic representation of $\gamma$ Cas one-armed oscillations, confined in its equatorial plane at 1.5 $R_*$. (Berio, with permission).

In this section we present high angular resolution observations of this star obtained with the Keck I telescope using aperture masking interferometry and we will discuss our results.

Aperture masking interferometry was performed by placing custom-made plates in front of the Keck I secondary mirror, reducing light from the 10 m primary mirror to a predefined set of subapertures. The introduction of such a mask creates a series of overlapping two hole interference patterns projected onto the detector array, allowing the Fourier amplitude and phase of each baseline to be recovered without the addition of 'redundancy' noise (Baldwin et al., 1986). By summing the atmospherically degraded Fourier phases around each closed triangle, we obtain the closure phase, an observable that is independent of the corrupting phase sheet above the subapertures (Jennison, 1958). In the bright source limit, nonredundant aperture masking delivers a maximal signal-to-noise ratio of the closure phases, which is critical for reconstructions of noncentrosymmetric images. For fainter sources, a doubly redundant mask geometry in the shape of an annulus was used (Haniff and Buscher, 1992), balancing the desire for stellar flux with that for limited baseline redundancy. Further engineering and performance details may be found in Tuthill et al. (1999).

HD 62623 was observed in both December of 1997 and April 1999 at Keck I using the Near-Infrared Camera (NIRC), a cryogenically cooled 256–256 pixel InSb array. The observations were performed at 1.24, 1.65, 2.26 and 3.08 $\mu$m and the obtained visibilty curves are shown in Figure 4.

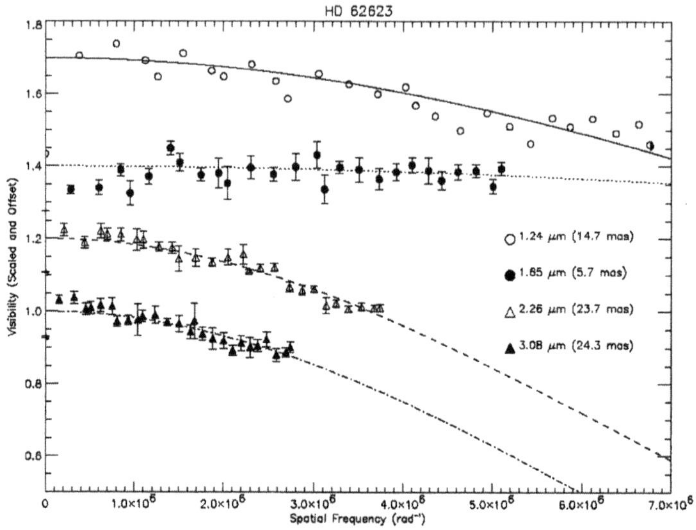


*Figure 4.* Visibility curves obtained at the Keck I telescope for the star HD 62623 and plotted as a function of baseline length by performing an azimuthal average. The graph corresponds to observations obtained in Decembre 1997 for four wavelengths. Constant offsets have been added to the variuos colors to separate them on the plot. The offset is a visibility of +0.3 for the 1.24 $\mu m$ and of +0.2 for the other wavelengths.

### 5.1. HD 62623: A STRANGE VISIBILITY BEHAVIOUR

From Figure 4 it is obvious that HD 62623 is resolved at 1.24, 2.26 and 3.08 $\mu$m wheareas at 1.65 $\mu$m it is unresolved (the corresponding visibility curve is constant with increasing spatial frequencies). Following the picture of a central hot star surrounded by an envelope of gas and at larger radii a circumstellar envelope of dust, the 1.24 $\mu$m emission may originates from both the central star and the hot circumstellar gas light that is scattered by dust grains. This implies the presence of a population of efficient scatterers at this wavelength. The resolved 2.26 and 3.08 $\mu$m curves can be interpreted as the thermal reemission of hot extended silicate dust grains at a temperature of about 1500 K. The presence of these grains is also supported by the low resolution spectrum of IRAS showing an emission 'bump' close to 10 $\mu$m mainly due to silicate dust. The most striking effect is coming from the 1.65 $\mu$m visibility curve (not very far in wavelength from the 1.24 $\mu$m observations) which seems to be unresolved, a result that was confirmed with the April 1999 observations. It means that HD 62623 is less extended at this wavelength and thus that the origin of the 1.65 $\mu$m emission is mainly photospheric. It also means that the chemical composition of the dust and its physical properties are unable to scatter the stellar light at this wavelength whereas it is efficient at 1.24 $\mu$m. One the other hand it is also possible to invoke the hot circumstellar gas emission as an extended source at 1.24 $\mu$m that is masking the central star, an effect already

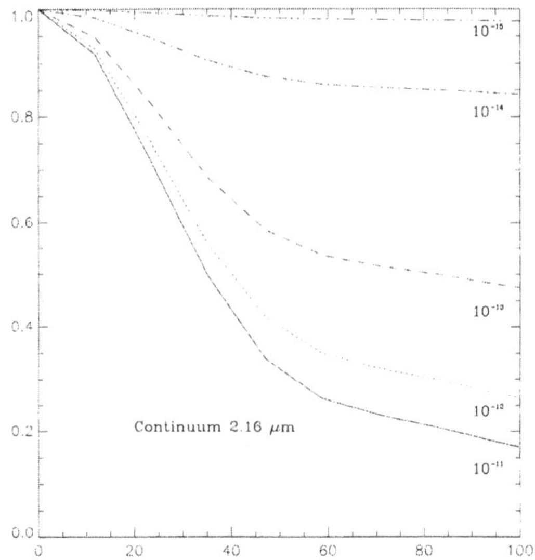


*Figure 5.* Visibility curves for a 25000 K, 10 R⊙ radius Be star seen at 200 parsecs as a function of baseline (in meters) for different wind densities at the base of the photosphere (in g cm$^{-3}$).

predicted by Stee (1998) for B[e] with mass loss larger than $10^{-5} M_{\odot}\mathrm{yr}^{-1}$. At 1.65 $\mu$m the hot circumstellar gas emission may be less opaque and thus most of the flux at this wavelength originates from the central unresolved star. Fitting the visibility curves from Figure 4 with an uniform disk model we obtain respectively diameters for the dust emission between 23.7 and 26.0 mas at 2.26 $\mu$m and between 24.3 and 30.35 mas at 3.08 $\mu$m.

## 6. Measuring Be Circumstellar Envelopes with the VLTI

In the following months, the MIDI instrument for the Very Large Telescopes Interferometer (VLTI) will provide the first IR measurements of the circumstellar disks at 10 $\mu$m whereas AMBER will perfom next year the first observations at 2 $\mu$m with angular resolution of a few mas and a spectral resolution up to 10000. Both instruments will allow a (u,v) plane coverage that will permit 2D analyses of Be disks with a detailled study of the kinematics within the disks thanks to the differential mode. Thus the question of the disk thickness will be hopefully cleared up especially for Be stars seen edge on (i.e. without sin $i$ ambiguity). A paper presented by Stee (2002) may help in the selection of the best VLTI targets since it appears that most of the Be stars, at least those in the Rinehart et al. (1999) sample, have wind densities between $10^{-13}$ and $10^{-12}$ g cm$^{-3}$ for K magnitudes within 2 and 6. In Figure 5 we have plotted visibilities for a 25000 K, 10R$_{\odot}$ Be star seen at 200 pc within the wind density range given in the previous section. It turns out that, for Be stars with a K magnitude between 2 and 6, i.e. a density range between $10^{-12}$

TABLE II

List of Be targets for the VLTI AMBER focal instrument with estimated visibilities, with a 60 meters baseline, for the two hypothesis of the wind density, i.e. $10^{-13}$ and $10^{-12}$ g cm$^{-3}$

**VLTI Be Target list**

| Name | RA (1950) | Dec (1950) | SpT | $m_K$ | d (pc) | $V_{max}$ | $V_{min}$ |
|---|---|---|---|---|---|---|---|
| HD 10144 | 01 35 51.307 | –57 29 24.92 | B3Vpe | 0.88 | 44.0 | 0.11 | 0.07 |
| HD 37795 | 05 37 50.211 | –34 05 58.71 | B7IVe | 2.84 | 82.2 | 0.22 | 0.14 |
| HD 50013 | 06 47 58.321 | –32 26 58.64 | B1.5IVne | 3.57 | 242.1 | 0.65 | 0.42 |
| HD 56139 | 07 12 46.89 | –26 41 05.0 | B2IV-Ve | 3.99 | 283.2 | 0.76 | 0.49 |
| HD 58715 | 07 24 26.357 | +08 23 29.89 | B8Ve | 3.04 | 52.2 | 0.14 | 0.09 |
| HD 63462 | 07 46 00.39 | –25 48 42.7 | B1IV:nne | 3.98 | 757.5 | 1.0 | 1.0 |
| HD 89080 | 10 12 33.047 | –69 47 21.33 | B8IIIe | 3.2 | 113.5 | 0.30 | 0.19 |
| HD 91465 | 10 30 14.487 | –61 25 39.66 | B4Vne | 3.05 | 152.4 | 0.41 | 0.26 |
| HD 105382 | 12 05 29.23 | –50 22 58.1 | B6IIIe | 4.8 | 115.2 | 0.31 | 0.20 |
| HD 105435 | 12 05 45.445 | –50 26 38.32 | B2IVne | 3.3 | 121.2 | 0.32 | 0.21 |
| HD 112091 | 12 51 39.65 | –56 53 51.0 | B5Vne | 5.6 | 110.7 | 0.29 | 0.19 |
| HD 127972 | 14 32 19.305 | –41 56 21.76 | B1.5Vne | 3.04 | 94.6 | 0.25 | 0.16 |
| HD 135734 | 15 15 02.58 | –47 41 33.4 | B8Ve | 4.3 | 89.1 | 0.24 | 0.15 |
| HD 148184 | 16 24 07.29 | –18 20 40.2 | B2Vne | 3.42 | 149.9 | 0.40 | 0.26 |
| HD 158427 | 17 27 58.353 | –49 50 19.59 | B2Vne | 3.8 | 74.3 | 0.20 | 0.13 |
| HD 212571 | 22 22 43.381 | +01 07 22.87 | B1Ve | 3.87 | 337.8 | 0.91 | 0.59 |

$V_{max}$ : Wind density $\rho_0 = 10^{-13}$ g cm$^{-3}$.
$V_{min}$: Wind density $\rho_0 = 10^{-12}$ g cm$^{-3}$.

and $10^{-13}$ g cm$^{-3}$, the disk must be very well resolved even with short baselines on the order of 40 meters. Nevertheless, the exact visibility will depend on the angle of inclination (sin $i$), the position of the envelope major axis projected into the sky plane compared to the projection of the interferometer baseline and of course, the distance of the star.

The star effective temperature is not a key parameter since the visibilities remain unchanged for $T_{eff}$ between 16000 and 40000 K (see Stee, 2002). Thus, in order to obtain an estimation of the visibility of a Be star disk seen at a distance d (in parsec), with a wind density of $10^{-13}$ g cm$^{-3}$ and a 60 meters baseline you may use the relation given by:

$$V_{max} = 2.7 \times 10^{-3} d_{pc} \quad (1)$$

whereas for the same star but with a wind density of $10^{-12}$ g cm$^{-3}$ the relation becomes:

$$V_{min} = 1.75 \times 10^{-3} d_{pc} \quad (2)$$

Using Eq. (1) and (2) it may help to chose a larger or shorter interferometric baseline in order to better constrain the disk size of a given Be star knowing its distance in parsec.

In Table II we present the list of our Be targets proposed for the AMBER VLTI focal instrument with, for each star, the estimated visibilities for the two wind density hypothesis, i.e. $10^{-13}$ and $10^{-12}$ g cm$^{-3}$. It comes clearly that except HD 63462 all our 16 targets will be resolved with a 60 meters baseline even for disks with a low density at the base of the stellar photosphere. It is also important to note that using differential interferometry we may use shorter baselines, especially for very well resolved Be stars (for instance HD 10144 or HD 58715). As already mentioned, in this case, the objective is to measure the photocenter $\vec{\epsilon}(\lambda)$ displacement as a function of wavelength ($\lambda$) within the hydrogen emission line Br $\gamma$. The relation between the phase of the visibility and the photocenter displacement remains valid as far as the source remains unresolved or 'slightly' resolved.

## 7. Conclusion

It is clear that the first VLTI observations of active hot star shells will put very strong constraints on the physics of Be and B[e] stars and will definitely clear the question about the disk size (predicted by Stee and Bittar (2001) to be very extended at near infrared wavelengths) and its thickness.

## References

Allen, D.A.: 1973, *MNRAS* **161**, 145.
Baldwin, J.E., Haniff, C.A., Mackay, C.D. and Warnier, P.J.: 1986, *Nature* **320**, 595.
Berio, Ph., Stee, Ph., Vakili, F. et al.: 1999, *A&A* **354**, 203.
Bjorman, J.E. and Cassinelli, J.P.: 1993, *ApJ* **409**, 429.
Castor, J.I., Abbott, D.C. and Klein, R.I.: 1975, *ApJ* **195**, 157.
Dachs, J. Engels, D. and Kiehling, R.: 1988, *A&A* **194**, 167.
Dougherty, S.M. and Taylor, A.R.: 1992, *Nature* **359**, 808.
Gehrz, R.D., Hackwell, J.A. and Jones T.W.: 1974, *ApJ* **191**, 675.
Haniff, C.A. and Buscher, D.F.: 1992, *J. Opt. Soc. Am.* **9**, 203.
Harmanec, P., Morand, F., Bonneau et al.: 1996, *A&A* **312**, 879.
Hirata, R. and Hubert-Delpace, A.-M.: 1981, in: C. Sterken (ed.), *G.E.V.O.N.*, Workshop on Pulsating B Stars. Nice Observatory, p 217.
Hoffleit, D. and Jaschek, C.: 1982, *The Bright Star Catalogue*, Yale University Observatory, New Haven, USA.
Hony, S., Waters, L.B.F.M., Zaal, P.A. et al.: 2000, *A&A* **355**, 187.
Jennison, R.C.: 1958, *MNRAS* **118**, 276.
Lamers, H.J.G.L.M., Snow, T.P. and Lindholm, D.M.: 1995, *ApJ* **455**, 269.
Marlborough, J.P.M.: 1987, in: Slettebak (?) and T.P. Snow (eds.), *IAU Colloquium 92*, Cambridge University Press, Cambridge.
Mourard, D., Tallon-bosc, I., Blazit, A., Bonneau, D., Merlin, G., Morand, F., Vakili, F. and Labeyrie, A.: 1994a, *A&A* **283**, 705.

Mourard, D., Tallon-bosc, I., Rigal, F., Vakili, F., Bonneau, D., Morand, F. and Stee, Ph.: 1994b, *A&A* **288**, 675.

Mourard, D., Thureau, N., Antonelli, P. et al.: 1998, *SPIE Conference on Astronomical Interferometry*, Vol **3350**, p. 517.

Okazaki, A.T.: 1997, *A&A* **318**, 548.

Plets, H., Waelkens, C. and Trams, N.R.: 1995, *A&A* **293**, 263.

Quirrenbach, A., Hummel, C.A., Buscher, D.F. et al.: 1993, *ApJ* **416**, L25.

Quirrenbach, A., Buscher, D.F., Mozurkewich, D. et al.: 1994, *A&A* **283**, L13.

Quirrenbach, A., Bjorkman, K.S., Bjorkman, J.E. et al.: 1997, *ApJ* **479**, 477.

Rinehart, S.A., Houck, J.R. and Smith, J.D.: 1999, *AJ* **118**, 2974.

Shao, M., Colavita, M.M., Hines, B.E. et al.: 1988, *A&A* **193**, 357.

Scuderi, S., Panagia, N., Stanghellini, C. et al.: 1998, *A&A* **332**, 251.

Stee, Ph. and Araújo, F.X.: 1994, *A&A* **292**, 221.

Stee, Ph., Araújo, F.X., Vakili, F., Mourard, D., Arnold, L., Bonneau, D., Morand, F. and Tallon-Bosc, I.: 1995, *A&A* **301**, 219.

Stee, Ph.: 1996, *A&A* **311**, 945.

Stee, Ph., Vakili, F., Bonneau, D. et al.: 1998, *A&A* **332**, 268.

Stee, Ph. and Bittar, J.: 2001, *A&A* **367**, 532.

Stee, Ph. and A. Domiciano de Souza Jr: 2002, in: *Science with Differential Phase*, EuroWinter School 'Les Houches', EDP Sciences, EAS Publications series, in press.

Stee, Ph.: 2000, in: M.A. Smith, H.F. Henrichs and J. Fabregat (eds.), *IAU Colloquium 175*, ASP, Conference Series Volume **214**, p. 129.

Stee, Ph.: 2002, *A&A*, submitted.

Thom, C., Granès, P. and Vakili, F.: 1986, *A&A* **165**, L13.

Trams, N.R., Waters, L.B.F.M., Lamers, H.J.G.L.M., Waelkens, C., Geballe, T.R. and Thé, P.S.: 1991, *A&AS* **87**, 361.

Trams, N.R., Lamers, H.J.G.L.M., Waters, L.B.F.M., Waelkens, C. and Groth, H.G.: 1990, in: M.O. Mennessier and A. Omont (eds.), *From Miras to Planetary Nebulae: Which Path For Stellar Evolution?*, Editions Frontières, France, p. 542.

Tuthill, P., Monnier, J.D., Haniff, C.A. and Danchi, W.C.: 1999, in preparation.

Vakili, F., Mourard, D., Stee, Ph. et al.: 1998, *A&A* **335**, 261.

Waters, L.B.F.M.: 1986, *A&A* **162**, 121.

Waters, L.B.F.M., Cot, J. and Lamers, H.G.J.L.M.: 1987, *A&A* **185**, 206.

Waters, L.B.F.M., Taylor, A.R., van den Hanvel, E.P.J., Habets, G.M.H.J. and Persi, P.: 1988, *A&A* **198**, 200.

Waters, L.B.F.M. and Marlborough, J.M.: 1992, *A&A* **253**, L25.

Yudin, R.V.: 2001, *A&A* **368**, 912.

# THE WOLF-RAYET STARS AT HIGH (SPATIAL) RESOLUTION

A. NIEDZIELSKI
*Toruń Centre for Astronomy, N. Copernicus University, Gagarina 11, 87-100 Toruń, Poland*
*E-mail: aniedzi@astri.uni.torun.pl*

**Abstract.** The Wolf-Rayet stars represent an advanced stage of evolution of the most massive stars. Their next immediate stadium will be supernova explosion. The most striking property of this very rare but exceptionally hot and bright objects is their extreme mass loss, of the order of $10^{-5}$ solar mass per year. In turn of evolution before and during the Wolf-Rayet phase such stars eject a lot of matter ($\sim$ 10 $M_\odot$) with velocity up to 3000 km/s that surrounds them in the form of gas and dust. In the case of binary Wolf-Rayet star such expanding envelope may interact with a companion (usually hot OB star) wind forming a tail extended for $\sim$ 100 AU. This spectacular phenomenon as well as some other connected with Wolf-Rayet stars that can be studied with high spatial resolution instruments (both astrometric and imaging) are reviewed.

**Keywords:** Wolf-Rayet stars, interferometry

## 1. Wolf-Rayet Stars

The Wolf-Rayet (WR) stars were discovered by C.J.E. Wolf and G. Rayet in 1867. On the contrary to the spectra of majority of stars which are dominated by narrow absorption lines, the spectra of WR stars show broad and bright emission lines. Classification of WR stars is based upon the appearance of optical emission lines of different ions of helium, carbon, nitrogen and oxygen (Beals, 1938). They are arranged into two main spectral types. The nitrogen or WN type is the one in which helium and nitrogen lines (He I-II and N III-V) dominate. In the carbon one, WC, helium, carbon and oxygen lines (CII-IV,He I-II and O III-VI) dominate. The emission lines seen in any WR star represent a wide range of ionizations and excitations therefore spectral types WN, WC are divided into subtypes defined by line ratios, yielding a classification mainly by ionization.

Chemical abundance anomalies follow the optical spectrum appearance of WR stars with hydrogen absent in WC and strongly depleted in WN. Average masses range from 12 $M_\odot$ (WC) to 22 $M_\odot$ (WN). More detailed studies show that the emission line spectra of WR stars are created in dense, expanding winds accelerated by these hot and luminous stars (T=40-100kK, $\log(L/L_\odot)$=5-6) up to over $v_\infty$=3000 km/s. The resulting mass-loss is of the order of $10^{-5}$ $M_\odot$ per year.

The Wolf-Rayet stars evolve from massive progenitors essentially in two ways, either via binary system evolution (Paczynski, 1967) or in the course of single star evolution with a strong wind (Conti, 1976; Chiosi et al., 1978). Both channels,

*Astrophysics and Space Science* **286:** 237–244, 2003.

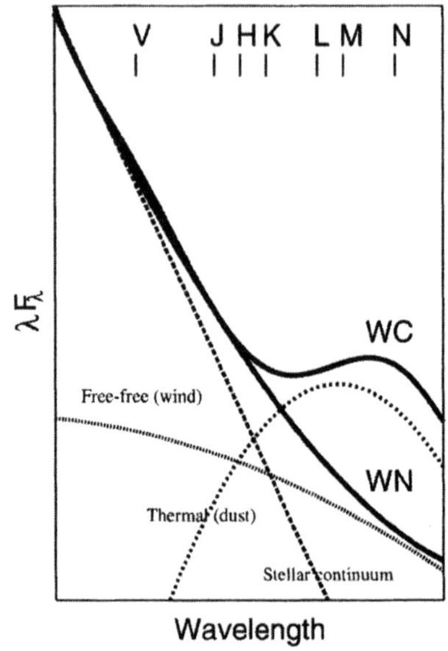


*Figure 1.* A simple sketch (not to scale) explaining the observed IR continuum excess in WR stars. Both free-free (wind) and thermal (due to dust emission) components are presented as well as the resulting WR spectra. Note the characteristic IR bump of WC stars. Approximate positions of visible/IR photometric bands are indicated at the top.

which relative efficiency is not well known, produce also large amount of circumstellar matter before and during the WR phase. The Wolf-Rayet (WR) stars are believed to be the last stable evolutionary phase of the most massive stars evolution before SN explosion.

Presently we know 234 WR stars in the Galaxy (van der Hucht, 2001a, b) 135 in LMC (Breysacher et al., 1999; Massey et al., 2000), 11 in SMC (van der Hucht, 2001b) and many other in galaxies of the Local Group.

Due to presence of stellar winds and dust shells WR stars are relatively bright in the NIR region (Figure 1). Strong stellar winds of WR stars produce extra radiation due to free-free emission (Hackwell et al., 1974). This process is most important for WN stars. In the case of many late type WC stars another mechanism, thermal radiation of the dust at temperature of T$\sim$1000K produces even stronger infrared radiation (Allen et al., 1972). In result WR stars are brighter in NIR than expected from their temperatures and many WR stars are accessible for current interferometric instruments (Figure 2).

## 2. Wolf-Rayet Stars at High Angular Resolution

At high angular resolution many phenomena connected with WR stars can be studied. The most promising subjects, organized roughly according to decreasing linear

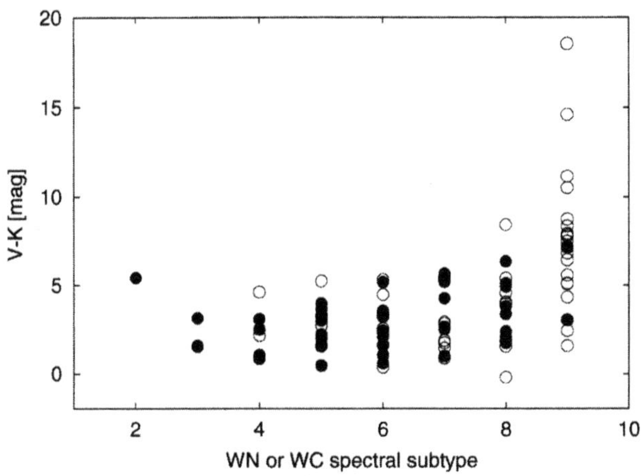


*Figure 2.* The Wolf-Rayet stars are much brighter in K band than for example in V due to extra emission in the near-infrared (filled circles represent WN stars, open ones – WC). The figure shows that an average WN star is about 5 magnitude brighter in K than in V while the dust producing late type WC stars up to 20 magnitude.

dimension are: star formation regions, ring nebulae associated with WR stars, dust shells around WCL stars, colliding-wind binaries, visual binary systems, spectroscopic binary systems, structure of WR winds, WR stars radii and their deviations from spherical symmetry. In this section we will review them in short.

## 2.1. Wolf-Rayet stars in regions of recent star formation

WR stars are good tracers of the star formation regions in galaxies. Being massive and extremely fast evolving objects they appear in regions of recent star formation, in massive-star clusters located in the center of HII regions. Typical linear sizes of nearby HII or starburst regions are of the order of 1kpc. Their central parts, containing young star clusters are of 1-5pc in diameter.

Analysis of these regions is of great importance for astrophysics since it yields information on the high mass end of the initial mass function. It also allows us to understand very early stages of evolution of very rare, extremely massive stars.

Lattanzi et al. (1994) using interferometric observations with HST FGS resolved the most powerful in the Local Group mini-starburst R136a located in the center of the 30 Dor nebula in LMC. With resolution of 15 mas they found a new component to this cluster. Another such cluster present in the Galaxy, NGC 3603 was resolved by speckle interferometry at 2.2 ESO-MPG by Hofmann et al. (1995). The central 6.3×6.3 arcsec (or 0.2×0.2pc) region was observed at 75 mas resolution.

Very high resolution imaging will resolve star formations regions in Galaxy into individual objects while additional astrometry will allow for dynamical studies and mass determination.

### 2.2. Ring nebulae

Very spectacular mass ejecta surrounding some stars, the Ring Nebulae (RN), were first observed by Johnson and Hogg (1965) (S 308 connected with WR 6, NGC 2359 related to WR 7 and NGC 6888 around WR 136). Approximately 25% of WR stars are associated with RN (Miller and Chu, 1993). These circumstellar objects, driven by stellar wind at different phases of stellar evolution and excited by luminous central objects are usually of 1–30 pc in diameter. In the Galaxy that translates to 1.2–80 arc min. Presently we know 63 RN in the Galaxy, ~60% of them associated with similar structures observed in IR – the IRAS shells (Marston, 1996).

Chu (1981) defined three general types of RN: radiatively excited HII regions, stellar ejecta and wind blown bubles. Detailed studies of RN yield information on: effective temperatures of central stars, interactions between stars and ISM, evolution with mass ejection. They also represent boundary conditions in which supernova explosions occur.

Due to large linear size the galactic RN are generally resolved with photographic or CCD observations. Very high resolution imaging will resolve RN in other galaxies of the Local Group and will allow to trace dynamical evolution of details of RN in the Galaxy.

### 2.3. Visual and spectroscopic binaries

About ~38% of the galactic WR stars are binaries (van der Hucht, 2001a) but only 13 visual binaries with separations below 1 arcsec are known. Niemela et al. (1998) resolved WR 146, WC 6, (r = 168 ±31 mas), visual binary known from radio observations, with HST WFPC2. Hartkopf et al. (1999) resolved WR 48 = Θ Mus (r = 46 ±9 mas) using speckle masking technique at 2.2 ESO-MPG.

Search for new binaries at high angular resolution is an relatively easy task in case of nearest WR stars but it may result in quite fundamental findings. Observational determination of relative efficiency of various WR stars formation channels requires large sample of well studied single and binary stars, unaffected by observational limitations. High resolution angular observations may help to create such a sample in solar vicinity.

Of 234 known galactic WR stars 24 spectroscopic binaries with WR component have RV solutions (~30% of WR binaries). Spectroscopically determined diameters of orbits range from $a \sin i = 2.9\ R_{\odot}$ to $a \sin i = 2750\ R_{\odot}$. Unfortunately for very few systems precise estimates of inclination angle - $i$ can be done. In such a situation high angular resolution observations, similar to those of Hummel et al. (1995), that in addition to RV curve can give relative positions of components in resolved binary systems will allow to obtain detailed masses of both components.

### 2.4. Colliding winds

An interesting phenomenon connected with physics of binary systems composed of hot luminous stars, the colliding winds, were noted by Cherepaschuk (1967). These areas of very high temperature plasmas (T$\sim 10^6$K) resulting from collision of hot, fast and dense WR wind with fast and hot OB star wind radiate mostly in X-rays but due to dust formation are also visible in NIR. Their location within system depends on momentum ratio of both winds, and is very close to the OB component usually. In the best studied example, WR 147 the angular distance between components is D = 640 mas (403 AU) while the colliding wind is located only 5.4 mas (36 AU) apart from the B component (Williams et al., 1997). Presently we know about 20 such systems with both WC- and WN-type WR components.

### 2.5. Dust shells

Dust shells around some of the late type WC stars were first detected in NIR by photometry (Allen et al., 1972). Follow-up studies at high angular resolution with instruments like speckle interferometry cameras resolved first of them. Allen et al. (1981) resolved the dust shell of Ve2-45 (WC 9) with 3.9 AAT speckle instrument and obtained its diameter as 0.13 $\pm$0.015 arcsec (at 2.2 $\mu$m). Later on this disc was resolved by Tuthill et al. (1999) with Keck I by use of the aperture masking technique and maximum entropy images were presented. They showed complicated, spiral-shaped, dust zone rotating with a period of 220$\pm$30$^d$. Similar study exists for WR 98a (Monnier et al., 1999) and for WR 140 (Monnier et al., 2002). Higher resolution observations are needed to resolve the binary system inside (within central 65 mas or 150 AU in the case of WR 104), to analyze dust production process in the colliding wind zone and dynamics of the tail.

### 2.6. Structure of the wind

Probably the most characteristic feature of WR stars is the strong stellar wind. One of the most interesting problems concerning WR winds is their (un)uniformity and geometry. Deviations from spherical symmetry were postulated by many authors. Interesting possibility of explanation of the observed line profile variations is rotation of the wind that induces co-rotating interaction regions, CIRs (Cranmer and Owocki, 1996; Owocki et al., 1997). Dessart and Chesneau (2002) proposed interferometric studies of CIRs in WR winds with AMBER. They suggest interferometric observations of visibility and differential phase as complementary to spectroscopic observations.

### 2.7. Radii and shapes

The WR radii are generally unknown. In the case of objects with thick winds the definition of radius is far from its intuitive meaning since we do not observe

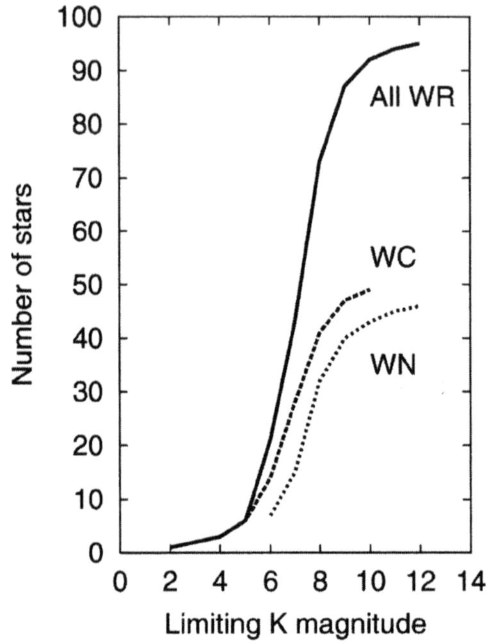


*Figure 3.* The number of Wolf-Rayet stars accessible to current and future interferometric optical/NIRinstruments is large. About 100 WR stars are brighter than K=10.

stellar surface at all. In model calculations of WR winds the radii are defined as sizes at certain optical depth, usually presented as $R_{2/3}$ or $R_\star$ (at Rosseland optical depth of 2/3 or 20, terminology according to Kiel-Potsdam models) which are inner boundaries of the wind. Unfortunately they are little or not at all connected with the hydrostatic core radii calculated in models of stellar evolution (Schaerer and Maeder, 1992), which are much smaller (Nugis & Lamers, 2002) nor with the radii determined from analysis of close binary systems that are also smaller (Moffat and Marchenko, 1996). One should note also that rotating WR stars may appear unspherical due to wind compression (Ignace et al., 1996) as shown in Langer and Hager (1999). The radii, $R_\star$, as we understand them now, range from 0.5 to 35 $R_\odot$ in the case of WN and from 1.2 $R_\odot$ to 5 $R_\odot$ in case of WC (Hamann and Koesterke 1998; Koesterke and Hamann, 1995). Given very large distances to WR stars we note that even the largest radii of the nearest WNL stars translate to very small angular sizes: $\sim$ 155 $\mu$as for WR 105 (WN9h) or $\sim$97 $\mu$as for WR 40 (WN8h).

## 3. Summary

Existing NIR photometry, that covers at present less then 50% of WR stars, shows that there are 6 WR stars brighter than $K=5^m$, and 92 brighter than $K=10^m$ – Figure 3. That makes substantial fraction of known galactic WR population accessible for

VLTI instruments. High angular resolution observations of WR stars will help to understand these objects better through exact masses determinations, radii measurements and analysis of the structure of winds, their interaction with ISM and the dust formation mechanism.

## Acknowledgements

Comments of Prof. A. Cherepaschuk, that resulted in large improvement of the manuscript of this paper are acknowledged with thanks.

## References

Allen, D.A., Swings, J.P. and Harvey, P.M.: 1972, *A&A* **20**, 333.
Beals, C.S.: 1938, *Trans. IAU* **6**, 248.
Breysacher, J., Azzopardi, M. and Testor, G.: 1999, *A&AS* **137**, 117.
Cherepaschuk, A.: 1967, *Perem. Zv.* **16**, 226.
Chiosi, C., Nasi, E. and Sreenivasan, S.R.: 1978, *A&A* **63**, 103.
Chu, Y.-H.: 1981, *ApJ* **249**, 195.
Conti, P.S.: 1976, *Mém. Soc. Roy. Liége.* **6**(9), 193.
Cranmer, S.R. and Owocki, S.P.: 1996, *ApJ* **462**, 469.
Dessart, L. and Chesneau, O.: 2002, *A&A* **395**, 209.
Hackwell, J.A., Gehrz, R.D. and Smith, J.R.: 1974, *ApJ* **192**, 383.
Hamann, W.-R. and Koesterke, L.: 1998, *A&A* **333**, 251.
Hartkopf, W.I., Mason, B.D., Gies, D.R., Ten Brummelaar, T., McAlister, H.A., Moffat, A.F.J., Shara, M.M. and Wallace, D.: 1999, *AJ* **118**, 509.
Hofmann, K.-H., Seggewiss, W. and Weigelt, G.: 1995, *A&A* **300**, 403.
van der Hucht, K.A.: 2001a, *New Astronomy Reviews* **45**, 145.
van der Hucht, K.A.: 2001b, in: D. Vanbeveren (ed.), *The Influence of Binaries on Stellar Population Studies*, Proc. Int. Conf., Kluwer, Dordrecht, p. 141.
Hummel, C.A., Armstrong, J.T., Buscher, D.F., Mozurkewich, D., Quirrenbach, A. and Vivekanand, M.: 1995, *AJ* **110**, 376.
Ignace, R., Cassinelli, J.P. and Bjorkman, J.E.: 1996, *ApJ* **459**, 671.
Johnson, H.M. and Hogg, D.E.: 1965, *ApJ* **142**, 1033.
Koesterke, L. and Hamann, W.-R.: 1995, *A&A* **299**, 503.
Langer, N. and Hager, A.: 1999, in: K.A. van der Hucht, G. Koenigsberger and P.R.J. Eenens (eds.), *Wolf-Rayet Phenomena in Massive Stars and Starburst Galaxies*, Proc. 193rd IAU symp., ASP, p. 187.
Lattanzi, M.G., Hershey, J.L., Burg, R., Taff, L.G., Holfeltz, S.T., Bucciarelli, B., Evans, I.N., Gilmozzi, R., Pringle, J. and Walborn, N.R.: 1994, *ApJ* **427**, 21.
Marston, A.P.: 1996, *AJ* **112**, 2828.
Massey, P., Waterhouse, E. and DeGioia-Eastwood, K.: 2000, *AJ* **119**, 2214.
Miller, G.J. and Chu, Y.-H.: 1993, *ApJS* **85**, 137.
Moffat, A.F.J. and Marchenko, S.V.: 1996, *A&A* **305**, 29.
Monnier, J.D., Tuthill, P.G. and Danchi, W.C.: 1999, *ApJ* **525**, 97.
Monnier, J.D., Tuthill, P.G. and Danchi, W.C.: *ApJ* **567**, 137.
Niemela, V.S., Shara, M.M., Wallace, D.J., Zurek, D.R. and Moffat, A.F.J.: 1998, *ApJ* **115**, 2047.
Nugis, T. and Lamers, H.J.G.L.M.: 2002, *A&A* **389**, 162.

Owocki, S.P., Cranmer, S.R. and Fullerton, A.W.: 1997, *ApJ* **453**, 37.
Paczyǹski, B.: 1967, *Acta Astronomica* **17**, 355.
Schaerer, D. and Maeder, A.: 1992, *A&AS* **96**, 269.
Tuthill, P.G., Monnier, J.D. and Danchi, W.C.: 1999, *Nature* **398**, 487.
Williams, P.M., Dougherty, S.M., Davis, R.J., van der Hucht, K.A., Bode, M.F. and Setia Gunawan, D.Y.A.: 1997, *MNRAS* **289**, 10.

# EXTRAGALACTIC ASTRONOMY WITH THE VLTI: A NEW WINDOW ON THE UNIVERSE

ALESSANDRO MARCONI and ROBERTO MAIOLINO
*INAF-Osservatorio Astrofisico di Arcetri, Largo E. Fermi 5, I-50125 Firenze, Italy*

ROMAIN G. PETROV
*Université de Nice Sophia Antipolis UMR 6525, Parc Valrose, 06108 Nice, France*

**Abstract.** Interferometry in the optical and near infrared has so far played a marginal role in Extragalactic Astronomy. Active Galactic Nuclei are the brightest and most compact extragalactic sources, nonetheless only a very limited number could be studied with speckle interferometry and none with long baseline interferometry. The VLTI will allow the study of moderately faint extragalactic objects with very high spatial resolution thus opening a new window on the universe. With this paper we focus on three scientific cases to show how AMBER and MIDI can be used to tackle open issues in extragalactic astronomy.

**Keywords:** interferometry, VLTI, extragalactic astronomy, active galactic nuclei, circumnuclear tori, broad line regions, massive black holes

## 1. Introduction

Optical and near infrared interferometry has so far played a marginal role in extragalactic astronomy and in particular no multiple aperture observations have ever been performed. The VLTI sensitivity will allow to explore spatial scales which have not been probed so far, thus opening a new window on extragalactic astronomy. Previous observations of extragalactic sources have been limited to speckle interferometry of a few of the brightest Active Galactic Nuclei like NGC 1068 (Meaburn et al., 1982; Chelli et al., 1987; Afanas'ev et al., 1992; Wittkowski et al., 1998; Weinberger et al., 1999), NGC 1386 (Mauder et al., 1992) and NGC 4151 (Ebstein et al., 1989; Ayers et al., 1990). Even with the use of the 8m telescopes, VLTI observations will be limited to the brightest sources. The objects with the highest flux within the diffraction limit PSF of the 8m telescopes are Active Galactic Nuclei, which, at least initially, will be the only extragalactic targets observed.

Active Galactic Nuclei are galactic nuclei showing evidence for non-stellar production of energy. The most stringent constraints come from the emitted luminosities (L$\sim 10^8 - 10^{13}$ L$_\odot$), the flat ($L_\nu \sim \nu^{-1}$) non-stellar spectra extending from radio to $\gamma$-rays, the high efficiencies of matter-energy conversion ($\simeq 0.1$), the rapid time variabilities (observed on scales as short as a few hours in optical and X-rays), the compact source sizes (directly measured in radio sources and smaller

*Astrophysics and Space Science* **286:** 245–254, 2003.

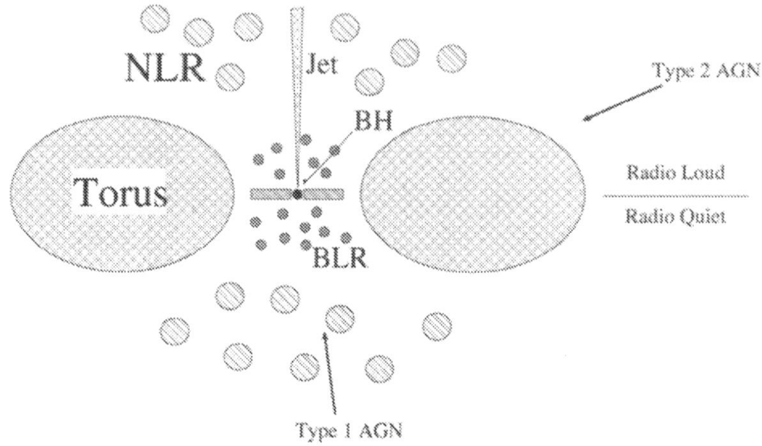


*Figure 1.* Schematic view of the unified model for AGNs. The components are not drawn in scale. The BLR (Broad Line Region) size ranges from a few light days in local Seyfert 1 galaxies to a few light years in bright quasars. Similarly the obscuring torus size ranges from $\sim 1$ to $\sim 100$ pc. Finally the NLR (Narrow Line Region) can range from $\sim 100$ pc up to several Kpc.

than a few light days) and the presence of relativistic jets (e.g. Blandford, Netzer and Woltjer, 1990). Observationally all AGNs can be roughly divided into two main classes, type 1 and type 2 (e.g. Seyfert 1 and 2 nuclei). The type 1 AGNs show broad ($FWHM > 1000$ km/s) permitted emission lines in their optical and near-IR spectra while type 2 AGNs only show narrow lines ($FWHM < 1000$ km/s). The most widely accepted model (see Figure 1) comprises a central black hole with mass in the range $10^6 - 10^{10}$ $M_\odot$ surrounded by an accretion disk that converts gravitational energy into radiation and outflows. The radiation emitted in the optical – UV – soft X-rays accounts for most of the AGN bolometric luminosity. Broad emission lines originate in small high density gas clouds ($N_e \sim 10^9$ $cm^{-3}$) orbiting around the nuclear source. Variability time scales of the broad line fluxes and other theoretical arguments suggest that the BLR size ranges from a few light days for nearby Seyfert 1 galaxies up to a few light years in the most luminous quasars. Plasma jets are emitted perpendicular to the disk. At large radii ($\sim 1 - 100$ pc), an obscuring torus of cold gas and dust surrounds the nucleus. The orientation of this torus relative to the line of sight naturally accounts for the differences between type 1 and type 2 AGNs (figure 1; see Antonucci, 1993 for a review). Up to now optical and near-IR observations have been able to probe spatial scales larger than the size of the obscuring torus. Interferometric radio observations are able to probe smaller scales, but very little information is available on many constituent of the AGN model, like the torus or the BLR. Thus, in the following, we will show how the VLTI, and in particular AMBER, can be used to tackle open issues on the obscuring torus, the BLR and the central supermassive Black Hole.

## 2. The Obscuring Torus

The nuclear near-IR luminosity of several AGNs is dominated by emission of hot dust, close to the sublimation limit, which traces the inner walls of the circumnuclear molecular torus. The inner radius of the torus is set by the minimum distance from the nuclear UV source at which dust can survive against sublimation, i.e. $R_{in} \simeq 0.2 - 4\,L_{46}^{1/2}$pc where $L_{46}$ is the optical-UV luminosity of the AGN in units of $10^{46}$ erg/s (Laor and Draine, 1993). The smallest radius is for large grains and the largest for small grains. Within the torus, the absorbing dust reprocesses the optical-UV primary radiation into the infrared, via heating of the dust grains. Several authors have modeled the infrared spectral energy distribution of AGNs and have shown that, indeed, the nuclear infrared emission can be explained with reprocessing of the nuclear radiation by dust (e.g. Pier and Krolik, 1993; Granato and Danese, 1994; Efstathiou and Rowan-Robinson, 1995). The obscuring dusty medium is expected to extend from the sublimation radius (less than a pc) up to about 100 pc (Granato et al., 1997; Maiolino and Rieke, 1995).

Until a few years ago, the only constraints on the models were the infrared spectral energy distributions and the upper limits on the size of the nuclear H and K band sources obtained from ground-based and HST/NICMOS observations (e.g. Thatte et al., 1997; Maiolino et al., 1998). Without strong constraints on the sizes and geometries of the tori, there is a degeneracy between the model parameters (like dust composition, distribution, etc.) meaning that the same spectra and size constraints can be adequately explained with very different torus models. The radiative transfer models have been recently compared with the IR SED of nearby Seyfert galaxies and with high angular resolution near- and mid-IR observations to try to constrain the physics and the geometry of the dusty torus (e.g. Alloin et al., 2000; Bock et al., 2000; Maiolino et al., 1998; Rouan et al., 1998) with the finding that the near- and mid-IR emitting dust is somewhat more extended than expected by the early models. The geometry and the physics of the dusty gas in the circumnuclear region of AGN are thus more complex than assumed so far. An additional complication is that the properties of dust in AGNs might be different with respect to the diffuse interstellar medium. More specifically, Maiolino et al. (2001a; 2001b) have shown that probably the dust in the circumnuclear region of AGNs is biased in favor of large grains, while in all models for the molecular torus a 'standard' Galactic dust mixture has been assumed. A distribution of dust biased in favor of large grains would change significantly the distribution of temperatures and also the location of the dust sublimation radius.

We can consider, for example, the case of the famous Seyfert 2 galaxy NGC 1068 whose luminosity, for an assumed distance of D=14.4 Mpc, is $L = 3.7\,10^{45}$ erg/s. The inner radius for a Galactic dust mixture is $0.5L_{46}^{0.5}$ pc, thus the expected inner radius for the NGC 1068 torus is 0.3pc corresponding to $\simeq$ 4 mas. Then, in the model by Granato et al. (1997), the torus should have a K-band size of $\simeq$ 14 mas FWHM (18 mas at 10$\mu$m). These sizes can be directly measured using AMBER

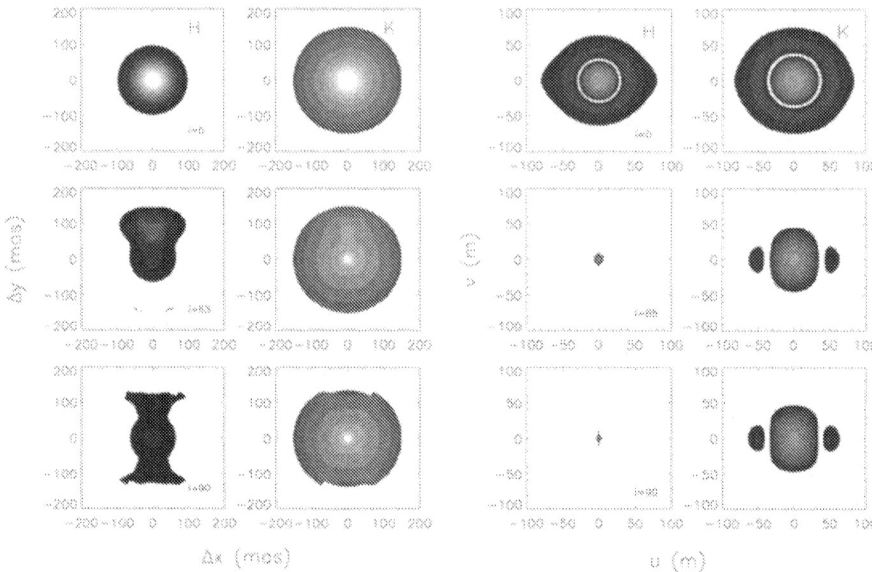


*Figure 2.* Left: Expected surface brightness distributions of H and K emission for the NGC 1068 torus model by Granato et al. (1997) seen at different inclinations with the line of sight. There are eight greyscale levels starting from -15 with steps of 0.5 in logarithm of erg $s^{-1}$ $cm^{-2}$ $hz^{-1}$ $sterad^{-1}$. Smaller intensity values are darker. (Model data courtesy of G.L. Granato). Right: Expected visibilities, $|V|$. Notations are as in the previous panel. There are eight greyscale levels with logarithmic steps of 0.2dex starting from 0.04.

and MIDI at the VLTI. As a zero order approximation, the observed visibilities at different points in the uv plane, in different spectral bands (H, K and 10$\mu$m) can be fit with a ring model to estimate the inner and outer radii. The measure of the inner radius will directly tell if the dust composition is as in our galaxy or biased toward larger or smaller grains. However one can perform more detailed studies and try to distinguish between different models. A library of models with different parameters (inner radius, thickness, inclination, density, optical depth, mixture of dust grains) can be computed. Then for each model one can compute the expected visibilities and phases for the points in the uv plane covered by the observations. By using AMBER with 3 telescopes and the low spectral resolution mode ($\mathcal{R} = 35$), a single observation will provide 3 visibilities and one closure phase in any of the $\sim$ 30 spectral channels present in J, H and K. It is important to remember that the large wavelength coverage (1.0 to 2.4 $\mu$m) substantially increases the instantaneous coverage of the uv plane which can be further increased taking advantage of Earth rotation. MIDI will similarly provide one visibility point per single observation. Thus, the best model should be able to reproduce at the same time the AMBER and MIDI observations together with the ir spectra. Figure 2 shows the expected spatial distributions and visibilities of H and K emission for the NGC 1068 torus model by

Granato et al. (1997) seen at different inclinations with respect to the line of sight. For a given geometry, different inclinations produce different morphologies which can be distinguished with VLTI.

In conclusion, the AMBER and MIDI data will remove the degeneracy among the model fitting parameters, at variance with previous studies which could compare models only with the IR spectral energy distribution.

## 3. The Broad Line Region of Active Galactic Nuclei

The BLR (Broad Line Region) is the region where the broad (FWHM> 1000 km/s) permitted lines observed in the spectra of type 1 AGNs originate (e.g. Blandford, Netzer and Woltjer, 1990). Given the small distance from the central super massive black hole (SMBH) the width of the broad lines is likely to originate from the gravitational motion of gas clouds around the SMBH. So far, the size of the BLR could not be directly measured and the only available information is provided by the so-called reverberation mapping technique (e.g. Peterson, 1994). The BLR size is estimated as $c\Delta\tau$ where $\Delta\tau$ is the time lag between the continuum and line variation. Clearly this represents an average size weighted over the BLR geometry and physical conditions. In principle the BLR geometry and kinematics can be derived from the detailed behavior of the light curves, but the inversion is not unique mainly because of the correspondence between the 1-dimensional nature of light curves and the 3-dimensional nature of the BLR. Also the non optimal time sampling of the observations strongly reduces the constraints which can be derived. Suggested spatial distribution of the BLR clouds are spherical, disk-like or conical. Dynamically, the BLR might be dominated by gravitational motions (either a virialized system with chaotic motions or a disk in keplerian rotation), or might be part of a radiation pressure driven outflow, or of an inflow. Obtaining a direct measure of the BLR size and constraining its morphology and kinematics is fundamental in order to understand its origins and relationship with AGN activity, and to measure the mass of the SMBH and verify reverberation mapping techniques. BLR sizes determined with the reverberation mapping have been found to correlate with the quasar luminosity. A recent estimate by Kaspi et al. (2000) gives:

$$R_{BLR} \sim 33 \left( \frac{\lambda L_\lambda(5100\text{Å})}{10^{44}\, erg\, s^{-1}} \right)^{0.7} \text{light days} \qquad (1)$$

where $L_\lambda$ is the rest-frame monochromatic luminosity at 5100 Å. Assuming a standard cosmology ($H_0 = 70$ km/s/Mpc, $\Omega_M = 0.3$ and $\Omega_\Lambda = 0.7$), $R_{BLR}$ can be translated into an apparent size in the plane of the sky obtaining for instance $R_{BLR} = 0.03$ mas for a $L = 10^{12}$ $L_\odot$ quasar at $z = 0.1$ (we have assumed $\lambda L_\lambda$(5100Å) = L/15). For $L = 10^{13}$ $L_\odot$ and $z = 0.5$, $R_{BLR} = 0.04$ mas. These numbers imply that BLRs are not resolvable with the VLTI. However, when a

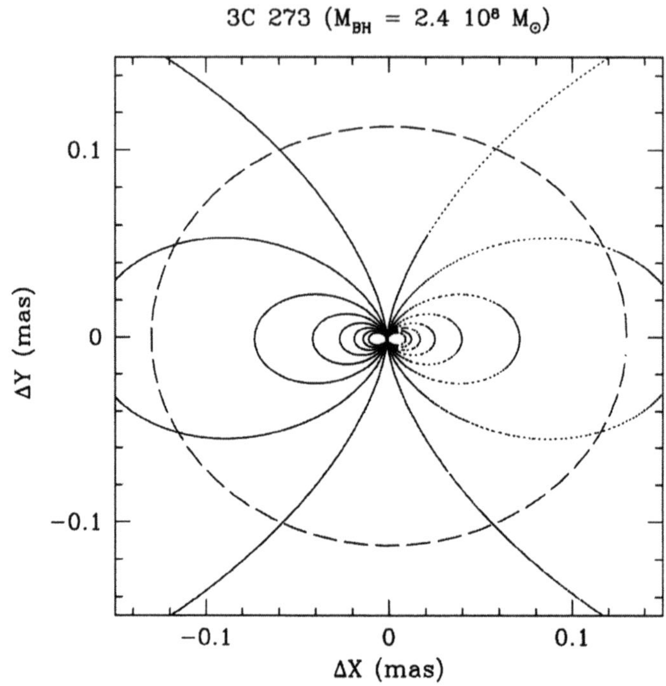


*Figure 3.* Isovelocity contours for the BLR of 3C273. 'Red' velocities are represented with a solid line, 'blue' velocities with a dotted line. Contours go from 400 to 3200 km/s with steps of 400 km/s. The dashed line is the BLR radius measured with reverberation mapping. The assumed disk inclination with respect to the line of sight is 30 deg.

source is not resolved by an interferometer, the differential phase measures the displacement of the photocenter with wavelength along the baseline direction. This is valid even when this displacement is much smaller than the standard resolution limit of the interferometer. The position of the photocenter with wavelength is a powerful constraint for the geometry and kinematics of the BLR and can be obtained by taking advantage of the AMBER spectral resolution.

For example, one can consider the case of the famous quasar 3C273. Its BLR radius measured with reverberation mapping by Kaspi et al. (2000), $R_{BLR} = 387 \pm 55$ lt-days corresponds to 0.13 mas (at the distance of 3C 273, 1 mas corresponds to 2.6 pc) and the inferred BH mass is $\sim 2.4 \times 10^8$ $M_\odot$. Figure 3 shows the expected isovelocity contours with steps of 400 km/s assuming that the BLR is in a Keplerian disk and that it is inclined of 30 deg with respect to the line of sight. Using AMBER with Medium spectral resolution ($\mathcal{R} = 750$), one can measure the differential phase of the broad Pa$\alpha$ in the K band and obtain information in 400 km/s velocity bins. At 2.2$\mu$m, with a projected baseline of 80 m, a photocenter displacement of 0.13 mas will result in a phase shift on the line of $\phi \sim 0.15$ rad. Since in each spectral channel we expect $\sim 300$ Pa$\alpha$ photons/sec, the expected accuracy which can be reached in 1 h of observation is $\sigma_\phi \sim 0.002$ rad ($\sim 2\,\mu$arcsec in K with a 80 m baseline), thus allowing to measure the photocenter displacement. Since a given measure with two telescopes yields mainly one component of the photocenter displacement, in order to obtain a full map of the photocenter displacement one should measure the differential phase using observations from two perpendicular baselines.

The differential phase in each velocity bin will provide the position of the photocenter of the part of the disk which is delimited by two contiguous contours in Figure 3. In this case, the photocenters will be aligned along the line of nodes of the disk and, combining positional and velocity information, one can directly measure the mass of the central BH. Conversely, if the BLR is composed by a spherically symmetric and isotropic ensemble of clouds the photocenters will all coincide with the position of the continuum photocenter and no differential phase will be measured. Other locations of the photocenters will result in more complicated geometries but, in any case, the amplitude of the photocenter displacement is directly related to the angular size of the BLR. If we are able to establish a relationship between the differential interferometry angular radius and the reverberation mapping linear radius, we can get a direct measurement of the AGN distance. Since, in the best case, this kind of measurement can be made on a magnitude $V \sim 20$ quasar when there is a nearby reference star, this could eventually lead to a new and independent technique to measure cosmological distances.

## 4. The Central Black Hole

It is now clear that a large fraction of local galaxies, if not all, contain a MBH (e.g. Kormendy and Gebhardt, 2001). Moreover, there is a suggestion in the data that the hole mass is proportional to the mass (or luminosity) of the host spheroid (e.g. Merritt and Ferrarese, 2001 and references therein). Recently Ferrarese and Merritt (2000) and Gebhardt et al. (2000) have shown that a tighter correlation holds between the BH mass and the velocity dispersion of the bulge. Clearly, any correlation of black hole and spheroid properties would have important implications for theories of galaxy formation in general, and bulge formation in particular. These results have prompted a frantic attempt to understand the physical origin of the observed correlations between BH mass and galaxy properties (e.g. Haehnelt and Kauffmann, 2000; Cavaliere and Vittorini, 2002) and to put constraints on galaxy and BH formation models.

However, there is a caveat which must be taken into account: the so called massive black holes in galactic nuclei are in reality massive dark objects because, at the moment, there is no conclusive proof that they are indeed black holes. The massive dark objects are point-like at the resolution of the observations, which, at most, are of the order of $0.1''$. Combining this with typical galaxy distances of 10–20 Mpc, one derives that the density of the massive dark objects are not

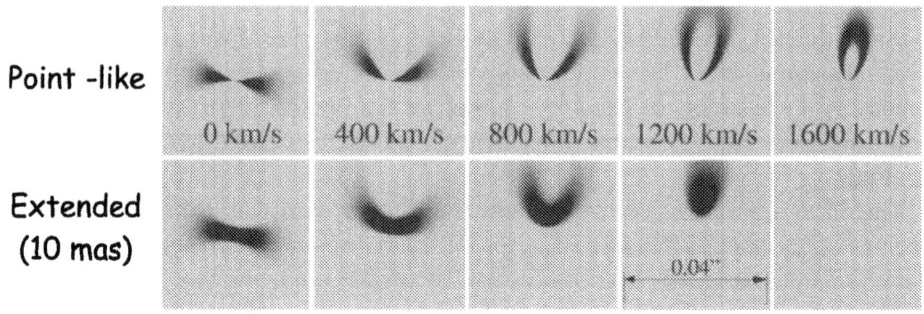


*Figure 4.* Intensity distribution from a rotating disk in various velocity ranges (toplabels; km/s) expected from a point like mass (top) and a mass extended over 10mas (bottom). The case considered is that of Centaurus A where a dark mass of $10^8\ M_\odot$ is detected from the rotation of a gaseous disk inclined by 35° with the line of sight. The images are $0.04'' \times 0.04''$. North is up and East is left.

so high that the only plausible alternative is a BH: indeed in most cases they are even smaller than the densities observed in core collapsed globular cluster ($\rho \sim 10^7\ M_\odot\ pc^{-3}$) which are the most dense star clusters known. Therefore the possibility that these dark masses could be clusters of 'dark' objects (e.g. neutron stars, planets etc.) cannot be excluded. Two notable exceptions are our galactic center ($\rho \sim 10^{12.6}\ M_\odot\ pc^{-3}$, Genzel et al., 2000) and NGC 4258 ($\rho \sim 10^{9.5}\ M_\odot\ pc^{-3}$, Miyoshi et al., 1995) where, given the high densities of the dark matter distribution, the only possibility is that of a massive black hole. In the first case the key point is the closeness (just 8 kpc) while in the second case the high spatial resolution is obtained with VLBI observations of water masers. Unfortunately our galactic center is unique and galaxies with water masers observable with VLBI are just a handful and none of them provides as good an evidence as NGC 4258 (Braatz et al., 1997).

Using the spectral resolution of AMBER it is possible to constrain the size of the dark mass distributions which are found at the galaxy centers to verify if they are indeed black holes. We can consider for example the case of Centaurus A (Marconi et al., 2001). In Centaurus A the dark mass is $\sim 10^8 M_\odot$ and if it can be constrained within 10mas (i.e. 0.17 pc) the density will be $2 \times 10^{10} M_\odot pc^{-3}$. Centaurus A will thus provide a better case for a BH than NGC 4258. Centaurus A has an ionized gas disk rotating around the putative BH. This disk is a strong emitter in H and Fe lines and can be observed with AMBER in Pa$\beta$ and [FeII] 1.26$\mu$m with medium spectral resolution ($\mathcal{R} = 750$, 400 km/s spectral channels). On the spatial scales accessible to AMBER, this disk will rotate with velocities of $\sim 0 - 3000$km/s and the regions at different velocities will have different morphologies if the mass is point like or extended (see Figure 4). The extreme case is for the velocity bin at 1600 km/s where no disk emission is expected if the mass is extended over 10 mas. The visibility points and differential phases obtained at several locations in the uv plane for each spectral channel can be compared with models of disks rotating around a point mass (BH) or around an extended mass distribution (several different distributions can be tested). By fitting the model to the data one can then determine the free parameters (total mass, PA of disk line of nodes, disk inclination) as well as constrain the region were the dynamical mass must be distributed.

## 5. Conclusions

The VLTI will allow the study of moderately faint extragalactic objects with very high spatial spatial resolution thus opening a new window on the universe. To show the potentiality of high spatial resolution interferometric observations we have presented three scientific cases which will be tackled with AMBER and MIDI. The combined use of AMBER and MIDI will make possible to spatially resolve the obscuring torus of Active Galactic Nuclei and to solve the degeneracies in torus models. The unique spectroscopic capabilities of AMBER will be able to spatially resolve the BLR and to constrain its geometry and kinematics. By relating the BLR sizes measured with reverberation mapping and those measured with differential interferometry one might directly measure cosmological distances. Finally, it will also be possible to constrain the size of the dark matter distributions found in galaxy nuclei to test if they are really supermassive black holes.

## References

Afanas'ev, V.L., Balega, I.I., Orlov, V.G. and Vasiuk, V.A.: 1992, *A&A* **266**, 15–20.
Alloin, D., Pantin, E., Lagage, P.O. and Granato, G.L.: 2000, *A&A* **363**, 926–932.
Antonucci, R.: 1993, *ARA&A* **31**, 473–521.
Ayers, G.R., Benson, J., Carels, K. et al.: 1990, *ApJ* **360**, 471–473.
Blandford, R.D., Netzer, H. and Woltjer, L.: 1990, Saas-Fee Advanced Course 20, Lecture Notes 1990, XII, 280 pp. 97, Springer-Verlag.
Bock, J.J., Neugebauer, G., Matthews, K. et al.: 2000, *AJ* **120**, 2904–2919.
Braatz, J.A., Wilson, A.S. and Henkel, C.: 1997, *ApJS* **110**, 321.
Cavaliere, A. and Vittorini, V.: 2002, *ApJ* **570**, 114–118.
Chelli, A., Cruz-Gonzalez, I., Carrasco, L. and Perrier, C.: 1987, *A&A* **177**, 51–62.
Ebstein, S.M., Carleton, N.P. and Papaliolios, C.: 1989, *ApJ* **336**, 103–111.
Efstathiou, A. and Rowan-Robinson, M.: 1995, *MNRAS* **273**, 649–661.
Ferrarese, L. and Merritt, D.: 2000, *ApJ* **539**, L9–L12.
Gebhardt, K. et al.: 2000, *ApJ* **539**, L13–L16.
Genzel, R., Pichon, C., Eckart, A. et al.: 2000, *MNRAS* **317**, 348–374.
Granato, G.L. and Danese, L.: 1994, *MNRAS* **268**, 235.
Granato, G.L., Danese, L. and Franceschini, A.: 1997, *ApJ* **486**, 147.
Haehnelt, M.G. and Kauffmann, G.: 2000, *MNRAS* **318**, L35–L38.
Kaspi, S., Smith, P.S., Netzer, H. et al.: 2000, *ApJ* **533**, 631–649.
Kormendy, J. and Gebhardt, K.: 2001, *Proc. of 20th Texas Symposium*, p. 363.
Laor, A. and Draine, B.T.: 1993, *ApJ* **402**, 441–468.
Maiolino, R. and Rieke, G.H.: 1995, *ApJ* **454**, 95.
Maiolino, R., Krabbe, A., Thatte, N. and Genzel, R.: 1998, *ApJ* **493**, 650.

Maiolino, R., Marconi, A., Salvati, M. et al.: 2001a, *A&A* **365**, 28–36.
Maiolino, R., Marconi, A. and Oliva, E.: 2001b, *A&A* **365**, 37–48.
Marconi, A., Capetti, A., Axon, D.J. et al.: 2001, *ApJ* **549**, 915–937.
Mauder, W., Appenzeller, I., Hofmann, K.-H. et al.: 1992, *A&A* **264**, L9–L12.
Meaburn, J., Morgan, B.L., Vine, H. et al.: 1982, *Nature* **296**, 331.
Merritt, D. and Ferrarese, L.: 2001, *ASP Conf. Ser.* **249**, 335.
Miyoshi, M., Moran, J., Herrnstein, J. et al.: 1995, *Nature* **373**, 127.
Peterson, B.M.: 1994, *ASP Conf. Ser.* **69**, 1.
Pier, E.A. and Krolik, J.H.: 1993, *ApJ* **418**, 673.
Rouan, D., Rigaut, F., Alloin, D. et al.: 1998, *A&A* **339**, 687–692.
Thatte, N., Quirrenbach, A., Genzel, R. et al.: 1997, *ApJ* **490**, 238.
Weinberger, A.J., Neugebauer, G. and Matthews, K.: 1999, *AJ* **117**, 2748–2756.
Wittkowski, M., Balega, Y., Beckert, T. et al.: 1998, *A&A* **329**, L45–L48.

# POTENTIAL OF THE VLTI FOR LINKING STELLAR FRAMES TO ICRF

GÉRARD DAIGNE, PATRICK CHARLOT and CHRISTINE DUCOURANT
*Observatoire Aquitain des Sciences de l'Univers – CNRS, 2 rue de l'Observatoire, F-33270 Floirac, France*

JEAN-FRANÇOIS LESTRADE
*Observatoire de Paris – CNRS, 77 Ave. Denfert-rochereau, F-75014 Paris, France*

**Abstract.** The question of positioning the optical counterparts of the ICRF quasars is outlined in the perspective of future space astrometry missions, which ultimately will bring a new realization of the ICRS in the optical range. Ground-based interferometry with a dual-field observing mode (PRIMA/VLTI), together with the missions DIVA and FAME, will have a key role in building an extragalactic reference frame in the optical/near-IR range with about the same accuracy as that of the present (VLBI) primary frame.

**Keywords:** astrometry, reference frames, interferometry, atmospheric effects

## 1. Introduction

The International Celestial Reference System (ICRS) adopted by IAU resolutions in the early 90's is a non-rotating system centered at the barycenter of the Solar System. It is realized by the most accurate positions of compact radio-loud quasars, measured through VLBI observing in the centimeter wavelength range (Arias et al., 1995). The primary realization of the ICRS is the International Celestial Reference Frame (ICRF), based on the positions of 212 defining objects, within a larger list of 608 potential objects (Ma et al., 1998). Position accuracy of the defining objects is 0.2–0.5 milliarcsec (mas), with the ICRF axes defined to an accuracy of 0.02 mas.

A celestial reference frame is a combination of compact objects and a measuring technique. In the optical range, the measuring technique has been the Hipparcos space astrometry mission, and the objects, more than one hundred thousand stars. Extragalactic sources could not be observed with Hipparcos due to its limited sensitivity, so that the position and proper motion of stars determined from the sphere Great Circle solution had to be linked to ICRF. This was achieved mainly through VLBI observations of stars (Lestrade et al., 1999). The actual frame for astrometric measurements in the optical range is now provided by most of the Hipparcos Catalogue objects. Its global uncertainty is 0.25 mas/yr in rotation, and 0.6 mas in the position of the origin at epoch 1991.25 (Kovalevsky et al., 1997).

Quasars and blazars have a huge potential for accurate realization of the ICRS in the optical range, with a source size for their time variable emission expected

*Astrophysics and Space Science* **286:** 255–260, 2003.

to be typically smaller than 0.1 pc, or 20 $\mu$as if the object is seen at a distance of 1000 Mpc.

Though 3 additional kinematic parameters (proper motion and parallaxe) are needed, stellar objects can materialize reference frames in the optical and near-IR ranges very accurately. This is due to their brightness, their sky density and their compactness. Indeed, the apparent diameter of a not too cold star with $V \geq 12$ is smaller than 100 $\mu$as.

In Section 2, we recall the present situation in positioning the optical counterpart of ICRF quasars. The impact of future space astrometry missions in building reference frames is outlined in Section 3. Then we show how the VLTI and all-sky survey missions like DIVA or FAME could contribute to materialization of the ICRF in the near-IR range, at such a precision level that the radio and optical emission sources could be closely related.

## 2. Optical Counterpart of the ICRF

To position the optical counterpart of the ICRF quasars, a two-step procedure was needed. An extended catalogue of star positions in the surroundings of ICRF quasars was first obtained with wide-field astrographs, then smaller fields were observed with larger telescopes and CCD imaging. Optical positions of quasars in the Hipparcos frame were thus obtained with an uncertainty of about 50 mas (Zacharias et al., 1999). Further analysis of these results has been performed by da Silva Neto et al. (2002). The ICRF quasars were separated into two sets according to the size of their radio structure on milliarcseconds scales, as derived from VLBI maps by Fey and Charlot (1997). Using the arc-length method for comparing different frame realizations (Arias et al., 1988), a statistical difference was found and explained as a non-coincidence between optical and radio positions for the set of extended quasars, the position departure being about 10 mas.

With a star density of about 700 objects per square degree, the USNO CCD Astrograph Catalog (UCAC) (Zacharias et al., 1999) will provide a significant improvement in the optical position accuracy of faint objects. This should allow us to tackle the relative positioning of optical and radio emission in ICRF quasars, and study in more detail the possible departure mentioned above.

Optical and radio source positions may differ if the emission is not dominated by the core of a central engine, powered by the accretion flow of a supermassive black hole. In the optical/near-IR, additional emission may come from the host galaxy, or from components with different spectra in the jet. Furthermore, the optical/near IR core emission may be extincted or beamed by a dust torus surrounding the central engine, and be observed with a displacement. Before high spatial resolution of these objects can be performed with long baseline interferometry, positioning the near-IR photocenter with an uncertainty smaller than 10 mas is important to elucidate the departure noted by da Silva Neto et al. (2002). Indeed, 10 mas is the typical size for the extended structures observed in radio maps.

## 3. Impact of Future Space Astrometry Missions

Future space astrometry missions like DIVA, FAME, SIM and GAIA will boost the quality of reference frames in optical astronomy: denser frames, fainter objects and more precise positions (and kinematic parameters). A global performance comparison of the 4 missions can be found in Mignard and Roeser (2002). Their expected contributions to the realization of reference frames are summarized below:

- DIVA should observe several tens of ICRF quasars (about 50?), with a precision in position measurements ranging from 0.3 mas to a few mas. Its determination of the proper motion of Hipparcos stars will greatly benefit from their first epoch of observation, about 15 years earlier. With the DIVA sphere solution aligned to ICRF, proper motion of more than $10^5$ stars will be measured with an accuracy better than 0.1 mas/yr. This will give a stringent control on the rotation rate of the DIVA sphere solution in the ICRS.
- FAME should observe more ICRF quasars than DIVA, with improved precision in position measurements. The same argument applies for constraining the zero in proper motion determination, with the observation of Hipparcos stars.
- With GAIA, thousands of QSOs will be observed, with much improved position accuracy (e.g. 40 $\mu$as for a magnitude G = 18), so that realization of the ICRS in the optical range could possibly be based on the GAIA measured QSOs positions (Mignard, 2002). Such a reference frame in the optical range would be much denser than the present ICRF, with a factor of about 10 in position accuracy improvement.
- SIM will bear a steering instrument, with a limited number of targets to be observed, and a sensitivity dependent on the exposure duration. SIM will build its own stellar frame, an astrometric grid made of about 3000 stars with magnitude V = 12–13. The SIM grid has to be placed into the ICRS with the observation of extragalactic objects; some of these objects have to be members of ICRF for the alignement of the two frames. The SIM project then addresses the question of selecting reference frame tie objects among ICRF quasars, and the question of relating the optical and radio emission sources, either galactic or extragalactic (Fey et al., 2001).

The DIVA or FAME sphere solution will have to be placed into the ICRS through the observation of extragalactic objects, and presumably of ICRF quasars in order to minimize any bias between optical and radio reference frames. A few ICRF quasars should be observed with DIVA at its nominal precision in position of 0.3 mas for R $\leq$ 12–13. More generally, any measuring technique of the angular distance between stars and quasars will be useful if highly accurate, that is able to take into account source structures. Ground-based interferometry is favoured, either in the radio range with ALMA, or in the near-IR range with the Keck Interferometer and the VLTI.

## 4. Potential of the VLTI in the DIVA/FAME Perspective

Due to their magnitude, quasars imaging with VLTI will most often require the PRIMA dual-field facility for tracking the interferometer delay on a bright star. The genuine observing mode of quasars in ground-based optical interferometry is thus based on simultaneous observation of stars in their surrounding. As a result, precise measurements of the angular distance between quasars and stars are expected.

The feasibility of linking frames through ground-based interferometry with phase-referencing will depend on the magnitude distribution of ICRF quasars in the near-IR range, and on the distribution of separation angles between quasars and bright enough stars. In a preliminary analysis based on the 2MASS* Second Incremental Data Release (covering half of the sky), we have searched for detected ICRF quasars in the K band, and studied the K band magnitude distribution of stellar objects in their vicinity.

With 153 detected quasars among 330 available fields, the detection rate is about 46%, which is larger than the detection rate (36%) obtained from the Veron-Cetty & Veron catalog (Barkhouse and Hall, 2001). These authors already noticed that radio-loud quasars were more likely to be detected in the 2MASS survey than radio-quiet quasars. The magnitude distributions of these ICRF quasars, in the K band and in the R band, are shown in Figure 1a. The magnitude distributions of stars in the fields surrounding these quasars are shown in Figure 1b with different field radii, from $20''$ to $1'$. The magnitude range is limited to K = 13, supposed to be the magnitude limit for fringe tracking with UTs (Derie et al., 2003).

Considering only differential optical path variations between the two observed directions, we have estimated the angular range for phase-referencing in a short-exposure mode, in terms of the faint object magnitude (Daigne and Lestrade, 2003). Our simulation results are based on an average atmospheric turbulence model at Paranal (PARSCA campaign). Differential optical path fluctuations could be followed from one sample to the next, and then averaged with separation angles exceeding $1'$. The faint object magnitude has to be $K \leq 17.5$ with UTs, and $K \leq 14.5$ with ATs. With the additional condition of a bright enough star for main delay tracking ($K \leq 13/11$ with UTs/ATs), dual-field interferometry with two UTs is much favoured for accurately positioning the near-IR counterpart of ICRF quasars.

* This publication makes use of data products from the the Two Micron All Sky Survey, which is a joint project of the University of Massachusetts and the Infrared Processing and Analysis Center/California Institute of Technology, funded by the National Aeronautics and Space Administration and the National Science Foundation.

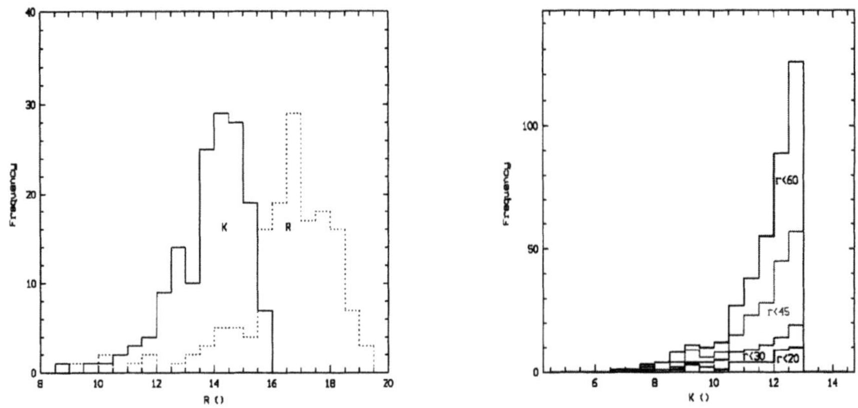


*Figure 1.* ICRF quasars and their surroundings in the 2MASS Second Incremental Data Release. (a) Magnitude distribution of the 153 detected quasars in the $K_S$ band from 2MASS (full line), and in the *R* band from USNO-A2.0 (dotted line). (b) $K_S$ magnitude distribution of stars in circular fields around these quasars, for radii 20", 30", 45" and 60".

A detailed comparison of positions and structures of ICRF quasars emission in the near-IR and radio ranges will be possible by combining the observations of VLTI/PRIMA and of a future all-sky survey mission, bringing dense stellar frames at the accuracy level of the primary (VLBI) reference frame.

## References

Arias, E.F., Lestrade, J.F. and Feissel, M.: 1988, Comparison of VLBI celestial reference frames, *Astron. Astrophys.* **199**, 357–363.

Arias, E.F., Charlot, P., Feissel, M. and Lestrade, J.-F.: 1995, The extragalactic reference system of the International Earth Rotation Service, ICRS, *Astron. Astrophys.* **303**, 604–608.

Barkhouse, W.A. and Hall, P.B.: 2001, Quasars in the 2MASS second incremental data release, *Astron. J.* **121**, 2843–2850.

Daigne, G. and Lestrade, J.-F.: 2003, Integration of the atmospheric fluctuations in a dual-field optical interferometer: a short exposure regime, *Astron. Astrophys.* **406**, 1167–1174.

Derie, F., Delplancke, F., Glindemann, A., Lévêque, S., Menardi, S., Paresce, F. and Wirenstrand, K.: 2003, PRIMA – Phase-referenced imaging and microarcsecond astrometry with the VLTI, this issue.

Fey, A.L. and Charlot, P.: 1997, VLBA observations of radio reference frame sources. II. Astrometric suitability based on observed structure, *Astrophys. J. Sup.* **111**, 95–142.

Fey, A.L., Boboltz, D.A., Gaume, R.A., Eubanks, T.M. and Johnston, K.J.: 2001, Extragalactic radio source selection for use in directly linking optical astrometric observations to the radio reference frame, *Astron. J.* **121**, 1741–1751.

Kovalevsky, J., Lindegren, L., Perryman, M.A.C. et al.: 1997, The HIPPARCOS catalogue as a realisation of the extragalactic reference system, *Astron. Astrophys.* **323**, 620–633.

Lestrade, J.-F., Preston, R.A., Jones, D.L., Phillips, R.B., Rogers, A.E.E., Titus, M.A., Rioja, M.J. and D.C. Gabuzda: 1999, High-precision VLBI astrometry of radio-emitting stars, *Astron. Astrophys.* **344**, 1014–1026.

Ma, C., Arias, E.F., Eubanks, T.M., Fey, A.L., Gontier, A.-M., Jacobs, C.S., Sovers, O.J., Archinal, B.A. and Charlot, P.: 1998, The international celestial reference frame as realized by very long baseline interferometry, *Astron. J.* **116**, 516–546.

Mignard, F. and Roeser, S.: 2002, Space Astrometry missions, in: O. Bienaymé and C. Turon (eds.), *GAIA: A European Space Project*, EAS Pub. Series, **2**, pp. 69–90.

Mignard, F.: 2002, Observations of QSOs and reference Frame with GAIA, in: O. Bienaymé and C. Turon (eds.,) *GAIA: A European Space Project*, EAS Pub. Series, **2**, pp. 327–339.

da Silva Neto, D.N., Andrei, A.H., Vieira Martins, R. and Assafin, M.: 2002, A pattern of noncoincidence between radio and optical positions of international celestial reference frame sources, *Astron. J.* **124**, 612–618.

Zacharias, N., Zacharias, M.I., Hall, D.M., Johnston, K.J., de Vegt, C. and Winter, L.: 1999, Accurate optical positions of extragalactic radio reference frame sources, *Astron. J.* **118**, 2511–2525.

Zacharias, N., Urban, S.E., Zacharias, M.I. et al.: 2000, The first USNO CCD astrograph catalog, *Astron. J.* **120**, 2131–2147.

# DISTANCES ON COSMOLOGICAL SCALES WITH VLTI

MARGARITA KAROVSKA, MARTIN ELVIS and MASSIMO MARENGO
*Harvard-Smithsonian Center for Astrophysics*

**Abstract.** We present here a new method using interferometric measurements of quasars, that allows the determination of direct geometrical distances on cosmic scales. Quasar Broad Emission Line Regions sizes provide a 'meter rule' with which to measure the metric of the Universe. This method is less dependent of model assumptions, and even of variations in the fundamental constants (other than c). We discuss the spectral and spatial requirements on the VLTI observations needed to carry out these measurements.

## 1. Introduction

Determining the 3-D geometry of the Universe involves measurement of the large 'cosmic' scale distances of high redshift sources such as distant supernovae and quasars. Cosmic distance scale determination methods include relative and absolute distance estimators. Absolute distance estimators include various applications of the: Baade-Wesselink method (eg. Sasselov and Karovska, 1994; Marengo et al., this workshop), and distances estimators using time delays (eg. SN1987A; Panagia et al., 2001). Relative distance estimating methods include period-luminosity relation for Cepheid star variability (used by the *Hubble* Key Project; Freedman et al., 2001), and supernovae of type SN1a as 'standard candles'. Most strikingly, recent distance measurements using the brightness of supernovae of type SN1a at $z{\sim}1.5$ as 'standard candles' suggest that that the universe has a non-zero cosmological constant, $\Lambda$, so that the expansion of the Universe is currently accelerating (Perlmutter et al., 1999).

Relative distance estimators often involve assumptions and correlations, and have inevitable model dependencies. Absolute methods on the other hand have the advantage of having lesser dependence on physical models and provide an independent way to determine the distance scale.

We describe here an absolute method for estimating distances to quasars at different $z$ using a time delay measurements of linear sizes of quasars Broad Emission Line Regions (BELRs, Peterson, 1997) and measurements of their angular sizes using large baseline interferometers such as the VLTI. This method could allow determining direct geometrical distances to radio-quiet quasars and therefore measuring the cosmological constant, $\Lambda$, with minimal assumptions.

*Astrophysics and Space Science* **286:** 261–266, 2003.

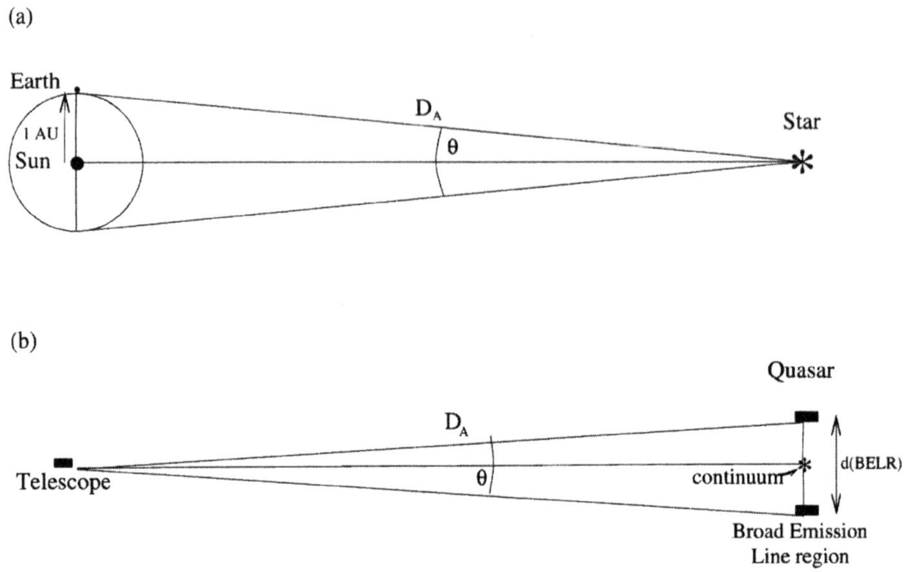


*Figure 1.* Measuring geometric distances to astronomical objects: (a) parallax to a star; (b) broad emission line region distance to a quasar.

## 2. Quasar Parallax

We propose to use a new geometric method to determine distances to quasars that is equivalent to the parallax method. However, in this case instead of using the Earth's orbit as a known 'standard length' (Figure 1a), we use the size of the quasar BELR (Figure 1b) as measured using reverberation mapping and interferometry (Elvis and Karovska, 2002).

The size of the BELR is known from light travel time measurements. BELRs emit a series of strong lines from permitted transitions, indicating Doppler velocities of 5000–10,000 km $s^{-1}$ – defining characteristic of quasars. The Broad Emission Lines (BELs) respond to changes in the continuum source in the center of the quasar by changing their intensity with a time-lag of a few days or more (for higher luminosity and redshift). The time-lag is induced by the light travel time from the continuum source in the center of the quasar to the gas clouds in the BELRs that produce the broad emission lines (Peterson, 1993, 2001). By studying the response of the BELs to continuum changes it is possible to map out the structure of the region emitting the BELs. This method is called 'reverberation mapping'.

Reverberation mapping exists for a few tens of active galaxies and quasars. The size of the broad emission line region depends on the line measured. Observations indicate that a single timescale dominate for each emission line (Nezter and

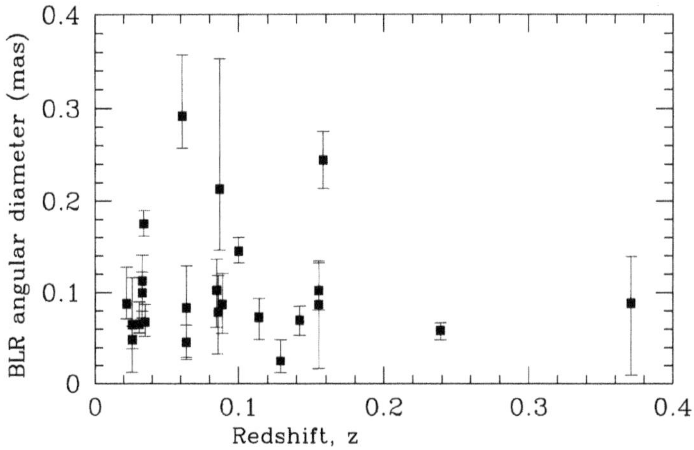


*Figure 2.* Angular diameters for the Hα and Hβ BELRs of nearby active galaxies, assuming $H_0$=65 km $s^{-1}Mpc^{-1}$ (Peterson et al., 1998; Kaspi et al., 2000). (The values of Ω and Λ are unimportant at these low redshifts.

Peterson, 1997). For low redshift quasars the size of the BELRs is ~10 light days. The timescale of BEL variations increases with redshift as $(1+z)^{2.5}$, since not only does the BELR size increase, but also the intrinsic variation timescale is dilated by $(1+z)$. Typical reverberation time for quasars at z = 3 requires observing programs spanning of 3–5 years.

The 'angular diameter distance' derived using the reverberation mapping and the interferometric measurements can be mapped as a function of redshift. This relation will describe the basic characteristic of the space-time metric of the Universe, and could eventually allow a derivation of Λ (Elvis and Karovska, 2002). This method of derivating Λ is much less dependent on physical models and of changes in the fundamental constants (other than $c$, the speed of light) because it uses a 'standard length' approach rather than a 'standard candle'.

## 3. Interferometry of Quasar BELRs

Measurements of the angular sizes of BELRs are not easy even for the nearby quasars and AGNs. For example, at z<0.2 the BELRs have measured lags that imply angular sizes ~0.1 mas (Figure 2.). This angular size is about 10 times smaller then the baselines of the existing ground based interferometers operating in optical and near-IR wavelengths (e.g. NPOI, IOTA). In addition, their sensitivity and the limiting magnitude are far bellow what is required for measuring the BELRs.

To measure accurately the sub-milliarcsecond scale BELR angular diameters therefore requires very long-baseline interferometers with large aperture telescopes, and with a capability of observing in selected quasar emission lines. In addition, the baseline of an interferometer that can resolve the BELR increases with $z$ for a fixed physical size, because the observed wavelength of any emission line increases as $(1 + z)$.

Reverberation mapping suggests that shorter wavelength emission lines (in the UV or X-ray bands) may be more favorable for BELR size measurements. However, the space-based interferometers operating at sub-optical wavelength range are currently only in planning stages, and the ground-based interferometry is at present the only mean to carry out these measurements.

Several long-baseline ground-based optical and near-IR interferometers that have a potential of measuring the sizes of the BELRs in nearby quasars and AGNs are currently being built or commissioned. These interferometers, including the VLTI, are employing large aperture telescopes (e.g. VLTI 1.8 m to 8-m) on baselines of several hundred meters and have the sensitivity required to observe BELRs in selected spectral lines used in the reverberation measurements.

Since lag-time and angular size measurements must be carried out using the same emission lines, we are currently exploring optical and near-IR lines that may be suitable for reverberation mapping studies as well as for interferometry. Our initial studies indicate that in the optical H$\beta$, H$\alpha$ lines could be appropriate for these measurements. The next generation VLTI instruments operating in the optical would be able to resolve the nearby AGN BELRs and study their geometry. This is very important, because the geometry of quasar BELRs is not well known. Possibilities range from a simple orderly wind (Elvis, 2000) to a maximally chaotic distribution of clouds (Baldwin et al., 1995).

In the near-IR the VLTI could measure the BELRs sizes of nearby (very low redshift) unobscured AGNs (eg. with AMBER; Petrov et al., 2000). Possible useful lines include HeII 10830Å, H Pa-$\alpha$ (18751Å), and H Pa-$\beta$ (12818Å). For the nearest AGNs (eg. Cen A) the expected BELRs sizes are significantly larger and therefore one could carry out imaging and study their geometry (Paresce, 2001). For more distant AGNs and quasars one can then use this information to estimate the size of their BELRs. Our simulations show that using high-signal to noise visibility measurements (less then 1%) and a model brightness distribution, one can significantly exceed the limiting resolution of the VLTI and estimate diameters much smaller then 1–2 mas. By assuming a model for the geometry of the BELRs the VLTI observations could therefore determine the distances to the nearby AGNs and quasars, and improve the $H_0$ accuracy to $\sim$5%.

However, there are a number of issues regarding the feasibility of the interferometric measurements that need to be addressed. For example, there is an inherent problem in relating reverberation lag times as currently measured to the angular sizes that an interferometer would measure. Even perfectly sampled data will produce a broadened cross-correlation function (CCF), and it is not obvious which delay time, e.g. the CCF peak, or the CCF centroid, corresponds to the angular size from an image. The reverberation mapping measures a 'responsivity weighted' size of the light delay surface, while interferometry measures an 'emission weighted' size.

The amplitude of variation of the emission lines is much less than that of the continuum (eg. factor 2 changes in the UV continuum produce only 20% changes in the emission lines). By monitoring a number of quasars for continuum changes, one could identify the onset of a significant continuum increase in a particular quasar, which will then trigger 'target of opportunity' VLTI measurements. The VLTI could observe the target quasar in both low and high continuum states, with appropriate lags included, so that the difference map will be a measure only of the 'responsive fraction' of the line, i.e. that part which responds to continuum changes. Evidently, the best approach would be to continuously monitor quasars with the VLTI and derive the lag times and sizes from the same data.

## 4. Conclusions

Quasar BELRs provide a meter rule for determining the cosmic distance scale independent of changes in the fundamental constants (other than $c$, the speed of light). We describe a new absolute distance estimating method using reverberation time-lag measurements of the linear BELR sizes, combined with the VLTI interferometric angular size measurements, to determine the distances to quasars. VLTI can measure the sizes of BELRs and explore their geometry at low redshifts. This will provide input to BELR models and allow measurement of $H_0$. Long-baseline interferometers with a resolution of 0.01 mas are needed to measure the size of the BELR in z = 2 quasars, and therefore the 3-D geometry of the Universe (and $\Lambda$) which appear plausible with near future extrapolations of current technology.

## Acknowledgements

This work was supported in part by NASA contract NAS8-39073 (Chandra X-ray Center).

## References

Baldwin, J.A., Ferland, G., Korista, K. and Verner, D.: 1995, *ApJL* **455**, L119.
Elvis, M.: 2000, *ApJ* **545**, 63.
Elvis, M. and Karovska, M.: 2002, *ApJ Letters*, in press.
Freedman, W. et al.: 2001, *ApJ* **553**, 47.
Kaspi, S., Smith, P.S., Netzer, H., Maoz, D., Jannuzi, B.T. and Giveon, U.: 2000, *ApJ* **533**, 631.
Netzer, H. and Peterson, B.M.: 1997, in: D. Maoz, A. Sternberg and E.M. Leibowitz (eds.), *Astronomical Time Series,* Kluwer, Dordrecht, p. 85.
Panagia, N., Gilmozzi, R., Macchetto, F., Adorf, H.-M. and Kirshner, R.P.: 2001, *ApJ* **380**, L23.

Paresce, F.: 2001, *Science with VLTI*, (http://www.eso.org/projects/vlti/).
Perlmutter, S. et al.: 1999, *ApJ* **517**, 565.
Peterson, B.M.: 1993, *PASP* **105**, 247.
Peterson, B.M.: 1997, *An Introduction to Active Galactic Nuclei*, CUP, Cambridge.
Peterson, B.M., Wanders, I., Betram, R., Hunley, J.F., Pogge, R.W. and Wagner, R.M.: 1998, *ApJ* **501**, 82.
Peterson, B.M.: 2001, in: *The Starburst-AGN Connection 2001*, World Scientific, Singapore, astro-ph/0109495
Petrov, R. et al.: 2000, *Proc. SPIE* Vol. **4006**, p. 68, in P.J. Lena and A. Quirrenbach (eds.), *Interferometry in Optical Astronomy*, http://buz.obs-nice.fr/amber/)
Sasselov, D. and Karovska, M.: 1994, *ApJ* **432**, 367.

# THE VLTI FUTURE

# LONG RANGE SCIENCE PERSPECTIVES FOR THE VLTI

A. ECKART, T. BERTRAM, N. MOUAWAD, T. VIEHMANN, C. STRAUBMEIER and J. ZUTHER
*University of Cologne, I. Physikalisches Institut, Zülpicher Str. 77, 50937 Köln, Germany*
*E-mail: eckart@ph1.uni-koeln.de*

**Abstract.** VLTI interferometry will allow imaging of galactic and extragalactic sources with milliarcsecond angular resolution. For moderately bright sources the spectral resolution will be of the order of 10000. These capabilities will allow detailed studies of solar system objects, stars, protoplanetary systems and the detection of hot extra-solar planets. The observations of galactic nuclei will allow unprecedented measurements of physical parameters in these systems. VLTI will be a prime instrument to study the immediate environment of the massive black hole at the center of the Milky Way. With the exception of a few 'self-referencing' sources the observations of extragalactic nuclei will benefit from an extended capability for simultaneous measurements of nearby reference sources for fringe tracking. With beam combination instruments like AMBER, MIDI, PRIMA, and GENIE the VLTI will reach full maturity at a time when other interferometric instruments at different wavelengths will be fully operational. Most important are ALMA (in the mm- and sub-mm-domain), LOFAR and SKA (in the radio meter to centimeter domain) and of course VLB-networks in the radio, and other – at that time – well developed interferometers in the optical. A major scientific potential of future scientific VLTI programs will lie in an efficient combination of these high angular resolution capabilities.

**Keywords:** VLTI, AMBER, MIDI, PRIMA, GENIE, KI, LBT, proto planetary disks, extra-solar planets, disks of galaxies, galactic nuclei, Galactic Center

**Abbreviations:** VLTI – Very Large Telescope Interferometry; AMBER – The VLTI Astronomical Multiple BEam Recombiner; MIDI – The VLTI MID-Infrared instrument; PRIMA – The VLTI instrument for Phase-Referenced Imaging and Microarcsecond Astrometry.; Darwin-GENIE – The ESA/ESO Ground based European Nulling Interferometer Experiment; KI – Keck Interferometry; LBT – Large Binocular Telescope; VLBI – Very Long Baseline Interferometry; ALMA – Atacama Large Millimeter Array; SKA – Square Kilometer Array; LOFAR – Low Frequency array

## 1. Instrumentation

The VLTI will allow sensitive interferometric observations with a very large collecting area and a suite of high-throughput beam combiner instruments. The fully developed VLTI site combined with other interferometric facilities will allow multi-frequency investigations of astrophysical sources in the angular resolution regime of 1 to few 100 mas. This will allow for comprehensive studies of scientific key objects like proto planetary disks and nuclei of distant galaxies.

In the NIR domain the **VLTI** will allow for baselines of the order of up to 200 m and an angular resolution of the order of 1 milliarcsecond (mas). A combined

*Astrophysics and Space Science* **286:** 269–276, 2003.

observation using the four 8.2 m diameter unit-telescopes (UT) as well as three 1.8 m diameter auxiliary telescopes (AT) will provide a total collecting area of approximately 220m$^2$ (e.g. Glindemann, 2002; Eckart et al., 1997). In total the site can host up to 8 telescopes with 8 delay lines. In addition to VLTI several high angular resolution, high sensitivity interferometric facilities will go online. Currently **AMBER** allows for three beam interferometry in the NIR (1–2.5$\mu$m) with expected limiting magnitudes of K = 14 and K = 20 with a bright reference source. Interferometric spectroscopic observations with R = 10000 will be possible. In the MIR **MIDI** is operational with an angular resolution of 20 mas on a 100 m baseline covering the the N-band (8–13$\mu$m, expandable to Q-band 17–26$\mu$m). Here limiting magnitudes are N = 4 and 5.8 (with reference source). For both instruments the field of view (FOV) is of the order of $\pm$1".

The effective FOV will be significantly increase with **PRIMA** within the next few years.The objective of PRIMA is to enable simultaneous interferometric observations of two objects – each with a maximum size of 2 arcsec – that are separated by up to 1 arcmin, without requiring a large continuous field of view. PRIMA will enhance the VLTI performance in several ways. The limiting magnitude will be shifted to K = 15–20, allowing imaging of faint objects with high angular resolution ($<$10 milli arcsec). In addition high precision astrometry ($<$10 $\mu$as) and therefore an improved planet detection will be possible.

The **KI** will operate over a maximum baseline of ~85 m with a resolution of the order of 6 mas in the NIR (e.g. Colavita, 2001). Its collecting area will be dominated by the two 10m class telescopes and will amount to about 160 m$^2$. Similar to the VLTI a two channel capability will allow measurements of a bright reference source for fringe tracking over angles of $\approx$2 arcminutes.

The **LBT** consists of two 8.4m diameter mirrors in one single mount over a fixed baseline of 14.4 m. The 110 m$^2$ collecting area samples virtually all baselines ranging from 0 to 22.8 m and the resolving power at 1.25$\mu$m wavelength will be ~9 mas. The NIR beam combination will be carried out by LINC (e.g. Straubmeier et al., 2002; Herbst et al., 2000). The LBT as a Fizeau interferometer will have an exceptionally large imaging FOV of the order of at least 1 arcminute. Using MCAO techniques (multi conjugate adaptive optics) this will be extended to at least 2 arcminutes.

The **CHARA** (McAlister 2001) and **OHANA** (Perrin et al., 2002) projects will have similar but complementary capabilities compared to the VLTI.

VLTI will also reach full maturity at a time when other interferometric instruments at different wavelengths will be fully operational (see Figure 1).

Most important is **ALMA** in the mm- and sub-mm-domain. At an altitude of 5000 m located in the Atacama desert an interferometer of 64 antennas with 12 m diameter each will allow for a typical resolution of 100 mas at 300 GHz in a 10" FOV. The frequency range will be at least 100 (possibly 30) to 1000 GHz.

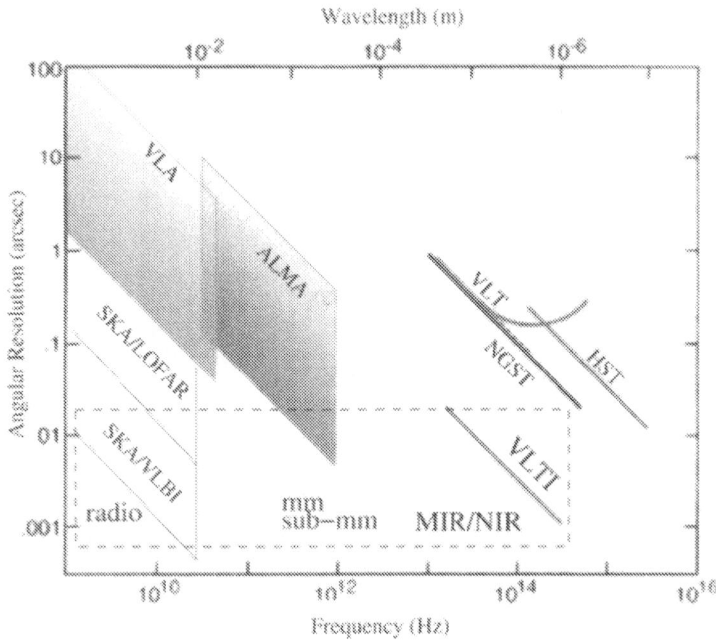


*Figure 1.* Frequency coverage and resolving power of existing of near future major interferometric facilities. A major science driver for future VLTI programs will lie in an efficient combination of these instrumental capabilities.

**LOFAR**, located in the Netherlands and neighboring countries, will operate over a frequency range of 0.15 to 1.4 GHz allowing for an angular resolution of 100 mas at 1.4 GHz and probably a simultaneous multi beam operation.

The **SKA** (at a not yet determined location) will operate over a frequency range of 0.15 to ~20 GHz allowing for an angular resolution of 10 mas at 20 GHz and a FOV of the order of 1 degree at 1.4 GHz.

In addition there will of course be **VLB**-networks in the radio, and other – at that time – well developed **optical interferometers**.

A major scientific potential of future scientific VLTI programs will lie in an efficient combination of these high angular resolution capabilities.

## 2. Proto Planetary Disks

An outstanding scientific question in modern astrophysics is the frequency at which solar systems form and what the nature of extra-solar planets is. Recently Vannier et al. (2002) have highlighted the possibilities of differential interferometry and astrometry using AMBER and MIDI for the detection of hot giant extra-solar planets around nearby stars. Detecting planets from the ground requires a high dynamic range and a low contrast between the central star and the planet. This can

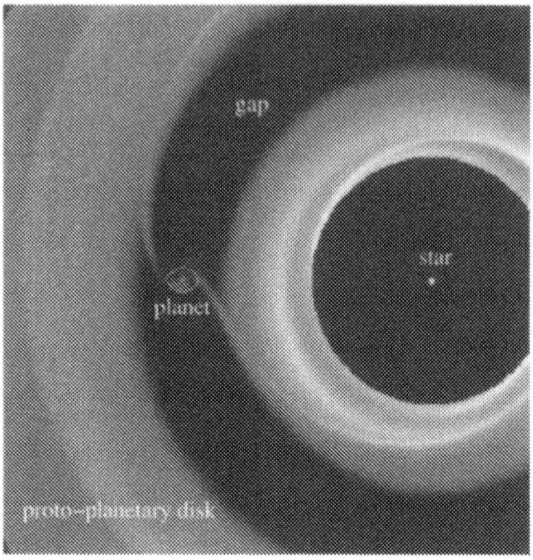


*Figure 2.* Schematic view of a gap in a proto planetary disk that is caused via the interaction of a forming Jupiter like planet. The central, bright portion of the disk close to the stars has been suppressed.

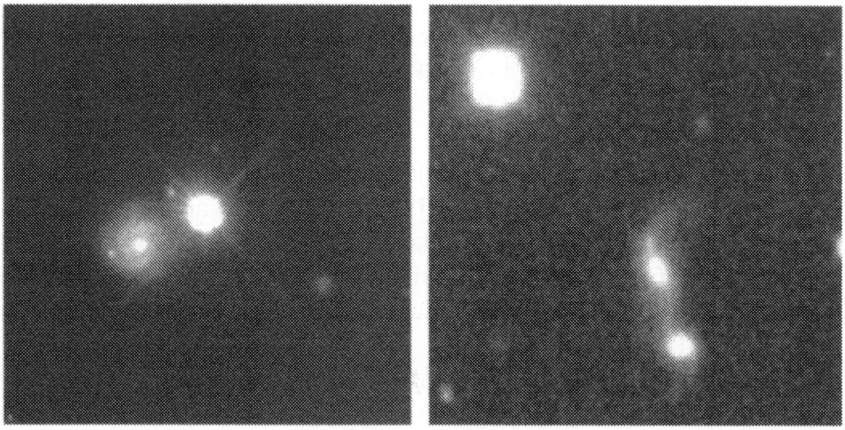

*Figure 3.* Two examples out of our catalog of galaxy guide star pairs. The guide stars are brighter than 14. magnitude in the visible. On the left the spiral galaxy contains several bright regions – possibly HII regions – areas of recent star formation. On the right an example for a strongly interacting galaxy pair is given.

best be achieved in the NIR/MIR wavelength regime. There the detection of at least young, hot giant or even terrestrial planets during their luminous final accretion can be detected (e.g. Stern, 1994). In foreseeable future the planned ESA/ESO ground-based nulling interferometric technology demonstrator, Darwin-GENIE using the VLTI, will also greatly contribute to the detection of hot planets and the dusty circum-stellar environments of nearby stellar systems (e.g. Absil et al., 2002).

Malbet and Bertout (1995) presented model visibilities and synthetic images of disks around T Tauri stars. The authors conclude that thermal emission from circumstellar disk around T Tauri stars is detectable with interferometric techniques at wavelengths longer than 2.2$\mu$m. Similarly, the work by Millan-Gabet, Schloerb and Traub (2001) points at future possibilities in the field of disks around Herbig Ae/Bs stars. Simulations of the interaction between giant planets and the gas disk out of which they formed have shown that the planet may open a gap and cause spiral density waves in the disk (see Figure 2 and e.g. Kley, 1999; and Kley, D'Angelo and Henning, 2001). The gap may extend up to several astronomical units in width and may in fact be used for the indirect detection of planets in such disks.

Recently Wolf et al. (2002) investigated the possibility of finding evidence for planets in circumstellar disks by infrared and submillimeter interferometry. They present three-dimensional simulations of a circumstellar disk around a solar-type star with an embedded planet of 1 Jupiter mass. The corresponding intensity maps the authors obtained provide the basis for the simulation of VLTI MIDI and ALMA measurements. The work shows that studies of proto planetary disks will greatly benefit from combined interferometric results obtained in different wavelength regimes. Measurements in several wavelengths regimes (i.e. NIR/MIR and sub-mm) will be necessary in order to derive physical parameters of the disk, the properties of the gap and hence the properties of the planet responsible for it.

## 3. Disks and Nuclei of Galaxies

A major corner stone for the future of ground-based astronomy is imaging and spectroscopy of the disks and nuclei of external galaxies at the diffraction limit of large telescopes using recent implementations of adaptive optics. While already with AMBER and MIDI several dozens of bright, nearby extragalactic objects (self-referencing for AO and fringe tracking) are within reach (among them NGC 1068, the brightest closest Sey 2 nucleus; see Alloin et al., 2001 and references therein) the statistical basis for extragalactic work will be greatly enlarged by objects with reference stars within their iso-planatic or iso-kinetic patch. Here the VLTI has an extraordinary advantage by providing a very large collecting area to allow interferometric measurements in the field of NIR/MIR extragalactic astrophysics.

A wavefront correction using a bright natural guide stars (NGS) close to the target galaxy will provide the best Strehl ratios that can be obtained. With the high resolving power provided by ground based AO systems on large telescopes it is possible to investigate star formation, nuclear activity, as well as the dynamics and structure of distant galaxies in the NIR at scales below 100 pc for $z < 0.05$ and at scales below 500 pc at $z < 2$. However, most of the deep extragalactic surveys conducted so far have been deliberately taken in regions devoid of bright reference stars. This strategy minimizes the usability and usefulness of AO and fringe tracking in studying distant extragalactic objects.

Infrared imaging surveys can now quite successfully be used to search for star-galaxy pairs that can be observed with AO. As an example about 300 pairs of galaxies and natural guide stars with V$\leq$14 and star-galaxies separations of less than 30 arcseconds have been extracted from the early data release of the SDSS (e.g. Hikage et al., 2002). Examples are given in Figure 3. These objects can then be used to efficiently investigate the distribution of (1) nuclear star clusters and (2) young, bright, massive star clusters in the disks of spirals and nuclear stellar clusters in spiral galaxies of different Hubble type with redshifts of less than z = 0.3.

Several galaxies in this sub-sample show clear signs of an interaction with a galaxy of comparable size or smaller satellites. Therefore the target galaxies can be used to derive characteristic interaction parameters in order to improve our understanding of merger senarii and to use them as input parameters (mass estimate, projected distances, orientations, structure etc.) for multi-particle and SPH simulations of merger processes.

Galaxy interactions are an essential process in the evolution of galaxies and the extent of star formation in these systems. Recent studies have shown that especially a large fraction of ultra-luminous IR galaxies (ULIRGs), with $L_{FIR} > 10^{12}$ $L_{\odot}$ are interacting or merging systems, and many contain active galactic nuclei. The NIR adaptive optics and especially interferometric imaging is ideal to study mergers at high angular resolution especially because the NIR bands are more sensitive to the mass dominating stellar populations, and at the same time less sensitive to the absorption effects of interstellar dust.

## 4. The Galactic Center

Measurements of stellar velocities and variable X-ray emission at the centre of the Milky Way have provided the strongest evidence so far that the dark mass concentrations seen in many galactic nuclei are likely supermassive black holes. The combined NTT/SHARP speckle and NACO aptive optics data (Figure 4) recently allowed us to trace two thirds of a complete orbit of the star S2 currently closest to SgrA* (Schödel et al., 2002; Eckart et al., 2002a). In particular, the authors succeeded in observing both peri- and apo-center passages. The data show that the star is on a bound, highly elliptical Keplerian orbit around SgrA*, with an orbital period of 15.4 years and a peri-center distance of only 17 light hours. The orbital elements require an enclosed point mass of $3.7 \pm 1.5 \times 10^6$ solar masses, which agrees well with the mass at much larger distances. The new data now also exclude with high confidence that the central mass can consist of a cluster of dark astrophysical objects or massive, degenerate fermions, and strongly constrain the central density structure.

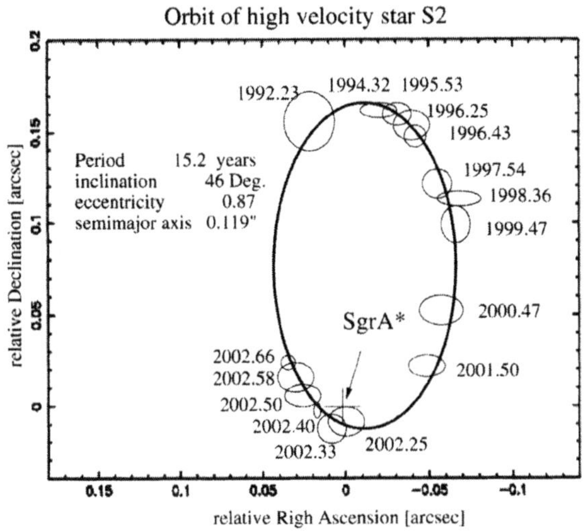


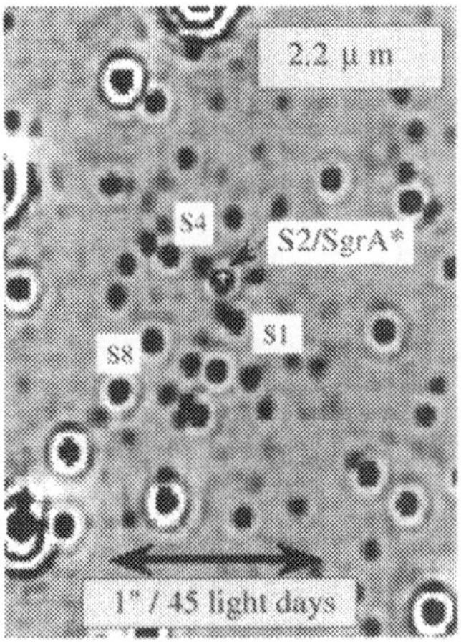


*Figure 4.* The orbit of star S2 close the the central black hole at the position of the compact radio source SgrA*. On the left the results obtained with the MPE SHARP camera at the NTT and the VLT UT4 NACO adaptive optics system are shown. On the right the central 1″ diameter cluster of high velocity stars is shown.

Future observation with the VLTI (e.g. Eckart et al., 2002b) will allow to closely investigate individual sources in the central stellar cluster. Here especially the interferometric investigation of individual sources in the MIR domaine will be possible (see e.g. AO results by Clénet et al., 2001). This will cover dust embedded objects like IRS 21, possible binary stars like IRS 16SW, as well as the position of SgrA* associated with the central massive black hole. Here it is especially the future dual-feed capability of the fully mature VLTI system via the PRIMA instrument that will allow us to obtain valuable results on the Galactic Center of unprecedented quality. The measurements can be combined with simultaneous large FOV observations or VLA/VLBI obervations of SgrA* and OH/$H_2O$ maser stars within the central cluster.

## 5. Summary

The VLTI will allow to carry out sensitive observations over a wide range of sources of astrophysical interest in the 1 to 100mas angular resolution domain. The large collecting area (more than 220 $m^2$ and in future up to 8 telescopes with 8 delay lines) and wide range of interferometric instrumentation (AMBER, MIDI, PRIMA, Darwin-GENIE) will make the VLTI the prime choise for conduction sensitive highest angular resolution inetferometric observations from the shortwavelength end of the NIR well into the MIR domaine. Since the VLTI will be fully operational at a time at which other sensitive interferometric facilities at different wavelengths are also available the scientific yield of the measurements will be greatly enhanced via an efficient combination of all of these high angular resolution capabilities.

## References


Absil, O., Gondoin, Ph., den Hartog, R., Erd, C., Fridlund, M. and Rando, N.: 2002, in: F. Combes and D. Barret (eds.), *SF2A-2002: Semaine de l'Astrophysique Francaise*, meeting held in Paris, France, June 24–29, 2002, EdP-Sciences (Editions de Physique), Conference Series.

Alloin, D., Galliano, E., Cuby, J.G., Marco, O., Rouan, D., Clénet, Y., Granato, G.L. and Franceschini, A.: 2001, *A&A* **369**, L33.

Clènet, Y., Rouan, D., Gendron, E., Montri, J., Rigaut, F., Lna, P. and Lacombe, F.: 2001, *A&A* **376**, 124.

Colavita, M.: 2001, *American Astronomical Society Meeting* **198**, #51.02.

Eckart, A. et al.: 1997, in: *Science with the VLT Interferometer*, Proceedings of the ESO workshop, held at Garching, Germany, 18–21 June 1996, Publisher: Berlin, New York: Springer-Verlag, ESO Astrophysics Symposia, ISBN: 3540632646, 1997., p. 259.

Eckart, A., Genzel, R., Ott, T. and Schödel, R.: 2002a, *MNRAS* **331**, 917.

Eckart, A., Mouawad, N., Krips, M., Straubmeier, C. and Bertram, T.: 2002b, *SPIE* **4835-03**, Conf. Proc. of the SPIE Meeting on 'Astronomical Telescopes and Instrumentation', held in Waikoloa, Hawaii, 22–28 August 2002.

Glindemann, A.: 2002, *Scientific Drivers for ESO Future VLT/VLTI Instrumentation Proceedings of the ESO Workshop held in Garching*, Germany, 11–15 June, 2001, p. 279.

Herbst, T.M., Rix, H.-W., Bizenberger, P. and Ollivier, M.: 2000, ŜPIE **4006**, 673.

Hikage, C. et al.: 2002, *PASJ* **54**, 707.

Kley, W.: 1999, *MNRAS* **303**, 696.

Kley, W., D'Angelo, G. and Henning, T.: 2001, *ApJ* **547**, 457.

Malbet, F. and Bertout, C.: 1995, *A&A Suppl.* **113**, 369.

McAlister, H.: 2001, *American Astronomical Society Meeting* **198**, #51.04.

Millan-Gabet, R., Schloerb, F.P. and Traub, W.A.: 2001, *ApJ* **546**, 358.

Perrin, G., Woillez, J., Guerin, J. and Lai, O.: in: F. Combes and D. Barret (eds.), *SF2A-2002: Semaine de l'Astrophysique Francaise*, meeting held in Paris, France, June 24–29, 2002, EdP-Sciences (Editions de Physique), Conference Series.

Schödel, R. et al.: 2002, *Nature* **419**, 694.

Stern, S.A.: 1994, *AJ* **108**, 2312.

Straubmeier, C., Eckart, A., Bertram, T., Zealouk, L. and Wang, Y.: 2002, *SPIE* **4835-52**, Conf. Proc. of the SPIE Meeting on 'Astronomical Telescopes and Instrumentation', held in Waikoloa, Hawaii, 22–28 August 200?.

Vannier, M., Petrov, R.G., Bensammar, S. and Lopez, B.: 2002, in: F. Combes and D. Barret (eds.), *SF2A-2002: Semaine de l'Astrophysique Francaise*, meeting held in Paris, France, June 24–29, 2002, EdP-Sciences (Editions de Physique), Conference Series.

Wolf, S., Gueth, F., Henning, T. and Kley, W.: 2002, *ApJ* **566**, L97.

# NEXT GENERATION INSTRUMENTATION FOR THE VERY LARGE TELESCOPE INTERFEROMETER

A. QUIRRENBACH
*Sterrewacht Leiden, Postbus 9513, NL-2300 RA Leiden, The Netherlands*

**Abstract.** The scientific capabilities of the VLT Interferometer can be substantially enhanced through new focal-plane instruments. Many interferometric techniques – astrometry, phase-referenced imaging, nulling, and differential phase measurements – require control of the phase to $\lesssim 1$ rad; this capability will be provided at the VLTI by the PRIMA facility. Phase-coherent operation of the VLTI will also make it possible to perform interferometry with spectral resolution up to $R \sim 100,000$ by building fiber links to the high-resolution spectrographs UVES and CRIRES. These developments will open new approaches to fundamental problems in fields as diverse as extrasolar planets, stellar atmospheres, circumstellar matter, and active galactic nuclei.

**Keywords:** interferometry, astronomical instrumentation, extrasolar planets, stellar atmospheres, circumstellar matter, active galactic nuclei

## 1. Introduction

The 'first fringes' observed with MIDI and two of the 8 m telescopes in December 2002 mark an important milestone for the VLTI: the first scientific instrument is now nearing completion. It will open the $10\,\mu$m window to observations with unprecedented spatial resolution and sufficient sensitivity to carry out studies of nearby active galactic nuclei and of circumstellar material (Leinert et al., 2000). An upgrade of MIDI that will also enable observations in the $20\,\mu$m window is planned for the near future. The near-infrared instrument AMBER is scheduled for delivery to Cerro Paranal in 2003; it will offer closure phase imaging with up to three telescopes in the J, H, and K bands, at spectral resolutions up to $R = 10,000$ (Petrov et al. 2000). The European astronomical community will thus soon gain access to powerful and flexible new tools for interferometric observations of many classes of objects.

The VLTI is unique among all interferometers world wide in that a large number of institutions are contributing their expertise to its instrumentation program, and in its ambition to provide an interferometric facility that is accessible to a large number of users and well-suited for diverse observing programs. (For an overview of the VLTI science program see Paresce 1997.) It is thus important to consider the question in which directions the VLTI should evolve after the completion of its first-generation instruments. There is clearly a large phase space to be explored by new instrument concepts. Sensitivity, calibration accuracy, wavelength coverage,

*Astrophysics and Space Science* **286:** 277–289, 2003.

spectral resolution, polarization capability, and number of telescopes that can be combined simultaneously are perhaps the most important parameters in which improvements are desirable and possible. In the following sections I will try to describe some techniques and concepts that could be considered for the second generation VLTI instrumentation.

## 2. Multi-Telescope Closure Phase Imaging

While it is certainly true that a lot of interesting science can be extracted from a limited number of visibility measurements, there are also many good reasons why one would like to obtain true images with milliarcsecond resolution. Since we do not have much a priori information about the distribution of spots on stellar surfaces, of the material surrounding young stars, of winds from AGB stars, or of dust and emission line regions in active galactic nuclei, imaging is clearly required to obtain objective information about these objects. A good case in point are intriguing interferometric observations indicating that the extended near-infrared emission regions of Herbig Ae/Be stars are not disk-like (Millan-Gabet et al., 2001) – but what exactly is their geometry? Or consider the dust spiral around the Wolf-Rayet binary WR104 observed with aperture masking on the Keck I telescope (Tuthill et al., 1999) – would it have been possible to infer its structure from observations with only a handful of baselines?

The principles of aperture synthesis imaging with closure phases are well-known from radio astronomy, and have also been successfully demonstrated with optical interferometers (Baldwin et al., 1996). The key is using Earth rotation and a sufficient number of baselines to sample the Fourier ($uv$) plane adequately, in a time that is short compared to the time scale on which the source varies. The combination of data taken in multiple spectral channels can further increase the $uv$ plane coverage, since the spatial frequency is inversely proportional to the wavelength. This technique obviously works best if the source structure is independent of wavelength, as for example in binary systems with two components of equal spectral type (Benson et al., 1997). A generalization is possible if the wavelength dependence can be easily parameterized, such as a spotted stellar surface, which can be described by brightness and temperature for each point.

The infrastructure of the VLTI allows for the simultaneous combination of up to eight telescopes; six delay lines, four 1.8 m Auxiliary Telescopes, and, of course, the four 8 m Unit Telescopes have been funded so far. Building an instrument that can combine the light from six or eight telescopes is therefore a logical next step; such an instrument would dramatically increase the scientific productivity of the VLTI. In my opinion, the main question is not whether, but how such an instrument should be built. The combination of multiple beams with bulk optics is a proven concept that offers a low-risk approach. On the other hand, integrated optics have many potential advantages which could include a lower cost (Berger et al., 2000), but the technology is less mature. The successful test of an integrated optics beam combiner at the Infrared Optical Telescope Array (Berger et al., 2001) raises the hopes, however, that integrated optics will be a viable alternative to the more conservative bulk optics approach for a multi-telescope closure-phase imaging instrument for the VLTI.

## 3. Fringe Scanning and Phase-Coherent Interferometry

The VLTI currently uses delay scans through the fringe packet for visibility measurements and to control the delay lines. This 'coherencing' is sufficient (but not optimal, from the signal-to-noise point of view) for single-baseline visibility measurements and for closure phase imaging. Most other interferometric techniques (astrometry, phase-referenced imaging, nulling, differential phase measurements) require phase-coherent operation, i.e., control of the residual phase to $\lesssim 1$ rad. The FINITO fringe sensor unit (Gai et al., 2002) will soon introduce this capability at the VLTI; the PRIMA facility (see Sect. 4) will develop phase-coherent methods in a more systematic way.

The implementation of co-phasing techniques requires a metrology system that can measure the optical pathlength difference (OPD) between the interferometer arms with a precision of a few nanometers (Quirrenbach et al., 1998). The main challenges here are keeping vibrations at a very low level, and eliminating differential errors between the metrology laser beams and the starlight, e.g. due to beam walk on the optical surfaces in the interferometer.* Anisoplanatism of atmospheric turbulence (i.e., differences in the OPD fluctuations in two different viewing directions separated by an angle $\theta$) causes a phase variance that scales with $\theta^{5/3}$ and restricts off-axis phase-referencing to angles of a few tens of arcseconds in the near infrared.

The on-axis co-phasing performance is ultimately limited by dispersion effects. The air-filled delay lines of the VLTI cause a systematic delay-dependent distortion of the fringe packet, and the zero-OPD point becomes a function of wavelength (Daigne and Lestrade, 1999). This leads to a decorrelation for wide bandpasses and systematic astrometric and phase errors. Moreover, random OPD fluctuations in the free atmosphere and in the delay lines are not completely achromatic. These fluctuations have zero mean, but can also lead to a fairly dramatic decorrelation between different wavelengths. The observed net dispersion is the sum of dry air dispersion and a term due to water vapor. Pressure balance limits the fluctuations of dry air; the relative water vapor fluctuations are therefore larger and dominate the dispersion fluctuations in the infrared (Meisner and le Poole, 2002). It would

* A metrology beam with a small diameter probes different areas of optical surfaces than a larger starlight beam. Optical elements have typical surface errors of order $\gtrsim 10$ nm. A small change of the beam footprints, e.g., when a mirror is actuated, therefore leads to a 'beam walk' error of similar magnitude.

be technically possible (but expensive) to retrofit the VLTI delay lines with vacuum tanks. This would simplify the fringe phase estimation, eliminate several terms in the astrometric and phase-referencing error budgets (due to temperature differences between the beams and dispersion between the metrology and starlight), and improve the nulling performance. Analyses carried out for the Keck Interferometer as well as the VLTI indicate that evacuated delay lines are required for spectroscopy of extrasolar planets with the differential-phase technique, unless a double-differential measurement with respect to a an external reference star is carried out (see Section 4).

## 4. PRIMA

The PRIMA (Phase-Referenced Imaging and Microarcsecond Astrometry, Quirrenbach et al., 1998; Delplancke et al., 2000) facility will substantially enhance the capabilities of the VLTI. Its astrometric mode will be used for an extensive extrasolar planet survey aimed at measuring their masses and orbits (Quirrenbach, 2000; Quirrenbach et al., 2002). The phase-referenced imaging mode of PRIMA will push the sensitivity limit of the VLTI by several magnitudes, because once the array is co-phased, its point-source sensitivity (within the isoplanatic patch) is essentially equal to that of a single telescope with equal collecting area. One should caution that the sensitivity of an interferometer to heavily resolved, fuzzy structures is much reduced, but many of the most interesting targets for studies with very high angular resolution are sufficiently compact. Among the scientific drivers for sensitive observations with milliarcsecond resolution are:

- Determining the size and geometry of obscuring tori and emission line regions in Active Galactic Nuclei.
- Resolving the stellar populations in dense clusters, for example in the cores of globular clusters and in the Galactic Center.
- Observations of individual stars in external galaxies, including the Magellanic Clouds.
- Characterization of faint binary systems, for example interacting binaries, X-ray binaries, cataclysmic variables.
- Watching the expansion of nova shells in the light of emission lines.

An important aspect of PRIMA is its ability to provide fringe tracking for observations with other instruments. MIDI will benefit tremendously from this capability, because many of its potential targets are too faint for fringe tracking at 10 $\mu$m, but can be self-referenced in the H and K bands.

The optimization of the fringe-tracking sensitivity is critically important for PRIMA, because this determines the availability of reference stars and therefore the sky coverage. An important consideration in this regard concerns the trade-off between white-light fringe tracking and techniques based on dispersed fringes (le Poole and Quirrenbach, 2002). White-light fringe tracking gives the highest sensitivity, because it allows for the simplest optical design (and thus for the highest optical throughput), and because it uses all available light with the smallest possible number of pixel-reads. On the other hand, white-light fringe tracking is also most sensitive to fringe mis-identification and fringe jumps. The best way to overcome this is by considering spectrally dispersed fringes. A compromise between these two conflicting requirements is to send part of the light to the white-light tracker, and part to a dispersed fringe sensor to locate the central fringe at a lower sampling rate. It is obvious that this works well only if fringe jumps are sufficiently rare, which in turn sets a limit on the minimum necessary signal-to-noise ratio and thus on the required brightness of the reference star. Detectors with intrinsic spectral resolution would be very advantageous to solve this dilemma; they could combine the advantages of white-light fringe tracking (efficiency and sensitivity) with a simultaneous capability to measure the group delay, and thus to ensure proper identification of the central fringe. Detector types that are currently under development include Superconducting Tunneling Junctions (STJs) and Superconducting Edge Sensors (Perryman and Peacock, 2000; Romani et al., 1999; Bruijn et al., 2000). A second-generation fringe tracker based on an adaptation of these detector types for near-infrared observations with a spectral resolving power of $R \approx 20$ would produce an almost optimum solution for fringe tracking on faint sources.

While there is a single goal for the reference channel, namely optimizing the fringe tracking sensitivity, the requirements for the science channel vary with the specific astrophysical goal. With co-phasing operational on the reference star, the integration time in the science channel can be arbitrarily long. The only limitation is set by the rate of change of the $uv$ coordinates of the Fourier components observed due to Earth rotation. Such smearing of course is not a serious limitation. One must only make sure that the detectors are read sufficiently often and that each measurement is put in its correct $uv$ bin. If required, for instance in order to enhance the signal-to-noise ratio of the results, one may come back the next night and put more integration time into the same $uv$ bin. (Unless, of course, the target source is variable on that time scale.) Thus the performance will only be limited by the available telescope time. It is important to realize that this potential for long integration times also enables high spectral resolution observations. This of course is very important for many astrophysical problems (see Sect. 5).

Another application of phase referencing is making differential phase measurements between different wavelengths. The near-infrared spectra of extra-solar planets should be characterized by extremely deep absorption bands of water and methane (e.g. Burrows et al., 1997). This opens the possibility of using wavelength-dependent astrometric information for the detection of extrasolar planets, and even to obtain their spectra. The photocenter of a star – planet system is slightly different between molecular bands, where the planet is relatively bright, and within bands, where it is much fainter. Actually, the shift of the photocenter is proportional to the planet / star brightness ratio and can thus be used as a proxy for the planet spectrum (Quirrenbach, 2000). The shift of the photocenter gives rise to a corresponding wavelength dependence of the interferometer phase, which can be measured if the signal-to-noise ratio is sufficient and systematic effects are kept small. In the case of 'Hot Jupiters' like 51 Peg B, which is quite favorable because the planet is close to the star and therefore hot and bright, the expected effect on the interferometer phase is $\approx 0.5$ mrad on the longest baselines of the VLTI. To measure such a small phase difference, a signal-to-noise ratio of $\approx 3000$ is needed, and it remains to be seen whether the systematic instrumental and atmospheric effects can be overcome at this level. For this technique, the dispersion in the air in the delay lines is a serious difficulty, but double differential measurements

$$\phi_{dd} \equiv \phi_t(\lambda_1) - \phi_t(\lambda_2) - [\phi_r(\lambda_1) - \phi_r(\lambda_2)] \quad , \tag{1}$$

where the subscripts $t$ and $r$ stand for 'target' and 'reference' respectively, could solve this problem (le Poole and Quirrenbach 2002).

Finally, it should be pointed out that referenced phases are better than closure phases for imaging of resolved sources. A Y-shaped imaging array optimized for good coverage of the $uv$ plane (such as the VLA) has a geometric progression of station locations along the arms of the Y. In contrast, the Navy Prototype Optical Interferometer (NPOI) has been designed with a linear arrangement of the telescope stations, in order to enable 'baseline bootstrapping' (Armstrong et al., 1998). This approach sacrifices $uv$ coverage for fringe tracking capability. If one can co-phase the array externally, one is again entirely free to choose the array configuration for the optimization of the image to be obtained. This has several important advantages:

- One can have many more pixels across the source, even with a relatively small number of telescopes.
- The image fidelity is higher, because the sampling of the $uv$ plane can be made more uniform.
- One gathers more information: $N(N-1)/2$ phases versus only $(N-1)(N-2)/2$ independent closure phases, which is a significant difference for arrays with a small number of telescopes $N$.

While closure phase imaging is frequently considered to be the most promising method to obtain images of stellar surfaces, phase-referenced imaging does in fact offer a very attractive alternative. For example, one could co-phase the VLTI on Sirius B (an 8.7 mag star), in order to image Sirius A at a separation of $\sim 7''$. In the visible, this would yield a picture with 10 resolution elements across the stellar disk. Cool stars with hot companions may present even better opportunities for phase-referenced imaging.

## 5. Interferometric High-Resolution Spectroscopy

Observations that combine high angular and spectral resolution will enable many completely new approaches to fundamental questions in stellar astrophysics. Examples of astrophysical problems that can be addressed by interferometric spectroscopy include:

– Measuring the variation of the stellar diameter with wavelength, or even better wavelength-dependent limb darkening profiles, provides a sensitive probe for the structure of strongly extended atmospheres of cool giant stars. Such data can be directly compared with predictions of theoretical models, and provide qualitatively new tests of stellar model atmospheres (Quirrenbach and Aufdenberg, 2002).
– Tracing the propagation of shocks in the atmospheres of Mira stars gives new information on their density structure, and on pulsation mechanisms.
– It is possible to perform direct measurements of the projection factors of Cepheid pulsations, which relate the true velocity of the pulsation to the observed radial velocity. Uncertainties in the 'p factor', which presently has to be computed from theoretical models, are a serious limiting factor in current estimates of Cepheid distances with the Baade-Wesselink method (Sabbey et al., 1995; Marengo et al., 2002).
– Generalized Doppler imaging provides detailed information on stellar surface structure. The reconstruction of stellar surface features from line profile variations is plagued with ambiguities. These can to a large extent be resolved by the additional phase information contained in interferometric data.
– Determining the rotation axis orientation of the components in binary systems can provide new test for binary formation theories. Stellar rotation induces a phase difference between the red wings and the blue wings of stellar absorption lines. Measuring the direction of the phase gradient on the sky allows determination of the orientation of the stellar axis (Petrov, 1989; Chelli and Petrov, 1995). One can thus check whether the rotation axes of binaries are aligned with each other, and with the orbital angular momentum.
– Velocity-resolved interferometric observations in emission lines can be used to determine the geometry and velocity field of outflows from pre-main-sequence stars and of disks around young stars and Be stars (Quirrenbach, 1997).

The required spectral resolution $R$ depends on the width of the stellar lines under consideration, of course. For cool stars with narrow lines, the information content increases dramatically with $R$, calling for an interferometric instrument providing $R \approx 100,000$.

Scanning through the fringe packet can in principle provide the desired spectroscopic information; the envelope of the fringe packet is the Fourier transform of the correlated source spectrum (multiplied by the bandpass of the instrument). This technique is known as double-Fourier interferometry (Mariotti and Ridgway,

1988; Mékarnia and Gay, 1990); it is appealing because it does not require any special hardware in addition to the scanning delay line. Reaching very high spectral resolution with this approach is difficult, however, since the delay range that must be scanned is proportional to $R$, according to the Nyquist sampling theorem. Atmospheric fluctuations pose a further problem, since phase coherence must be preserved during the whole delay scan. This can be achieved by monitoring the phase in a separate channel, and applying a suitable on-line or off-line correction. This method should work quite well for observations of strong isolated emission lines, but it is unsuitable for measurements of stellar absorption line spectra. In this case, the spectral resolution and the integrity of the observed spectrum are most certainly limited by calibration problems, because the shape of the whole fringe envelope has to be determined very precisely to allow the extraction of spectral information by Fourier transforming.

It thus appears that standard dispersive spectroscopy is much better suited for most of the projects delineated above. But building high-resolution spectrographs is an expensive proposition, and they would take up a lot of precious space in the crowded interferometry laboratory. It seems thus very appealing to take advantage of the existing high-resolution spectrographs of the VLT, UVES and CRIRES, for interferometric observations. I will call the suggestion for an interferometric mode of these instruments UVES-I and CRIRES-I, respectively. The idea behind this concept is fairly straightforward. Let's assume that the interferometer phase is stabilized by a fringe tracking unit (e.g., FINITO or the PRIMA fringe sensor) to $\lesssim 1$ rad at the desired observing wavelength. Then the output from a second beam combiner can be fed via a fiber to the input of the high-resolution spectrograph. Because the OPD is being stabilized by the fringe tracker, the integration time can be very long (many minutes). Note also that the pathlength of the optical fiber is irrelevant, because it is located after the beam combiner. Consequently, there is no need to use a single-mode fiber; multi-mode fibers will do just as well.

To obtain the full information on the visibility phase and amplitude, one has to measure the four fringe quadratures, i.e., one has to count the photons at phases 0, $\pi/2$, $\pi$, and $3\pi/2$ (e.g., Shao et al., 1988). It has been proposed to generate these for outputs by taking advantage of the phase shift between the two polarization states upon total internal reflection (Cassaing et al., 2000). An alternative would be a two-stage beam combiner, which sends 50% of the light from each interferometer arm to a 'standard' beam combiner, giving the 0 and $\pi$ outputs, and 50% through an achromatic $\pi/2$ phase shifter to a second beam combiner, which thus produces the $\pi/2$ and $3\pi/2$ outputs. The achromatic phase shifter can be realized with adjustable dispersive elements (Mieremet et al., 2000). In any case, the output of the beam combiner are four fibers carrying the four quadrature signals.

The UVES spectrograph already has a fiber feed, which is used for multi-object spectroscopy in the visible (Pasquini et al., 2002). It can accept eight objects, and could thus be used to perform interferometric observations on two baselines simultaneously. The spectrograph part of UVES-I would operationally be almost identical to the standard multi-object mode; the origin of the light in the fibers from the interferometer is transparent to the instrument. With a 1″ slit width, the spectral resolution is 40,000; it should be possible to reach $R = 100,000$ with a smaller slit at the cost of substantial light losses. UVES-I would benefit tremendously from adaptive optics systems that could provide high Strehl numbers in the visible. In the absence of such systems, it is still possible to perform the science programs listed above. One would simply take the 1.8 m Auxiliary Telescopes and stop their apertures down to $\approx 3\,r_0$, i.e., about 50 cm. This means that UVES-I would be about 8 magnitudes less sensitive than UVES with an 8 m aperture, but this would still allow observations of bright stars (to $V \approx 7$) in a completely new part of the instrumental parameter space.

The infrared high-resolution spectrograph CRIRES can also accommodate multiple fibers, with some additional design required for the feed into the spectrograph (U. Käufl, priv. comm.). CRIRES-I would work in the same way as UVES-I, but in the J and H infrared bands. It is unlikely that operation at K will be possible because of the exceedingly large thermal background in this band. With a slit width-resolution product of 100,000, CRIRES-I would reach a spectral resolution similar to that of UVES-I. Together, these instruments would provide a very powerful facility for stellar astrophysics.

## 6. Wide-Field Interferometry

In a Michelson stellar interferometer such as the VLTI, the object-image relationship can not be described as a convolution, because the rearrangement of the apertures rearranges the high-spatial frequency part of the object spectrum in the Fourier plane (Tallon and Tallon-Bosc, 1992). This has an important consequence for off-axis objects: the image position does not coincide with the white-light fringe position (see Figure 1 in Tallon and Tallon-Bosc, 1992). For a finite spectral bandwidth this means that the fringe contrast decreases with field angle and the field-of-view is limited; the maximum size of an image from a Michelson interferometer is $\sim R \equiv \lambda/\Delta\lambda$ resolution elements in diameter.

To obtain a wider field-of-view, the output pupil of the interferometer has to be configured such that it is an exact scaled replica of the input pupil. This ‘homothetic mapping’ is quite complicated because the input pupil (as seen from the direction towards the target star) changes continuously due to Earth rotation. Nevertheless, it would be interesting to explore the possibility of building a homothetic imaging instrument to take advantage of the full 2″ unvignetted field of the VLTI delay lines. Such an instrument would form images of all objects in the field, each one with a fringe pattern given by the Fourier transform of the VLTI input pupil. This would allow simultaneous observations of hundreds of stars in the core of a globular cluster or in the cluster surrounding the center of our Galaxy. Unfortunately, the tolerances of a homothetic imager are extremely tight (Beckers, 1990), and the two-dimensional arrangement of the output pupil makes it difficult to implement a spectroscopic mode. It may in fact be easier to build a separate homothetic beam combiner for each individual baseline, and to combine the information from the various baselines after detection in a computer (Schöller, 2000).

Similar considerations apply to astrometric observations. Many astrometric programs, including the search for extrasolar planets, can be carried out with a dual 'zero' field, i.e., with two fields that are much smaller than the Airy disk of a single telescope. However, programs such as measuring proper motions and orbits in clusters could be carried out much more efficiently with an interferometric field-of-view of $\sim 1''$ in one of the two fields, which would allow the simultaneous measurement of many stars (either with respect to one another, or referenced to the guide star in the other field). This option appears attractive from the scientific point of view, but it places even more stringent requirements on many of the instrumental tolerances than wide-field imaging, and it makes the design of the focal plane and the data reduction procedures much more complicated.

For a beam compression factor of $1/100$, the tolerance for lateral pupil alignment is as stringent as $\sim 5\,\mu$m for a $1''$ field; the requirements on pupil rotation and magnification are similarly demanding. However, the main problem with a non-zero field-of-view are imperfections of optical surfaces located close to an image plane. A 5 nm goal on the OPD accuracy implies that these imperfections have to be calibrated to that same level over the 'footprint' of a $1''$ field on each relevant surface. In this context, the critical component in the VLTI design is likely the variable curvature mirror (VCM), which is located in the focal position of the delay line trolley's Cassegrain telescope. The VCM consists of a thin membrane, which forms the front surface of a pressure cell and whose curvature can be controlled by changing the pressure in the cell (Ferrari et al., 2000). It is the VCM which allows for proper pupil relay in the VLTI, but any slight deviations from its nominal shape cause field distortions which spoil the astrometry.* At present these errors seem prohibitively large, but implementing simultaneous astrometry over a 'wide' ($\sim 1''$) field would certainly be an attractive option for future upgrades of PRIMA, which could be considered when the actual performance of this facility for 'on-axis' observations has been fully characterized.

## 7. Conclusions

The range of astrophysical problems that are being addressed with optical and infrared interferometry has grown considerably over the past few years (Quirrenbach 2001, and references therein). The VLTI is poised to advance all of these, and to open new vistas through the PRIMA facility and the second generation of VLTI

* Note that this is also relevant for some applications that are normally not considered 'astrometry', e.g., the determination of orbits of binaries with separations of order $0.''1$.

instruments. In my view, the most promising concepts for these instruments are (in no particular order):

- instruments (or upgrades to the current instruments) adding coverage of the visible, L, and Q bands,
- a multi-telescope (i.e., six-way or eight-way) near-infrared closure-phase instrument, implemented either with bulk or with integrated optics,
- a four-way mid-infrared instrument, which would succeed MIDI and take advantage of the four large Unit Telescopes to obtain images in the 10 $\mu$m and 20 $\mu$m bands,
- an interferometric polarimeter (e.g., Rousselet-Perraut et al., 1997a, 2000),
- a nuller for the thermal infrared (L-band or N-band), which could help detect debris disks around main-sequence stars and hot extrasolar planets (for more information see the workshop web page `http://www.strw.leidenuniv.nl/~nevec/GENIE/index.html`),
- fiber couplers to the UVES and CRIRES spectrographs,
- an optimized fringe tracker based on detectors with intrinsic spectral resolution (for a detailed discussion see le Poole and Quirrenbach 2002),
- a homothetic imaging and wide-field astrometric instrument.

It should perhaps be emphasized again that the viability of most of these advanced instrument concepts depends critically on the development of phase-coherent methods, and on continued improvements to the VLTI infrastructure. We certainly hope that ten years after the first fringes between two UTs we will see a full VLT array complete with 8 Auxiliary Telescopes, 8 (vacuum?) delay lines, dual-star modules and adaptive optics at all telescopes, and with a suite of powerful second-generation instruments. This, of course, will require substantial further financial resources. Producing excellent scientific results with the first-generation VLTI instruments should be our highest priority, because they are the best argument that future investments should actually be made.

## References

Armstrong, J.T., Mozurkewich, D., Rickard, L.J., Hutter, D.J., Benson, J.A. et al.: 1998, *ApJ* **496**, 550–571.

Baldwin, J.E., Beckett, M.G., Boysen, R.C., Burns, D., Buscher, D.F. et al.: 1996, *A&A* **306**, L13–L16.

Beckers, J.M.: 1990, in: L.D. Barr (ed.), *Advanced Technology Optical Telescopes IV,* SPIE **1236**, Bellingham, WA, pp. 379–389.

Benson, J.A., Hutter, D.J., Elias, N.M., Bowers, P.F., Johnston, K.J. et al.: 1997, *AJ* **114**, 1221–126.

Berger, J.P., Benech, P., Schanen, I., Maury, G., Malbet, F., Reynaud, F.: 2000., in: P.J. Léna and A. Quirrenbach (eds.), *Interferometry in Optical Astronomy,* SPIE **4006**, Bellingham, WA, pp. 986–995.

Berger, J.P., Haguenauer, P., Kern, P., Perraut, K., Malbet, F. et al.: 2001, *A&A* **376**, L31–L34.

Bruijn, M.P., Tiest, W.B., Hoevers, H.F., van der Kuur, J., Mels, W.A. and de Korte, P.A.: 2000, in: J.E. Trümper and B. Aschenbach (eds.), *X-Ray optics, Instruments, and Missions III,* SPIE **4012**, Bellingham, WA, pp. 145–153.

Burrows, A., Marley, M., Hubbard, W.B., Lunine, J.I., Guillot, T. et al.: 1997, *ApJ* **491**, 856–875.

Cassaing, F., Fleury, B., Coudrain, C., Madec, P.Y., Di Folco, E. et al.: 2000, in: P.J. Léna and A. Quirrenbach (eds.), *Interferometry in Optical Astronomy,* SPIE **4006**, Bellingham, WA, pp. 152–163.

Chelli, A. and Petrov, R.G.: 1995, *A&AS* **109**, 401–415.

Daigne, G. and Lestrade, J.F.: 1999, *A&AS* **138**, 355–363.

Delplancke, F., Lévêque, S., Kervella, P., Glindemann, A. and d'Arcio, L.: 2000, in: P.J. Léna and A. Quirrenbach (eds.),*Interferometry in Optical Astronomy,* SPIE **4006**, Bellingham, WA, pp. 365–376.

Ferrari, M., Lemaître, G., Mazzanti, S., Lanzoni, P., Derie, F. et al.: 2000, in: P.J. Léna and A. Quirrenbach (eds.), *Interferometry in Optical Astronomy,* SPIE **4006**, Bellingham, WA, pp. 104–115.

Gai, M., Corcione, L., Delage, L., Gardiol, D., Gennai, A. et al.: 2002, in: E. Vernet, R. Ragazzoni, S. Esposito and N. Hubin (eds.), *Beyond Conventional Adaptive Optics,* European Southern Observatory, Garching, pp. 329–334.

Leinert, C., Graser, U., Waters, L.B.F.M., Perrin, G., Lopez, B. et al.: 2000, in: P.J. Léna and A. Quirrenbach (eds.), *Interferometry in Optical Astronomy,* SPIE **4006**, Bellingham, WA, pp. 43–53.

le Poole, R.S. and Quirrenbach, A.: 2002, in: W.A. Traub (ed.), *Interferometry for Optical Astronomy II,* SPIE **4838**, in press.

Marengo, M., Sasselov, D., Karovska, M., Papaliolios, C. and Armstrong, J.T.: 2002, *ApJ* **567**, 1131–1139.

Mariotti, J.M. and Ridgway, S.T.: 1988, *A&A* **195**, 350–363.

Meisner, J. and le Poole, R.S.: 2002, in: W.A. Traub (ed.), *Interferometry for optical astronomy II,* SPIE **4838**, in press.

Mékarnia, D. and Gay, J.: 1990, *A&A* **238**, 469–474.

Mieremet, A.L., Braat, J., Bokhove, H. and Ravel, K.: 2000, in: P.J. Léna and A. Quirrenbach (eds.), *Interferometry in Optical Astronomy,* SPIE **4006**, Bellingham, WA, pp. 1035–1041.

Millan-Gabet, Rafael, Schloerb, F.P. and Traub, W.A.: 2001, *ApJ* **546**, 358–381.

Paresce, F. (ed.): 1997, *Science with the VLT Interferometer,* Berlin/Heidelberg, Springer-Verlag, 404 pp.

Pasquini, L., Alonso, J., Avila, G., Barriga, P., Biereichel, P. et al.: 2002, in: *SPIE* **4841,** in press.

Perryman, M.A.C. and Peacock, A.: 2000, in: W. van Breugel and J. Bland-Hawthorn (eds.), *Imaging the Universe in Three Dimensions, ASP Conf. Ser.* **195**, pp. 487–494.

Petrov, R.G.: 1989, in: D.M. Alloin and J.M. Mariotti (eds.), *Diffraction-limited Imaging with Very Large Telescopes,* NATO ASI **274**, Kluwer, Dordrecht, pp. 249–271.

Petrov, R.G., Malbet, F., Richichi, A., Hofmann, K.H., Mourard, D. et al.: 2000, in: P.J. Léna and A. Quirrenbach (eds.), *Interferometry in Optical Astronomy,* SPIE **4006**, Bellingham, WA, pp. 68–79.

Quirrenbach, A.: 1997, in: F. Paresce (ed.), *Science with the VLT Interferometer,* Springer-Verlag, Berlin/Heidelberg, pp. 163–170.

Quirrenbach, A.: 2000, in: J. Bergeron and A. Renzini (eds.), *From Extrasolar Planets to Cosmology, the VLT Opening Symposium,* Springer-Verlag, Berlin/Heidelberg, pp. 462–467.

Quirrenbach, A.: 2001, *ARAA* **39**, 353–401.

Quirrenbach, A. and Aufdenberg, J.: 2002, Wavelength-dependent diameters of cool giant stars, in: *IAU Symp.* **210,** in press.

Quirrenbach, A., Coudé du Foresto, V., Daigne, G., Hofmann, K.H., Hofmann, R. et al.: 1998, in: R.D. Reasenberg (ed.), *Astronomical Interferometry,* SPIE **3350**, Bellingham, WA, pp. 807–817.

Quirrenbach, A., Frink, S. and Mitchell, D.: 2002, in: *Bioastronomy 2002, ASP Conf. Ser.*, in press.

Romani, R.W., Miller, A.J., Cabrera, B., Figueroa-Feliciano, E. and Nam, S.W.: 1999, *ApJ* **521**, L153–L56.

Rousselet-Perraut, K., Chesneau, O., Berio, P. and Vakili, F.: 2000, *Astron. Astrophys.* **354**, 595–604.

Rousselet-Perraut, K., Mourard, D. and Vakili, F.: 1997a, *Opt. Eng.* **36**, 980–987.

Sabbey, C.N., Sasselov, D.D., Fieldus, M.S., Lester, J.B., Venn, K.A. and Butler, R.P.: 1995, *ApJ* **446**, 250–260.

Schöller, M.: 2000, in: P.J. Léna and A. Quirrenbach (eds.), *Interferometry in Optical Astronomy,* SPIE **4006**, Bellingham, WA, pp. 116–123.

Shao, M., Colavita, M.M., Hines, B., Staelin, D., Hutter, D.J. et al.: 1988, *A&A* **193**, 357–371.

Tallon, M. and Tallon-Bosc, I.: 1992, *A&A* **253**, 641–645.

Tuthill, P.G., Monnier, J.D. and Danchi, W.C.: 1999, *Nature* **398**, 487–489.

# SECOND GENERATION INSTRUMENTATION FOR THE VLTI: THE FRENCH VLTI CONNECTION

DENIS MOURARD[1], OLIVIER LARDIERE[2], BRUNO LOPEZ[1], FABIEN MALBET[3] and PHILIPPE STEE[1]
[1]*Observatoire de la Côte d'Azur*
[2]*Observatoire de Haute Provence*
[3]*Laboratoire d'Astrophysique de l'Observatoire de Grenoble, France*
*E-mail: mourard@obs-azur.fr*

**Abstract.** We describe the current French ideas for the instrumentation of the second generation of the VLTI. Instruments concepts addressed include: integrated optics beam combiner, extension of MIDI to a four beam facility, extension of AMBER to the visible and a densified pupil direct imaging beam combiner.

**Keywords:** VLTI, instrumentation

## 1. Introduction

After more than ten years of development, the European optical long baseline interferometer (the VLTI) is gradually reaching its full performances with the scheduled commissioning in 2003 of a large number of main subsystems: focal instrument MIDI for the 10 $\mu$m band, focal instrument AMBER for the near infrared and the three telescopes mode, adaptive optic devices MACAO for the UTs, fringe sensor unit FINITO.

As such, the VLTI will become one of the very first interferometric facility in the world. A very high quality and quantity of astrophysical results could be certainly expected in the coming years. New extensions are soon foreseen like a dual field facility (PRIMA) that would allow astrometric measurements with a spatial resolution of the order of a few tens of micro arcseconds. This will also allow to increase the limiting magnitude of the first generation focal instruments by the use of the technique of the phase reference imaging. Also, a demonstrator of nulling interferometry (GENIE) is under study in the framework of a collaboration with the European Space Agency concerning the preparation of the Darwin mission.

During the June 2001 workshop about the second generation of instruments for the VLT/VLTI, ESO has concluded about two main priorities for the development of the VLTI: an imaging mode with 6 to 8 beams and an extension to the shortest wavelengths (visible band). In both cases, the French community has brought some original ideas that should be discussed in the European astronomical community. Among these, we will describe in more details the following actions: **IONIC-**

*Astrophysics and Space Science* **286:** 291–296, 2003.

**VLTI** (integrated optics beam combiner) **APRES-MIDI** (extension of MIDI to a four beams facility) **VINI** (extension of AMBER to the visible band) **VIDA** (densified pupil direct imaging beam combiner). Other actions have been started during the last two years in the context of the 'Action Spécifique Haute Résolution Angulaire' from INSU/CNRS and of the 'Jean-Marie Mariotti Center'. Different working groups have started to work on systems or software aspects especially in the domain of interferometric imaging.

## 2. IONIC-VLTI

With milli-arcsecond spatial resolution, many scientific topics like dust torus around AGN, disks around young stars, envelopes around AGBs, require imaging. The current VLTI instruments (AMBER and MIDI) give access to information at high angular resolution, yet without actual imaging. Since as a rule of thumb, the image quality accessible with an optical interferometer is roughly proportional to the number of telescopes used simultaneously, the more the apertures, the better the reconstructed image. We propose (Malbet et al., 2002) the interferometric beam combination of N telescopes (N between 4 and 8) thanks to the integrated optics technology: four 8-m telescopes in the short term and/or eight 1.8-m telescopes in the long term. The goal is to have access to an instrument whose image quality is equivalent to the one achieved with millimeter radio interferometers with a spatial resolution better than the one foreseen with ALMA, in the visible and/or in the near infrared. This instrument would be able to acquire routinely images with 1 mas resolution with a 13- to 20-mag sensitivity in the NIR with a spectral range from 0.5 to 2.5 microns. The high accuracy achievable with integrated optics on visibility amplitude and phase measurements allows to reach high dynamic range even on faint objects.

Components able to recombine up to 4 beams have been tested on the sky, first at the IOTA interferometer and recently on the VINCI test instrument on the VLTI. The main competitors of the VLTI are the Keck interferometer with two 10-m telescopes and four 1.8-m outriggers and CHARA with six 1-m telescopes. We could also cite the experimental program OHANA that will combine the large telescopes at the Mauna Kea summit. However only the VLTI will be able to combine 4 telescopes of 8-m class placed in the best configuration for imaging to achieve accurate investigation of extra-galactic objects, if a imaging instrument is developed soon. Following our experience in interferometry and the VLTI already existing infrastructure, we propose a strategy in two stages.

In the short term, an imaging instrument for the main array (4 UTs) working from 0.8 to 2.5 $\mu$m. The required infrastructure is limited: 4 adaptive optics already funded, 4 delay lines among the six that will be installed, 1 additional fringe sensor.

In the long term, an imaging instrument for the extended array with possibly 8 telescopes (additional ATs or combination of ATs with some UTs) working from

0.6 to 5 $\mu$m. The required infrastructure is bigger: additional telescopes with a total of 6 to 8 ATs, a total of 8 delay lines, adaptive optics for the ATs if extension to the visible domain, and a cophasing instrument (IO technology?). A full system study is needed to determine the best development strategy including the fringe tracking.

## 3. APRES-MIDI

This proposition is not really a second generation instrument. It could be developed as an upgrade of the MIDI beam combiner in order to extend its astrophysical and imaging capabilities. A group in France is currently studying an optical concept aiming at recombining 4 telescopes beams (Lopez et al., 2002). Interference fringes are sampled in the pupil plane. Such a principle is perfectly adapted for reconstructing images by aperture synthesis at 10 $\mu$m with the VLTI. This principle could be used for building a new generation 10 $\mu$m instrument, but instead of doing a totally new instrument, we propose the design of an optical module that can supply the current MIDI-VLTI instrument with 4 beams.

One of the major interest of the mid infrared band like the 10 $\mu$m domain concerns the study of the dusty environments in a wavelength domain where warm dust grains radiate efficiently. The research in several currently relevant astrophysical program will strongly benefit of interferometers operating in the mid infrared and aiming at producing images. These programs concern: the study of the geometry of dust torii in active galactic nuclei, the study of dust disc characteristics surrounding the young stellar objects, the study of the morphology of the circumstellar environments embedding late type stars. The 10 $\mu$m window being very well adapted to the mapping of the aerial surrounding various type of astrophysical sources and being not much employed by astronomers up to now due to a lack of facilities, many new results could be expected in the fundamental programs mentioned above. Aside from the study of dusty media, the observation of gaseous envelopes like those surrounding hot Be stars, as well as the observation of stellar atmospheres and the study of low mass companions in binary system are astrophysical topics that will strongly benefit also from 10 $\mu$m interferometric observations.

The expected sensitivity of APreS-MIDI is equivalent to the one of the MIDI instrument (Lopez et al., 2002). MIDI is presently under commissioning at ESO at its effective sensitivity will be known after the full operability of the VLTI and of MIDI will be achieved. The theoretical sensitivity of MIDI corresponds to a coherent flux of N = 4 (about 1 mJy – 10 sigma detection above noise) using UTs and without the assistance of the external fringe tracker (Leinert et al., 2002).

# 4. VINI

AMBER is the near infrared instrument installed in the VLTI focal laboratory. There, it benefits from the general interferometric infrastructure on Paranal. Three telescopes are selected among the 4 UTs and the 3 ATs available. For each telescope the wavefront errors are partially corrected by the MACAO Adaptive optics on the UTs and to tip-tilt correction on the ATs. The beams are transported through delay lines used to equalize the optical path difference, which is stabilized by a fringe tracker. A star separator and differential delay lines allow to use an off axis reference star, up to 1 arc minute away from the science source, for wavefront and fringe sensing . Originally, one of the (lower priority) design criteria for AMBER was an easy extension to the red wavelength, at least around the H$\alpha$ line. The key idea was that Adaptive Optics and a focal instrument efficient in the K band with 8 m telescopes ($D/r0 = 11$) would be well adapted to observations in the red with 1.8 m telescopes ($D/r0 = 12$). The original proposal included adaptive optics system in the focal laboratory usable in the infrared with the UTs and in the red with the ATs. After the Concept Design Revue, the work was concentrated on the infrared instrument, but at least at a preliminary analysis level, the extension to the R band remains possible.

In general terms the interest of a visible instrument compared to an infrared one is that (Stee et al., 2002) visible and near-IR interferometry are complementarity. For example, both infrared and optical interferometry are often required for detailed radiative transfer modelling. Also the spatial resolution is 2 times better than in J and 4 times better than in K, which is decisive for many sources. Then the astrophysical signal produced by many processes is substantially stronger in the visible. For exempla, the optical emission lines of Be star envelopes and of the AGN BLR are much stronger than the corresponding IR lines. It is also much easier to have a higher spectral resolution in the visible. If one accepts the same size constraints than for the infrared AMBER spectrograph, the limiting spectral resolution in the visible is of the order of 40000, which is the kind of resolution needed for real progress in stellar activity studies. And last, the detectors in the visible remain substantially more efficient. In particular, photon-counting detectors with fair quantum efficiency now exist and further progress is expected.

The main scientific cases for a visible extension of the VLTI concern the following programs: distance scale in the Universe with the Cepheids stars, Active Galactic Nuclei, Stellar activity, Stellar rotation and orientation axes, Binary and binary formation and evolution, Circumstellar material, Miras stars.

Originally the idea was an extension of the AMBER focal instrument but discussions are currently in progress in order to merge the different propositions and to design a full capabilities imaging second generation instrument working from the visible to the near infrared wavelengths.

## 5. VIDA

Only in the recent years did it has been realized that multi-aperture interferometric arrays could provide direct snapshot images and coronagraphic images in a non-Fizeau mode. Whereas homothetic mapping of entrance pupil to exit pupil is useless when the aperture is highly diluted, a 'densified-pupil' or 'hypertelescope' imaging mode can concentrate most light into a high-resolution Airy peak. In addition to the luminosity gain, there is a contrast gain particularly valuable for stellar coronagraphy and exoplanets finding. The current VLTI is able to combine light from two telescopes coherently. In subsequent phases, a combiner is planned for applying closure phase with up to eight telescopes (UT and AT). The small number of apertures currently considered at the VLTI, does not take full advantage of hypertelescope imaging, but still performs significantly better than other observing modes (+3.8mag gain in comparison with Fizeau mode (Lardière et al., 2002)). We propose some possible optical scheme for a densified-pupil combiner for the VLTI (Lardière et al., 2002). Beyond its science value, the proposed instrument can serve as a precursor for many-element post-VLTI hypertelescopes.

The pupil densification consists in enlarging each sub-aperture by the same factor, their centers pattern being kept invariant. This densification factor is defined as $\gamma_d = (do/Do)/(di/Di)$, where di and do are the diameter of each aperture element before and after the pupil densification respectively and Di and Do are the array size before and after the pupil densification respectively. After pupil densification, the central interference peak of the focal image is intensified by a factor ${\gamma_d}^2$, whereas the FOV diameter is divided by $\gamma_d$. Indeed, the usable field of the hypertelescope, called 'Zero Order Field' (ZOF), has an angular radius of $\lambda/((\gamma_d - 1).di)$ on the sky, where $\lambda$ is the wavelength.

The optimal amount of pupil densification is thus a trade-off between the image intensity and the FOV and this trade-off should be considered with respect to the scientific cases that could be covered by such a direct imaging instrument.

It appears also possible to increase the accessible field of view by using an array of pupil densifiers spaced by $\lambda/di$. The use of integrated optics systems and/or optical fibers should simplify a lot the design and fabrication of such complex beam combiner.

## 6. Conclusion

There is a strong interest of the French community for an imaging second generation instrument for the VLTI. The development of such an ambitious project is conditioned by a certain number of steps. The problematic of interferometric imaging could progressed a lot with short terms studies like IONIC-4T and APRES-MIDI.

We are now starting the definition phase of the second generation instrument taking into account the scientific potential of the current instruments, the unique potential of the VLTI infrastructure and of course the most promising scientific cases as well as new conceptual ideas in the field of long baseline interferometry.

## References

Lardière, O., Labeyrie, A., Mourard, D., Riaud, P., Arnold, L., Dejonghe, J. and Gillet, S.: 2002, VIDA (Vlti Imaging with a Densified Array), a densified combiner proposed for snapshot imaging with the VLTI, *SPIE Conf.* **4838** Hawaii.

Leinert, Ch. et al.: 2002, Ten micron instrument MIDI getting ready for observations on the VLTI, *SPIE Conf.* **4838** Hawaii.

Lopez, B., Mathias, P., Mékarnia, D., Thiébaut, E., Dutrey, A., Bresson, Y., Lagarde, S., Antonelli, P., Roussel, A., Dugué, M. and Gay, J.: 2002, APRES-MIDI, APerture Synthesis in the MID-Infrared with the VLTI, *SPIE Conf.* **4838** Hawaii.

Malbet, F., Kern, P., Berger, J.P., Haguenauer, P., Perraut, K. and Perrier, C.: 2002, VLTI imaging instrument using the full potential of the interferometric array, *SPIE Conf.* **4838** Hawaii.

Stee, Ph., Petrov, R., Weigelt, G., Mourard, D., Jankov, S., Mathias, P., Vakili, F., Lagarde, S., Robbe, S. and Foy, R.: The potential of an extension of AMBER to the visible, *ESO Conf.* Munich.

# THE VLTI: CHALLENGES FOR THE FUTURE

FRANCESCO PARESCE
*European Southern Observatory, D-85748 Garching b. Muenchen, Germany*

**Abstract.** ESO's VLT Interferometer is quickly ramping up to its full potential as the first instruments covering the near and mid IR are deployed and the 8m telescopes are equipped with state-of-the-art adaptive optics systems. I will review the expected scientific benefits of such improvements and outline the next steps in the continuing quest for higher spatial resolution and sensitivity.

## 1. Introduction

Figure 1 shows graphically the real challenge for the VLT Interferometer (VLTI). On the left, the best currently available 2D image taken by an optical interferometer is represented while, on the right, is shown a typical 2D image taken routinely nowadays by a radio interferometer.

The comparison is striking as it shows better than the proverbial one thousand words the long road optical long baseline interferometry (OLBIN) has to go before it can rival its cousin and reach the scientific objectives it has set for itself. The latter, of course, include obtaining pictures in the optical and near and mid infrared of astronomical targets such as planets, stellar accretion disks and jets, supernovae ejecta and the nuclei of nearby AGN with the resolution, clarity and fidelity of the type shown in the right panel of Figure 1.

Going from left to right in this figure is no easy task from the technical standpoint for a number of reasons that I will describe in the next paragraphs but some outstanding results obtained recently such as the one shown in the left panel of Figure 1 raise the distinct possibility that OLBIN will soon be able to fulfill its promise and we may soon see a period coming where images of the caliber routinely obtained in the radio are available in the optical at similar if not better spatial resolutions. I will use the case of the VLTI, the only currently operating very large telescope array designed from the start as an interferometer, as an example of the type of OLBIN facility that will carry the burden of transforming the potential into reality by the end of this decade.

*Astrophysics and Space Science* **286:** 297–309, 2003.

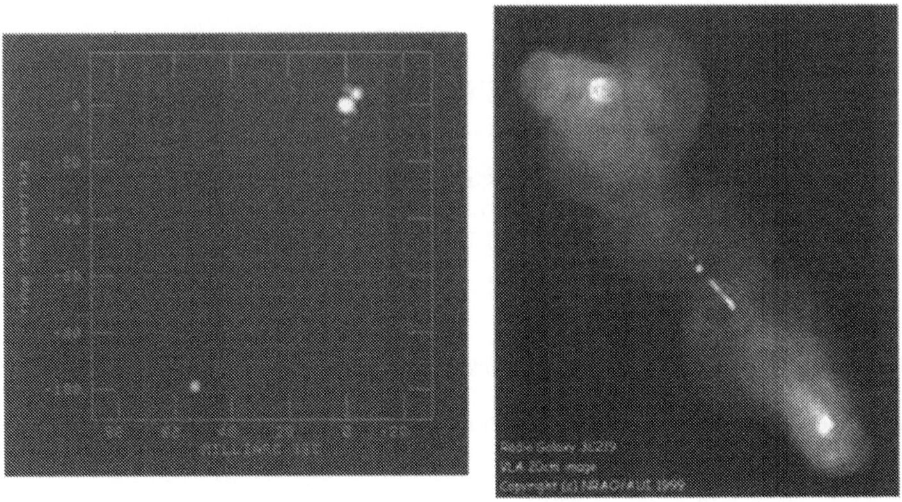

*Figure 1.* The triple star Eta Virginis observed in the optical with the Navy Prototype Optical Interferometer's 6-station array (left panel) and the radio galaxy 3C219 observed with the VLA at 20 cm (right panel).

## 2. Radio Versus Optical Long Baseline Interferometry

In order to understand the basic reasons behind the evident lag in performance between the two techniques let us consider some fundamental facts. In the radio range (10–600 THz), the atmospheric coherence time is typically of order $\sim$20 minutes and the isoplanatic patch $\sim$1–2° on the sky while the same parameters appropriate to the optical are typically 1-10 ms and 15–30″. This clearly means that high precision phase referencing in the radio range can be done by simply off-pointing to a calibrator star. Just as obviously, this is simply impossible in the optical.

Moreover, the physics of the detection process in the radio allows one to use heterodyne combination wherein detection is carried out at antenna level, the signal amplified, mixed with a reference signal of high coherence, digitized and combined in a correlator later at one's leisure. In contrast in the optical, this is not possible and one must resort to homodyne combination. In this case, one must directly detect the interference pattern only after correction for the optical path difference (OPD) and, since only the intensity of the field can be reliably detected the fringes must be modulated in phase to measure both the phase and amplitude of the complex visibility.

Some other practical considerations that have limited up to now the usefulness of OLBIN are associated with the precision and sensitivity of the combination process itself. For example, the maximum OPD allowed in an interferometer is the coherence length of the radiation used namely $\lambda_0^2/\Delta\lambda$ where $\lambda_0$ and $\Delta\lambda$ are the central wavelength and bandpass of the device, respectively which, in the optical at 1 $\mu$, corresponds to $\sim$ 10 $\mu$ over the baseline of $\sim$200 m or more or $\sim$ a part in $10^9$. If that were not challenging enough, the optical path jitter during the observation must not smear the fringes i.e. the dynamical stability of the OPD compensation must be of order of 10 nm. No wonder the technique has been so difficult to apply until recently.

The situation on the sensitivity was equally bleak up until the availability of accurate fringe trackers and very large telescopes equipped with adaptive optics systems. To understand this problem, consider the fact that the signal to noise ratio S/N in an optical interferometer is proportional to $NV^2$ where N corresponds to the number of photons detected per subaperture and per integration time and V is the measured visibility of the resulting fringe pattern. Thus, paradoxically, the detection accuracy decreases significantly with increasing source complexity in the sense that the correlated magnitude of a resolved object $m_{corr}$ is related to its unresolved magnitude m by $m_{corr} = m - 2.5 \log V^2$. This implies, for example, that an object of magnitude m whose visibility at a certain baseline is 10% (i.e. that the ratio of coherent to total radiation is 0.1) will have a correlated magnitude $m_{corr}$ = m+5 or, in other words, a sensitivity 100 times less.

Considering, finally, the twin facts that, without a fringe tracker, integration times are limited to $\sim$10 ms at best, that the background due to detector noise, sky and thermal emission from the telescope is high (especially at 10 $\mu$), an optical interferometer's sensitivity or limiting magnitude is shockingly low. Even with telescopes of $\sim$1m in diameter, the limiting uncorrelated magnitude of typical current interferometers is only K $\sim$ 5-7 in the best conditions.

## 3. Fundamental OLBIN Requirements

So, what do we need to reach our objectives of 2D imaging in the optical/IR at 1 mas resolution and K = 20 sensitivity that are both a must for astronomical competitiveness in the next few years?

We need an efficient array of the largest possible number n of telescopes with diameter D. The number of independent baselines increases rapidly with n (as n(n-1)) and, with it, the u,v plane coverage that is essential for accurate reconstruction of complex sources. This requirement is best shown in Figure 2 where the effect of u,v plane coverage on the shape of the synthesized beam is illustrated clearly. Starting from the situation shown in the top right where the sparse coverage of the u,v plane corresponds to a very dirty and essentially unusable beam, progressing down to the bottom right where the situation gets considerably better mainly due to the effect of earth rotation synthesis and, finally, to the left where a combination of more and/or mobile telescopes and earth rotation synthesis provides the ultimate resolution in a compact, dominating core of milliarcsecond (mas) size using a baseline of $\sim$100 m projected length.

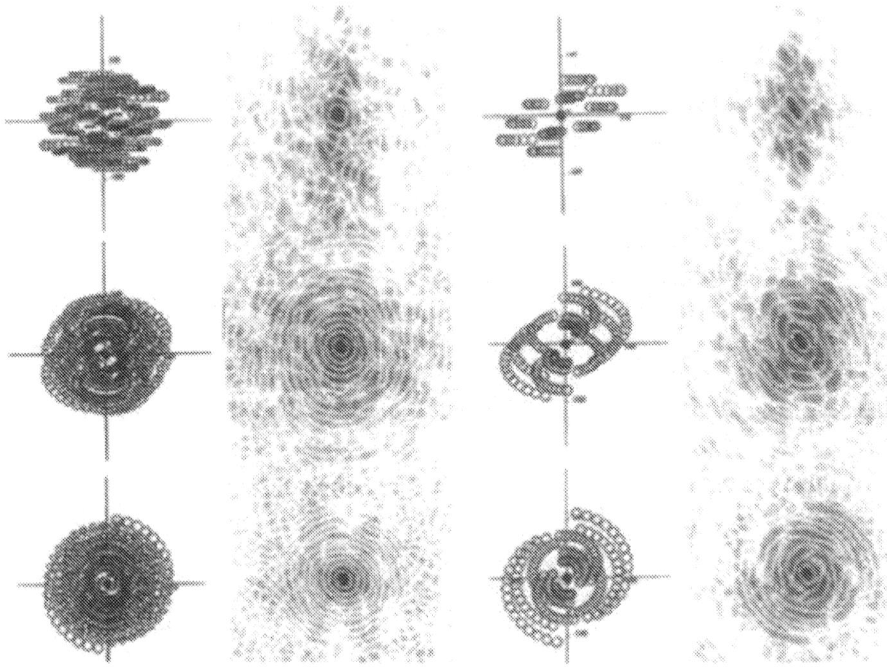

*Figure 2.* The effect of u,v plane coverage on the synthesized beam. Each panel consists in the u,v points corresponding to the particular configuration considered on the left and the corresponding synthesized beam on the right. Coverage increases from the top right to the bottom left.

We also need the largest possible baselines lengths B coupled with the largest possible D to maximize resolution and sensitivity. All telescopes of D>$\sim$1m need to be AO equipped to be used to their full potential. Extremely stable delay lines (DL) and accurate fringe trackers that allow the fringe packet to be frozen and, therefore, to allow integration times to grow hundred fold are other essential desiderata. n-way (with n$\geq$4) beam combiners covering the optical/IR range with spectral resolution of up to $10^4$ will also have to play an important role to overcome the inherent limitations of the technique brought about by the effects of the turbulent atmosphere by allowing the efficient use of standard closure phase techniques. Finally, the ultimate performance of OLBIN for faint object imaging and high precision astrometry will be attained when dual feed systems permitting the use of bright stars as phase reference for much fainter targets will be developed in the near future.

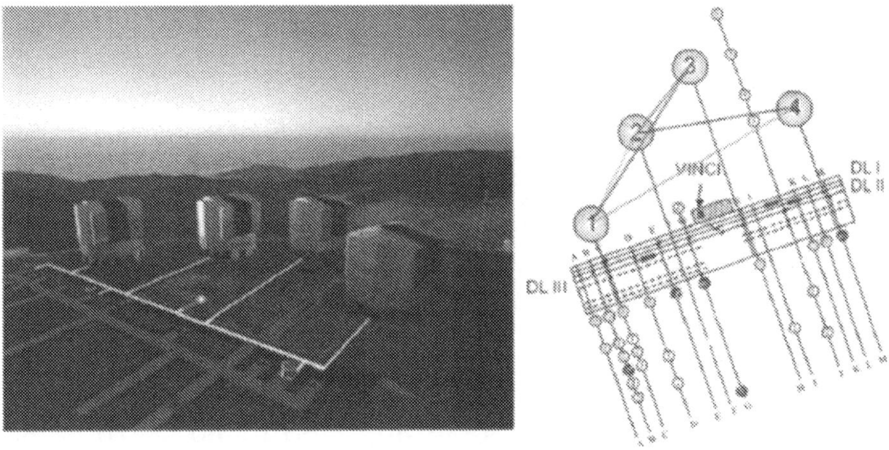


*Figure 3.* The observing platform on Cerro Paranal with the 4 UTs, the 120 m long delay line tunnel and the beam combination laboratory at the center of the VLTI array where the beams from the various telescopes converge and the stations for the auxiliary telescopes (left panel). The schematic configuration of the same array is shown on the right panel. The lines indicate the baselines used so far between the four UTs and the two siderostats located on the AT stations. DL I,II,III correspond to the 3 delay lines.

## 4. Where Are We? Current Status of VLTI

How, then, does the current instrumentation compare to these requirements? The situation for what regards the VLTI at the end of 2002 is that all four very large telescopes (UT1-4) have been integrated into the VLTI array in the configuration shown in Figure 3. All four telescopes have been combined in pairs to observe a number of targets successfully. Five baselines have been used now out of the available six, UT3/UT4 being the only one not used yet as there is only one delay line on the required side of the tunnel.

In particular, the UT1 and 3 telescopes have been equipped with tip/tilt sensors and used extensively to commission the array and obtain preliminary science results with the combination's 103m baseline. In addition, two 40 cm diameter siderostats in the positions indicated in Figure 3 with 16, 66 and 140 m long baselines were also used as the workhorses for system verification and are being used to carry out shared risk observations of astronomical targets proposed by community astronomers.

Currently, three 60 m long-stroke delay lines are operating in the 120 m-long tunnel at the center of the observation platform of Cerro Paranal shown in the left panel of Figure 3. These delay lines are the system's key feature that allows it to reach the stringent beam compensation and tracking requirements alluded to earlier. The measured flatness of the rails is now better than 25 $\mu$m over 65m with an absolute position accuracy of 30 $\mu$m. The relative position error of the carriages

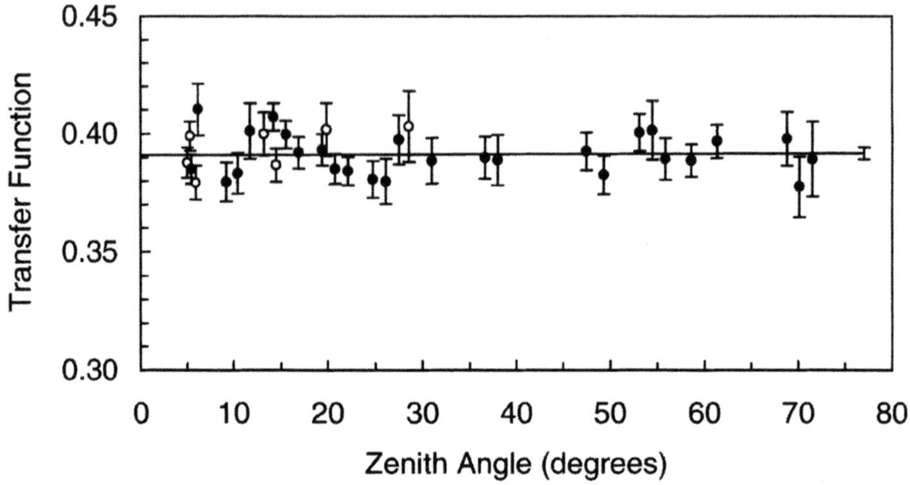


*Figure 4.* The transfer function of the VLTI as a function of zenith angle.

has been measured to be ~20 nm over a 50 ms integration time. The beams are combined in the pupil plane with the test instrument VINCI operating in the K band.

The available facility is described in much greater detail in the proceedings of the conference Interferometry for Optical Astronomy II, ed. Traub, W., Proc. SPIE 4838, 2003 and in the relevant ESO website:
`http://www.eso.org/projects/vlti/`.

To appreciate how well this facility is working in practice, Figure 4 shows the instrument's transfer function obtained with observations of Fomalhaut used both as the target and calibrator in a series of consecutive measurements with VINCI during a whole night in October, 2002 with the siderostats and the 140 m baseline (Davis et al.: 2003, submitted to *A&A*). The transfer function is defined as the ratio of the observed visibility squared to the expected visibility squared. The true visibility squared for a given object is then obtained by dividing its observed visibility squared by the transfer function for its associated calibrator.

As can be seen from this figure no significant variation in the transfer function was found for the zenith angle range 5° to 70°. The projected baseline varied between 139.7 m and 49.8 m during the observations and, as an integral part of the determination of the transfer function, a new accurate limb-darkened angular diameter for Fomalhaut of $2.112 \pm 0.011$ mas has been established. This has led to improved values for the emergent flux of $3.42 \pm 0.10 \times 10^8$ Wm$^{-2}$, an effective temperature of $8813 \pm 66$ K, a radius of $1.214 \pm 0.011 \times 10^9$ m ($R/R_\odot = 1.746 \pm 0.016$) and a luminosity of $6.33 \pm 0.22 \times 10^{27}$ W ($L/L_\odot = 16.5 \pm 0.6$).

The scientific objectives of this first phase of the VLTI in this configuration as it is offered to the community in the year starting on October 1, 2002 include high precision visibility measurements in the K band of the type shown above in order to determine the distance and mass of spectroscopic binaries, the precise distance of Cepheid variables, the diameter and $T_{eff}$ of stars across the HR diagram and a first look at the brighter complex systems like pre-main sequence stellar and debris disks and the envelopes and shells of evolved stars. In general, the observational philosophy in this first phase where we are restricted to measuring visibilities and not phases and, therefore, to objects having an axially symmetric intensity distribution is to obtain several hundred measurements at a 1% precision or better rather than several thousand measurements at 10% precision that would leave theoretical models unconstrained.

## 5. VLTI: Immediate Prospects (>2003)

Two new instruments will become available in the very near future namely in the period 2003–2004. They will extend the capabilities of the VLTI in three very significant ways: extended wavelength coverage, 3 way beam combination allowing phase closure imaging and increased spectral resolution. One of these is MIDI that will combine two beams in the N band (8-12 $\mu$) with a spectral resolution of 100. Its limiting magnitude (uncorrelated) is expected to reach N = 5 with the UTs with a field of view in the range 0.26-1.14″ and a minimum fringe spacing of 10 mas. MIDI obtained first fringes with the UTs in December, 2002 (see `http://www.eso.org/outreach/press-rel/pr-2002/pr-25-02.html`). The other will be AMBER operating in the J,H,K′ bands (1-2.4 $\mu$) at selectable spectral resolutions of 35, 1000, and 10,000. Its field of view will be in the range 0.06-0.24″ with a minimum fringe spacing of 1–2 mas and reaching a limiting uncorrelated magnitude of K=13 with the UTs. The exceptional aspect of this instrument will be its ability to combine 3 beams simultaneously.

But much more is on the way in this period. The FINITO fringe tracker will allow on axis fringe tracking in the H band of bright sources (H<12) observed at longer wavelengths thus significantly increasing the visibility measurement accuracy of these objects and the limiting magnitude of both instruments. Of similar importance will be the arrival of two adaptive optics systems on two UTs at the second half of 2003. This will allow removing almost all aberrations except piston with a Strehl ratio of 50% in K for a V<13 reference star and 25% for a V<16 star. The expected increase in limiting magnitude will be considerable. Two more AO systems will be added to the remaining UTs by 2004. Three more delay lines will also be installed in the next year allowing coverage of more than 90% of the AT stations.

Finally, the first two 1.8 m moveable ATs will be integrated in the VLTI facility by mid-2004 and account for an enormous increase in the critical u,v plane cov-

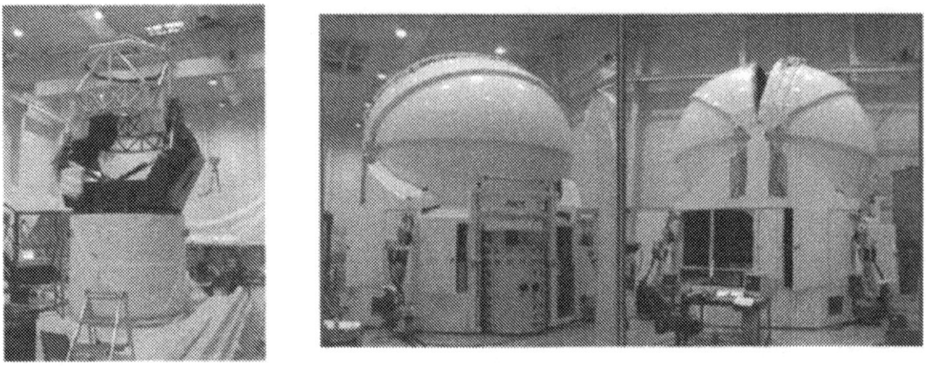

*Figure 5.* The AT being assembled in Europe (leftmost panel) and two views of the transporter completely assembled (right panels).

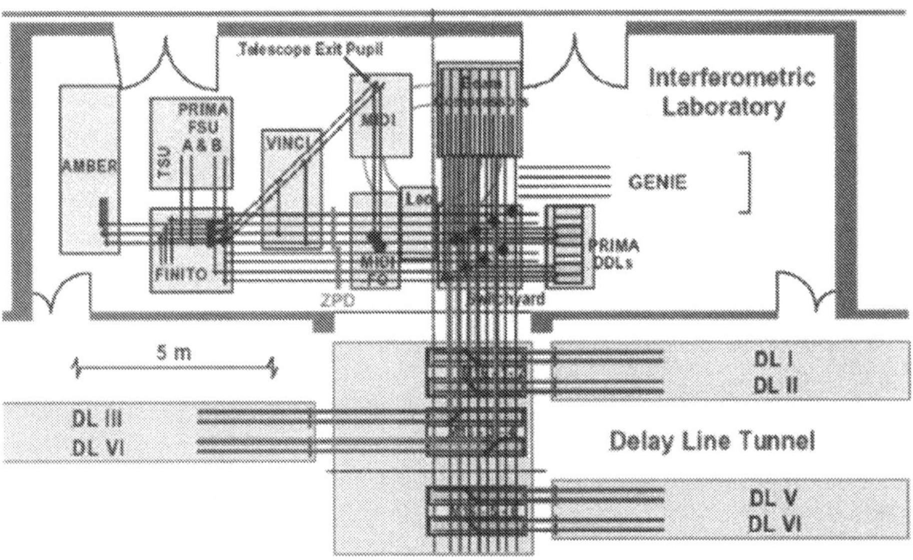


*Figure 6.* Beam location and distribution at the beam combination lab at the center of the array. ZPD indicates the Zero Path Difference position while DL is a delay line. The beam compressors convert the beam sizes from the various feeding telescopes to a standard value required for the instruments.

erage as discussed in the previous section. They are already now at an advanced stage of construction as shown in Figure 5.

They represent a special plus for the facility in that they can be used exclusively for interferometry without having to resort to the already heavily subscribed UTs. Of course, all this equipment will create a serious traffic jam at the combination point as shown in Figure 6.

The scientific objectives of this second phase of the VLTI development beyond $\sim$2003 are all centered on the exploitation of the increased sensitivity and precision of the facility with the various devices described above. In particular, the capability to perform real i.e. model-independent imaging on moderately bright complex sources (K<14) by determining the phase of the fringe packet in addition to its visibility opens up the possibility of accurate and faithful reconstruction of astronomical scenes. These would include, for example, the morphology of dust torii in nearby AGN, the structure of the circumstellar envelopes of mass losing giants, and long lived stellar surface features such as spots and faculae in the star's magnetic network.

Closure phase techniques with the 3-way beam combiner in AMBER and the mid IR capability of MIDI will be very useful in studying and, hopefully, understanding the complex and currently uncertain relationship between planets and the stellar accretion and debris disks from which they originate. The size, temperature structure and possible features in these disks due to the presence of protoplanets should be accurately measurable especially, possibly for the stars that are already known to possess planets via the radial velocity technique. Already the measurement of the binary fraction in nearby clusters and star forming regions will go a long way in elucidating the basic mechanisms underlying the complex process of star formation.

## 6. VLTI: Far Future Prospects (> 2005)

There are two effective ways to measure both the visibility and phase of the fringe packet formed at the beam combination point of the array. The first is closure phase that requires n-way beam combiners where $n \geq 3$. The larger n the better, of course, but the technique is restricted for the moment to relatively bright and stable objects of K<14 at best with the UTs. A better way is to use phase referencing using a bright reference star in the field of view in the fashion of adaptive optics. This allows high fidelity image reconstruction of faint complex sources (K<20 with the UTs) situated close to a bright reference. Close in the K band means within a radius of 30″ approximately. The way this works is indicated schematically in Figure 7.

Controlling all optical path lengths of the reference star and of the science star inside the interferometer ($OPD_{int}$) with a laser metrology system introduces the capability of imaging faint objects and of determining the precise angular separation between the two stars. The measurement has to be repeated for up to 30 min in order to average out the variations of the differential OPD caused by atmospheric turbulence ($OPD_{turb}$). With the two OPD terms being determined, the measurable is the sum $\Delta SB + \phi$. If both stars are point-like, the phase $\phi$ of the visibility function is zero, and if the baseline B is known with high accuracy, one obtains a high precision astrometric measurement of the angular separation $\Delta S$. If only the reference star is point-like and the science object is an extended object with a non-symmetric structure, the (non-zero) phase $\phi$ of the visibility function depends on the baseline

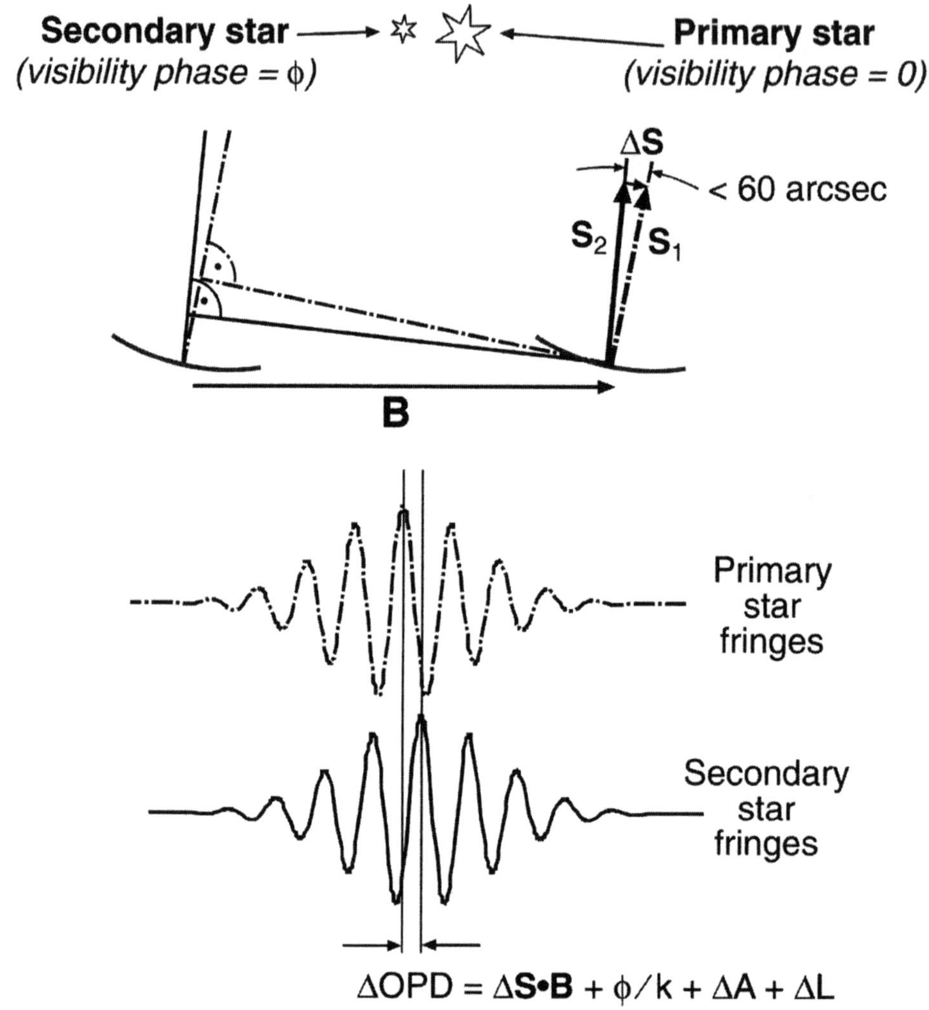


*Figure 7.* Schematic of the process of phase referenced imaging and astrometry. The stabilization of the fringe packet of the faint science object (secondary star, solid line) is accomplished by using a bright guide star (primary star, dot-dash line) as a reference. The difference in the positions of the two fringe packets are determined by the OPD given by the product of $\Delta S$ – the angular separation vector of the stars – and B – the baseline vector – by the phase $\phi$ of the visibility function of the science object, and by the OPD caused by the atmospheric turbulence and by the internal OPD.

vector B. Then, the measured sum $\Delta SB + \phi$ can be disentangled by repeating the measurement for several different baselines.

The facility being currently built is called PRIMA for Phase Referenced Imaging and Microarcsecond Astrometry and can be subdivided in four sub-systems: 1) a star separator i.e. an opto-mechanical system in the Coudé focus of the UTs and ATs to pick two objects within the 2 arcminute field of view and send the light to the Delay Line tunnel, 2) a fringe sensor unit to provide the signal for the fringe tracker, 3) a laser metrology system to measure the $OPD_{int}$, and 4) differential Delay Lines to correct the differential OPD for two objects that are separated up to 2 arcmin with a baseline up to 200 m. In this case, the maximum differential OPD is 130 mm.

PRIMA will enable VLTI to perform high precision narrow-angle astrometry down to the atmospheric limit of 10 microarcseconds ($\mu$as), real imaging of objects fainter than K$\sim$14 and nulling at contrast levels of $\sim 10^{-4}$. These capabilities will, in turn, enable VLTI to address directly a number of extremely important scientific issues that are currently at the top of the list of challenges for future astronomical high resolution instrumentation.

The first is the detection and characterization of extra-solar planets and their birth environment. This objective requires application of both the astrometric and the imaging capability of PRIMA. The astrometric capability will determine the main physical parameters of planets orbiting around nearby stars already found by radial velocity (RV) techniques including their precise mass, orbital inclination and a low resolution spectrum.

In addition and most importantly, astrometry will allow VLTI to extend the search for planets to stars that cannot be properly covered by the RV technique, in particular pre-main sequence (PMS) stars. The crucial exploration of the initial conditions for planetary formation in the stellar accretion disk as a function of age and composition will be finally possible. The simultaneous image at mas resolution or better of the accretion disk from which any particular planet is born will add a new and exciting dimension to our understanding of both planetary and stellar formation mechanisms since the complex accretion disk is expected to be the cradle of these objects.

Another key objective for VLTI with PRIMA will be the exploration of the nuclear regions of galaxies including our own. The resolution of the VLTI at 2 $\mu$ corresponds to 15 AU at the galactic center or about 1500 times the Schwarzschild radius of a $10^6$ $M_\odot$ black hole (BH). The first and most important goal will be to test for the presence of a central massive BH by measuring the three-dimensional velocity field of the star cluster centered on IRS16. The VLTI will be able to probe the 3D space motions of stars in the nuclear cluster down to $10^{-4}$ pc of the central source or approximately two orders of magnitude better than currently. This data will certainly provide very precise information on the central mass without the a-priori assumption of isotropic motions. High precision astrometry with PRIMA might even go so far as to probe the central BH to a few times its Schwarzschild radius ($2.5 \times 10^{-7}$ pc).

At this point, one can finally say that OLBIN will have reached the major leagues and be able to compete mas per mas with the radio facilities like the VLA and images such as the one shown on the right panel of Figure 1 obtained routinely in the near O/IR.

The VLTI as presently designed can accomplish this ambitious but absolutely critical task by relying on its capability to support high precision phase referencing.

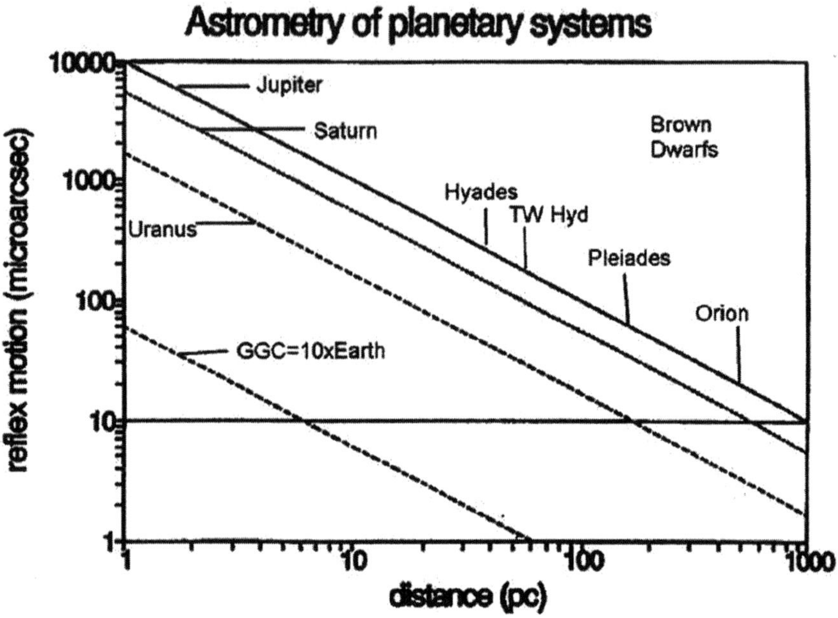


*Figure 8.* The expected astrometric reflex motion of objects of different masses orbiting a star as a function of the latter's distance. The VLTI is expected to measure reflex motions > 10 $\mu$as.

This characteristic will be critical in order to perform the high precision, narrow angle astrometry required to reach the predicted atmospheric limit of 10 $\mu$as accuracy at Paranal. Attainment of this performance will allow the reflex motion induced by an orbiting planet on a star located at the distances shown in Figure 8.

This figure shows how the VLTI in its astrometric mode would cope with a planet of a Jupiter, Saturn, Uranus or Gas Giant Core (GGC) of $\sim$10 earth masses placed at either the Jupiter radius of 5 AU as expected for a solar-like star or closer for a cooler companion. Since the closest star forming regions (SFR) are located between $\sim$50 (TW Hyd) and $\sim$140 pc (Taurus-Auriga) from the Sun, the VLTI would be in the enviable position of easily finding Jupiter-sized planets around stars of any spectral type and, most significantly from the physical point of view, of any age up to the MS. The VLTI then will be able to reach all the way out to Orion to study planets in a wide variety of environments.

The VLTI, by exploiting its dual mode of operation with phase referencing, can, at the same time, probe in considerable detail not only the planet itself through its effect on the motion of the parent star but also the evolution and structure of the proto-planetary/proto-stellar disk from which it eventually must emerge and interact. This affords another unprecedented opportunity to finally understand and pin down the presently uncertain connection between the two phenomena. At the typical distance of the MS early type star IR excess disks, for example, the VLTI IR spatial resolution of a few mas ensures a clear probe of these disks down to the level of a few tenths of AU resolution at a few AU radii where most of the interaction between a possible planet and the edge of the small particle disk is expected to occur.

Other research areas that the VLTI can contribute to significantly and deeper development of many of the topics covered in this document can be found in the proceedings of the ESO workshop 'Science with the VLTI', ESO Astrophysics Symposia Series, ed. F. Paresce, Springer-Verlag, Berlin, 1997 and in the document 'Scientific Objectives of the VLTI' by F. Paresce that can be found at `http://www.eso.org/projects/vlti/science/VLTIscienceMarch2001.pdf`.

## Acknowledgements

This document could not have been written without the fundamental scientific input of all the members of ESO's Interferometry Science Advisory Committee who contributed in the early years to the establishment of the scientific objectives of the VLTI and to the success of its efforts to get VLTI restarted in 1996 after its near demise in 1993. A particular debt of gratitude is due to Jean-Marie Mariotti whose noble example of dedication to and competence in the field of interferometry inspired many of the ideas in this document and infused the spirit of this new exciting endeavor. The VLTI would not be where it is today without critical contributions from many people too numerous to properly thank. A partial list would include C. Cesarsky, S. Correia, J. Davis, F. Delplancke, F. Derie, E. DiFolco, Duc Thanh Phan, A. Gennai, R. Giacconi, P. Gitton, A. Glindemann, P. Kervella, B. Koehler, S. Menardi, S. Morel, A. Richichi, M. Schoeller, J. Spyromilio, M. Tarenghi, A. Wallander, M. Wittkowski, R. Van Boekel.

# THE KLUWER LATEX STYLE FILE

Kluwer Academic Publishers has developed a special style file for authors who want to submit LaTeX articles. KLUWER.STY is a general LaTeX style file which is used for all Kluwer journals, irrespective of the publication's size or layout. (The specific journal characteristics are added later during the production process.) Authors are kindly requested always to use KLUWER.STY when creating a LaTeX article for a Kluwer journal.

*Instruction File*
Although KLUWER.STY is very similar to the ARTICLE.STY and uses many of the standard LaTeX commands, there are some differences. These are explained in the accompanying instruction file - KAPINS[number].TEX

## Getting the Kluwer Style File

KLUWER.STY is offered at a number of servers around the world. Unfortunately, those are unauthorized copies and authors are strongly advised not to use them. Kluwer can only guarantee the integrity of files obtained directly from Kluwer.

*WWW-site, E-mail or Air Mail*
Authors can obtain KLUWER.STY and the instruction file from Kluwer Academic Publishers' information system KAPIS. This free service is available at:

WWW.WKAP.NL

The stylefile may be requested from:

Kluwer Academic Publishers, Editorial Department,
P.O. Box 990, 3300 AZ Dordrecht, The Netherlands.
Telephone: (0)78-6576392, Fax: (0)78-6576555
E-mail: EDITDEPT@WKAP.NL

The files can be sent either by e-mail or on diskette. Please indicate the journal's name, your e-mail number, and postal address.

## Submitting Manuscripts

Please send your completed LaTeX article on diskette, together with the appropriate number of hard copies to the address listed in the *Instructions to Authors* of *Astrophysics and Space Science.*

*Via E-mail*
Experience has shown that sending articles via e-mail can sometimes result in lacunas appearing in the article. This is why Kluwer prefers to receive LaTeX articles on diskettes, accompanied by the hard copies. If you must send the LaTeX article via e-mail, do not forget to send the requisite number of hard copies by air mail.

## Questions

Should you have any questions or encounter problems using KLUWER.STY, please contact Kluwer Academic Publishers for assistance.

GPSR Compliance
The European Union's (EU) General Product Safety Regulation (GPSR) is a set of rules that requires consumer products to be safe and our obligations to ensure this.

If you have any concerns about our products, you can contact us on

ProductSafety@springernature.com

In case Publisher is established outside the EU, the EU authorized representative is:

Springer Nature Customer Service Center GmbH
Europaplatz 3
69115 Heidelberg, Germany

www.ingramcontent.com/pod-product-compliance
Ingram Content Group UK Ltd.
Pitfield, Milton Keynes, MK11 3LW, UK
UKHW020135160926
475607UK00003B/54

*9781402015182*